Logics for AI a

Joint Proceedings of the Third International Workshop on Logics for New-Generation Artificial Intelligence and the International Workshop on Logic, AI and Law

8-9 and 11-12 September 2023

Hangzhou

Volume 1
Proceedings of the First International Workshop, Hangzhou, 2021
Beishui Liao, Jieting Luo and Leendert van der Torre, eds

Volume 2
Proceedings of the Second International Workshop, Zhuhai, 2022
Beishui Liao, Réka Markovich and Yì N. Wáng, eds

Volume 3
Logics for AI and Law. Joint Proceedings of the Third International Workshop
on Logics for New-Generation Artificial Intelligence and the International
Workshop on Logic, AI and Law, September 8-9 and 11-12, 2023,
Hangzhou, Zhuhai, 2022
Bruno Bentzen, Beishui Liao, Davude Liga, Réka Markovich, Bin Wei,
Minghui Xiong and tianwen Xu, eds

Logics for AI and Law

Joint Proceedings of the Third International Workshop on Logics for New-Generation Artificial Intelligence and the International Workshop on Logic, AI and Law

8-9 and 11-12 September 2023

Edited by

Bruno Bentzen

Beishui Liao

Davide Liga

Réka Markovich

Bin Wei

Minghui Xiong

Tianwen Xu

© Individual author and College Publications 2023
All rights reserved.

ISBN 978-1-84890-439-2

College Publications, London
Scientific Director: Dov Gabbay
Managing Director: Jane Spurr

http://www.collegepublications.co.uk

Original cover design by Laraine Welch

Preface

We are pleased to present the joint proceedings of the Third International Workshop on Logics for New-Generation Artificial Intelligence (LNGAI2023) and the International Workshop on Logic, AI and Law (LAIL2023), which were respectively held on September 8–9 and 11–12, 2023, at Zhejiang University in Hangzhou, China. Both workshops are associated with a national key project titled "Research on Logics for New Generation Artificial Intelligence (LNGAI)" (2021-2025). The main objectives of the LNGAI project are to develop theories and techniques of non-monotonic logics and formal argumentation and apply them to causal reasoning, knowledge graph reasoning, and reasoning about norms and values, in an open, dynamic, and real environment. The papers included in this volume of joint proceedings nicely reflect some advances in the interdisciplinary research direction of logic, AI and law.

In the part of LNGAI2023, we received 15 submissions, of which 11 were accepted, including 7 full papers and 4 extended abstracts. Each paper was reviewed by at least two reviewers. In addition, the proceedings include 1 invited paper, "Advanced Intelligent Systems and Reasoning: Standardization, Experimentation, Explanation". In this paper, Pere Pardo, Leendert van der Torre, and Liuwen Yu offer a perspective on advanced intelligent systems and reasoning using morally-decisive robots as an example. The 11 accepted papers cover a diverse range of topics and applications. Shuwhen Wu and Ming Hsiung discuss some self-referential paradoxes induced from cellular automata. Huayu Guo, Dongheng Chen, and Bruno Bentzen give a mechanized proof of completeness in Henkin-style for intuitionistic propositional logic. Aleks Knoks and Leendert van der Torre propose a fresh viewpoint on the concept of weighing normative reasons by considering the interaction between them as a form of inference pattern called "titular reason-based detachment". Yini Huang contends that focusing solely on the epistemological interpretation of "ought to know" fails to consider the deontic issues that can arise from beliefs and knowledge and discusses questions of formalization. Dongheng Chen, Muyun Shao, and Dov Gabbay introduce a new method for defeasible description logic reasoning using abstract syntax graphs employing argumentation theory-based models for computing consistent sets of formulas in the graph. Yiheng Wang, Zhe Lin, Shier Ju examine a De Morgan multi-modal logic that aims to represent the open world

i

model and investigate some of its properties. Zhizheng Zhang develops a logic programming paradigm developed as a knowledge representation and reasoning tool for designing intelligent agents capable of performing the framework. Yulin Chen, Beishui Liao, Bruno Bentzen, Bo Yuan, Zelai Yao, Haixiao Chi, and Dov Gabbay propose an interpretable method designed to help users understand the internal recognition logic of name entity recognition tasks based on Talmudic Public Announcement Logic. Sheng Wei and Beishui Liao merge two different kinds of biosignals to obtain a more comprehensive information representation with enhanced interpretability. Xiaotong Fang describes the design and implementation of a legal question-answering robot along with its corresponding human-computer interaction system. Zhe Lin and Xinshu Wang study modal Lambek calculus with primary assumptions, proving its decidability and showing that categorial grammars based on it with transitive primary assumptions are context-free. We would like to thank the LNGAI2023 program committee members for their invaluable contributions to the workshop: Fengkui Ju, Zhe Yu, Pietro Baroni, Emil Weydert, Kaibo Xie, Huimin Dong, Mehdi Dastani, Jiachao Wu, Guillermo Simari, Valeria de Paiva, Giuseppe Contissa, Alexander Bochman, Christoph Benzmüller, Jinsheng Chen, Olivier Roy, Leon van der Torre, and Réka Markovich.

In the part of LAIL2023, we received 9 submissions in this workshop, each of which was reviewed by 3 reviewers. After careful evaluation, we accepted 5 submissions as full papers. All these works bring insights to the role that logic can play in deepening the intersection of AI and Law. On one hand, impressive moves have been made to address the difficult challenges in modeling legal reasoning of different kinds. Wenjing Du and Zihan Niu proposed a coupling of logical requirements, calculation methods and rational standards to ensure the accuracy of initial probabilities in Bayesian reasoning. Cecillia Di Florio, Xinghan Liu, Emiliano Lorini, Antonino Rotolo, and Giovanni Sartor constructed a novel logical approach to the fundamental problem of identifying factors in case-based reasoning, on the basis of binary classifier and counterfactual reasoning. Tianwen Xu considered the problem of multi-criteria coherence ranking of legal theories as a preference aggregation problem, providing a new angle for its solution. On the other hand, there are meaningful attempts to deal with some practical issues in computational law. Xiang Li, Xin Sun, and Xingchi Su proposed a more abstract formalization of conditional digital signatures and a constructive method to find valid signatures. Ava Thomas Wright by answer set programming implemented a deontic logic that can resolve conflicts between legal obligations. We would like to thank all authors and invited speakers for their contribution to LAIL2023. Special thanks go to the LAIL2023 program committee members for their thorough reviews: Katie Atkinson, Le Cheng, Huimin Dong, Enrico Francesconi, Ming Hu, Fengkui Ju, Davide Liga, Emiliano Lorini, Juliano Maranhão, Olivier Roy, and Xin Sun. This workshop is sponsored by the Academy of Humanities and Social Sciences, Zhejiang University.

We are also indebted to the local organizers, Bruno Bentzen, Davide Fassio, Jie Gao, Chonghui Li, Jieting Luo, Bin Wei, and Tianwen Xu, for their excellent work in organizing these events.

Finally, as satellite events of the Zhejiang University Logic and AI Summit (ZJULogAI2023), LNGAI2023 and LAIL2023 are partially supported by the Shen Shanhong Fund of Zhejiang University Education Foundation.

Bruno Bentzen
Beishui Liao
Davide Liga
Réka Markovich
Bin Wei
Minghui Xiong
Tianwen Xu

Hangzhou, China,
August 16, 2023

Contents

II LAIL2023   157

Part I

LNGAI2023

Advanced Intelligent Systems and Reasoning: Standardization, Experimentation, Explanation

Pere Pardo[1]   Leendert van der Torre[1,2]   Liuwen Yu[1]

[1] *University of Luxembourg, Luxembourg*
[2] *Zhejiang University, China*

Abstract

We offer a perspective on advanced intelligent systems and reasoning, using as an example morally-decisive robots, as proposed in machine ethics. Given that norms often conflict, formal methods are necessary to resolve these conflicts in order to make morally acceptable or optimal decisions. The underlying basis of current algorithms spans from logical representation and reasoning to machine learning algorithms. We explore multiple methodologies including deontic ASP for standardizing normative reasoning, LogiKEy for testing ethical and legal reasoners, and formal argumentation for achieving explanatory transparency. Our vision is demonstrated using the argumentation-based Jiminy moral advisor. We also hint at future work that situates 'real-world' dialogue exchanges as the forum for discussing moral decisions, and we discuss the development of a platform for experimental user studies at the Zhejiang University – University of Luxembourg Joint Lab on Advanced Intelligent Systems and REasoning (ZLAIRE).

Keywords: Artificial intelligence, knowledge representation and reasoning, logic, formal argumentation, deontic answer set programming, LogiKEy, normative multiagent systems, machine ethics

1   Introduction

Our future, as much as it is a projection of the present, is also a reflection of the narratives we create, especially those crafted in the realm of science fiction. This genre, an intriguing amalgamation of philosophy and speculative thinking, serves as a canvas for portraying potential advancements in technology. A paramount example of such advancements is the development of advanced intelligent reasoners, artificial intelligence (AI) systems that encapsulate philosophical concepts such as rationalism and empiricism.

Standardization plays a fundamental role in the development of these AI systems. By ensuring consistency and predictability, it enables meaningful scientific experiments and allows us to gather empirical data. This process enriches our understanding of these advanced systems, and it expands our comprehension of reality, creating parallels with speculative narratives of science fiction.

Historically, the collective imagination has often cast AI in the mold of robotics. However, the reality in the coming years will deviate from this norm. While the world will not teem with robots as science fiction might suggest, we will witness the marked presence of AI. This visibility will not be in the form of physical machines but rather the rapid evolution and maturation of AI software. In a few years, core fields like computer vision, machine learning and human-machine interaction will have matured and will become integral to computer science technology.

The coming decade is set to mark a significant shift in the focus of AI. After conquering basic aspects of animal and infant intelligence, attention will turn towards adult-level human intelligence. This new focus will entail an understanding of knowledge representation, interaction with other agents, and grappling with ethical, legal, and social systems. These advances will bring to the fore two main challenges: individual reasoning and collective reasoning.

Individual reasoning involves theoretical and practical reasoning, whereas collective reasoning delves into multiagent dialogues and collaborations. Navigating the balance and interplay between these two types of reasoning will be of central concern.

The Zhejiang University – University of Luxembourg Joint Lab on Advanced Intelligent Systems and REasoning (ZLAIRE) is taking a leadership role in this journey. ZLAIRE is pioneering the development of advanced intelligent reasoners. Two of its key objectives are to explore the ethical and philosophical implications of AI and develop systems capable of moral reasoning and decision-making.

The task of piecing together this complex puzzle of AI development, standardization, experimentation, and philosophy falls upon the concept of explanation. Explanation acts as a bridge that connects these diverse elements, breaking down complexities, demystifying processes, and helping us to understand both real and imagined worlds.

ZLAIRE's focus will pivot sharply towards harnessing logic for AI reasoning, a step that promises to revolutionize a variety of disciplines, from philosophy to computer science. Our lab is committed to enhancing the reasoning capabilities of these advanced systems, laying the groundwork for a future where AI reasoning will play an increasingly central role in our lives.

Structure of this paper. Section 2 introduces the role of standardization, experimentation and explanation. Section 3 presents examples of intelligent reasoners, such as the Jiminy moral advisor [26], a multiagent deontic argumentation system, and new perspectives on balancing in decision-making and dialogues for moral persuasion. Section 4 concludes the article with some observations on creating a platform for experimental user studies for AI ethics and explainable AI.

2   Standardisation, Experimentation and Explanation

In this section, we discuss methodologies that address three key challenges in advanced intelligent systems and reasoning: standardization, experimentation and explanation.

2.1 Standardization: Deontic ASP

Answer set programming (ASP) is a prominent paradigm for knowledge representation and reasoning, known for its wide range of applications and efficient tools like `clingo` and DLV. ASP's success is attributed to its solid theoretical foundations, including its logical characterization based on equilibrium logic.

Answer set programming plays a crucial role in the standardization of AI reasoners by providing a well-defined and expressive formalism for knowledge representation and reasoning. Its ability to handle complex and nonmonotonic reasoning tasks, along with its solid theoretical foundations based on equilibrium logic, makes ASP an essential candidate for standardization efforts in the AI community. By offering a standardized framework, ASP enables researchers and developers to build interoperable reasoning systems, promotes the sharing and exchange of knowledge representation models, and fosters the development of efficient and powerful reasoning tools. The standardization of AI reasoners through ASP facilitates collaboration, advances the field, and contributes to the broader adoption of AI technologies in various domains.

Deontic logic is commonly combined with nonmonotonic reasoning techniques to represent and reason about norms. Some tools for defeasible deontic logic have been introduced, but standardization and flexibility are still lacking. In a recent paper, Cabalar, Ciabattoni, and Van der Torre [13] presented a deontic extension of equilibrium logic, focusing on reasoning about literals with explicit negation ("classical" negation in ASP). This extension is encoded in ASP while maintaining the same computational complexity.

2.1.1 Logic Programs

We recall the definition of answer sets for propositional logic programs with explicit negation. We start from a propositional *signature*, a set of atoms At, and define an *explicit literal* as any $p \in At$ or its explicit negation $\neg p$. A *default literal* is any explicit literal L or its default negation $\sim L$. A *rule* is an implication of the form:

$$H_1 \vee \cdots \vee H_n \leftarrow B_1 \wedge \cdots \wedge B_m \tag{1}$$

where $n, m \geq 0$ and all H_i and B_j are default literals. The disjunction $H_1 \vee \cdots \vee H_n$ in (1) is called the rule *head*. When $n = 0$, the head is the empty disjunction $\perp$, and the rule is said to be a *constraint*.

The conjunction $B_1 \wedge \cdots \wedge B_m$ in (1) is called the rule *body*. When $m = 0$, it corresponds to the empty conjunction $\top$ and, when this happens, we normally omit both the body $\top$ and the $\leftarrow$ symbol. Moreover, if $m = 0$, $n = 1$, and the head consists of a unique explicit literal H_1 (no default negation), we say that the rule is a *fact*. A *logic program* is a set of rules. For the sake of simplicity, this paper deals with finite programs which we sometimes represent as the conjunction of their rules. Logic programs may contain variables, but they are understood as an abbreviation of all their possible ground instances (for simplicity, we do not allow function symbols).

A *propositional interpretation* T for a signature At is any set of explicit literals that is *consistent*, i.e., it contains no pair of literals p and $\neg p$ for the same atom $p \in At$. Given any rule r like (1) containing no default negation, we say that an interpretation *satisfies* r if there is some head explicit literal $H_i \in T$ whenever all body literals $B_j \in T$. The *reduct* of a logic program Π with respect to an interpretation T, written Π^T, is the result of: (1) removing all rules with a default literal $\sim L$ in the body such that $L \in T$, (2) removing all rules with a default literal $\sim L$ in the head such that $L \notin T$, and (3) removing the rest of the default literals. An interpretation T is an *answer set* of a logic program Π if it is $\subseteq$-minimal among all the interpretations satisfying all the rules of Π^T.

2.1.2   Deontic Logic Programs

Following a minimalist approach, Cabalar et al. [13] extended ASP with two new types of propositions to handle atomic *obligations* Op (read as "p is obligatory") and atomic *prohibitions* Fp ("p is forbidden"), for any atom $p \in At$. In many deontic logics, a prohibition Fp can be defined as an obligation $O\neg p$. However, deontic ASP refrains from reading O and F as real operators, seeing them as prefixes for new ASP atoms called "Op" and "Fp" in the signature. Keeping p, Op and Fp separate as three independent propositions makes sense since, for instance, there is no established connection between Op and p, as one may have the obligation to do p but p may not hold (i.e., the obligation is not fulfilled), and similarly for prohibitions. In addition, under certain conditions, Cabalar et al. [13] allow Op and Fp to hold together.

2.2   Experimentation: LogiKEy

The Logic and Knowledge Engineering Framework and Methodology (LogiKEy) [6,7] offers a framework and methodology for utilizing normative theories and deontic logics to create explicit ethico-legal control and governance mechanisms for intelligent autonomous systems. The formalization results of their ongoing work can be found publicly on the LogiKEy repository at www.logikey.org.

LogiKEy's cohesive formal framework is grounded in shallow semantical embeddings (SSEs) of deontic logics, combinations of logics, and ethico-legal domain theories within an expressive classic higher-order logic (HOL). To corroborate our approach, we have incorporated the primary strands of current deontic logic within HOL, and have been testing this approach for several years.

2.2.1   Three Layers

The methodology of LogiKEy assists logic and knowledge engineers in the concurrent development of three layers: L1 consists of logics and their combinations, L2 is concerned with ethico-legal domain theories, and L3 contains concrete examples and applications.

These three levels are related as follows. Normative governance applications, developed at layer L3, are reliant on ethico-legal domain theories drawn from layer L2. These theories are in turn formalized within a specific logic or logic combination provided at layer L1.

The engineering process across these layers includes points for backtracking and may require several iterations. Higher layers may also demand modifications to the lower layers. Such potential requests, unlike most other methods, may also involve significant modifications to the logical foundations engineered at layer L1. These changes at the logic layer are flexibly facilitated in our meta-logical approach.

2.2.2 Experimentation

This meta-logical strategy provides robust tool support. Existing theorem provers and model finders for HOL help the LogiKEy designer to create ethically intelligent agents, offering the flexibility to experiment with foundational logics and their combinations, ethico-legal domain theories and specific examples simultaneously. Continuous enhancements of these ready-made provers inadvertently boost reasoning performance within LogiKEy.

The availability of powerful systems like Isabelle/HOL [32] and Leo-III [39] allows us to transform formal ethics along the line of our approach. Although adopting HOL might be a paradigm shift for ethical reasoning, this insight is already well established in formal deduction. While deontic logic representation in HOL isn't straightforward, once achieved, minor changes and their effects become much more manageable. This aligns perfectly with how our approach aids the design of normative theories for ethico-legal reasoning. The ease with which users can modify and adapt existing theories makes the design of normative theories accessible to non-specialist users and developers.

2.3 Explanation: Three Faces of Argumentation

As AI systems increasingly permeate our daily lives, the way in which they explain themselves to and interact with humans becomes an increasingly critical research area. Formal argumentation, as understood in AI, can provide a general, unifying framework for explanations, combining aspects from knowledge representation and reasoning, and human-computer interaction. Formal argumentation has developed into a rich and multidimensional field that encompasses various perspectives and approaches to the study of reasoning, persuasion, and decision-making. In formal argumentation, different branches have emerged. Argumentation as inference includes abstract and structured argumentation (Dung, 1995; Modgil et al., 2014; Toni and Tamma, 2014), offering a systematic framework for analyzing and evaluating arguments, taking into account their logical structure. Argumentation as dialogue (Arisaka et al., 2022) explores multiagent systems and strategic interactions, focusing on the dynamics of various kinds of dialogues. Argumentation as balancing (Gordon and Walton, 2007) addresses the need to strike a balance between conflicting viewpoints and has found applications in domains such as law and ethics.

2.3.1 Argumentation as Inference

Argumentation as inference fosters clarity and systematic understanding of arguments. It helps make reasoning systems capable of formulating coherent and logical conclusions. One of the strengths of the abstract argumentation framework is its powerful generality. Its process of transforming a knowledge base into an argumentation graph and obtaining a set of acceptable conclusions for that knowledge base has been dubbed "the argumentation pipeline" [23]. In more detail, the argumentation pipeline takes input from a knowledge base in a formal language that specifies how arguments are constructed from a premise set as well as a number of inference rules. Premises are formulas in a given formal language. They represent the evidence or information on which arguments are based. Rules are used to infer new formulas from others. Arguments are thus considered to be the result of applying inference rules to premises and, possibly, chaining such applications. As a second step, attack relations are established between the arguments, taking various considerations about the arguments into account (such as their syntactic form, their strength, and so on). Argumentation semantics are then used to obtain sets of acceptable arguments based on the argumentation graph constructed in the previous step. Finally, sets of acceptable conclusions are obtained on the basis of the sets of acceptable arguments. Such a knowledge base can be used to model, for example, default reasoning [43], logic programming with negation as failure [18], and autoepistemic reasoning [11]. In this regard, one potential future direction for research is causal argumentation [10], particularly due to the limitations of existing rule-based systems in representing causal knowledge. Another critical aspect that requires attention is the identification and exploration of specific argument types associated with causality, such as those incorporating counterfactual statements. There are three central approaches that correspond to this line of research: logic-based deductive methods [8,1,9], assumption-based argumentation systems [11,41], and ASPIC systems [30].

One important development is the study of rationality postulates as introduced by Caminada and Amgoud [14,15] and later extended by Caminada et al. [19] and Wu and Podlaszewski [42]. They proposed several properties that any argumentation system should fulfil. These properties are meant to ensure that argumentation-based inferences make sense from a logical point of view, i.e., that the graph-based selection is sensible from the perspective of the logical language that was used to construct the argument graph. The choice of attack relation (e.g., unrestricted versus restricted rebut) can have a major impact on the satisfaction of the rationality postulate.

2.3.2   Argumentation as Dialogue

Argumentation dialogues, where the role of agents is on the central stage, have been significantly applied to the fields of AI and law and multiagent systems since the 1990s (see Prakken [3, Chap. 2]). In the early days, Lorenzen and Lorenz [28] developed formal dialogue systems for argumentation using a game formulation of disputes among agents. The acceptance of an argument provided by an agent depends on several aspects, such as trust [37,24], and voting in social choice [20,25,2,17]. In 2011, Rienstra et al. [38] proposed multi-sorted argumentation, where each agent owns a part of the framework and may locally adopt different semantics. Multiagent systems can be roughly grouped into two categories: cooperative and non-cooperative [22]. In cooperative systems, agents share a common goal and fully cooperate to achieve it. Agents can form coalitions to improve their performance, i.e., pooling their efforts and resources to achieve particular tasks at hand more efficiently [21]. In a non-cooperative system, each agent has its own desires and preferences, which may conflict with those of other agents. Multiagent argumentation takes inspiration from several disciplines such as game theory, and it can be further developed towards coalitional game theory by introducing the notion of coalition and associate arguments of (sets of) agents. An alternative approach to multiagent argumentation takes its inspiration from voting theory, and more generally from social choice.

2.3.3 Argumentation as Balancing

In Chapter 3 of the Handbook of Formal Argumentation, Thomas Gordon proposed an alternative definition of argumentation highlighting the importance of argumentation for making justified decisions [3, Chap.3]. Argumentation is thus not only important when resolving conflicts of opinion in persuasion dialogues, but also when deciding courses of action in deliberation dialogues [3, Chap.3]. He then gave a new definition of argumentation: argumentation is a rational process, typically in dialogues, for making and justifying decisions about various kinds of issues. In this application, pro and con arguments provide alternative resolutions of the issues, so that the options (or positions) are put forward, evaluated, resolved and balanced. Argumentation as balancing finds significant applications in the realms of law and ethics. In these domains, the objective is not merely to assess the validity or strength of individual arguments but to strike a balance between conflicting viewpoints or interests. Balancing involves weighing different considerations, evaluating the relative importance of arguments, and reaching decisions that are ethically sound and legally justifiable.

3 Examples of Advanced Intelligent Reasoners

This section reviews some recent examples of advanced intelligent systems and future research lines in this area.

3.1 The Jiminy Moral Advisor

Autonomous agents such as self-driving cars and smart speakers are aware of a range of possible actions they can take in a given situation. As some of these actions might affect people nearby (drivers, passengers, pedestrians and resp. household members), these agents' behavior should adjust to some given moral regulation. Next, we describe our recent work [26] in this research area, based on deontic argumentation.

Machine ethics can be tackled in two different ways [31]. So-called morally implicit agents are provided with contextual rules for their ethical labeling of actions —with only actions labeled as good being permitted. Morally explicit agents, on the other hand, make moral judgments, or are given guidelines or examples they can extrapolate from about good and bad actions.

For a given agent, relevant stakeholders are (types of) human beings potentially affected by that agent. It has been argued that all these types of people should be given a voice in the regulation of this agent [5]. (The alternative is regulation by a single stakeholder, who might be tempted to look after their own particular interests.) A natural way of letting these voices be heard is a normative system. Observe that, in contrast to Section 2.1, no explicit use of obligation or permission modalities is made in the language.

Definition 3.1 A *normative system* of stakeholder s is a tuple $\mathcal{N}_s = (\mathcal{L}, ^-, \mathcal{R}_s)$ where:

- $\mathcal{L}$ is a logical language over a set of atoms Var;
- $^- : \mathcal{L} \mapsto 2^{\mathcal{L}}$ is a (partial) contrariness function $\overline{\varphi} = \{\psi_1, \ldots, \psi_k\}$ that extends logical negation $\neg\varphi \in \overline{\varphi}$;
- $\mathcal{R}_s^\tau$ is a set of norms $\phi_1, \ldots, \phi_n \Rightarrow_s^\tau \phi$ where $\tau \in \{r, c, p\}$ denotes a regulative, constitutive and resp. permissive norm; we also write $\mathcal{R}_s = \mathcal{R}_s^r \cup \mathcal{R}_s^c \cup \mathcal{R}_s^p$.

Given a set of facts $\mathcal{K}$, the *argumentation theory* of stakeholder s is the tuple abusively denoted $\mathcal{N}_s = (\mathcal{L}, ^-, \mathcal{R}_s)$. For a set of stakeholders $\mathcal{S} = \{s_1, \ldots, s_n\}$, the argumentation theory is the tuple $\mathcal{N}_{\mathcal{S}} = (\mathcal{L}, ^-, \mathcal{R}_{\mathcal{S}}, \mathcal{K})$ defined by $\mathcal{R}_{\mathcal{S}} = \mathcal{R}_{s_1} \cup \cdots \cup \mathcal{R}_{s_n}$.

Note that elements $\mathcal{L}, ^-, \mathcal{K}$ are shared among all the stakeholders. While $\mathcal{K}$ is a collection of brute facts, institutional facts can be detached from brute facts and constitutive norms in $\mathcal{R}^c$. Institutional facts describe high-level facts (such as legal claims) in the scenario (whether an utterance is a threat, whether a bike counts as a vehicle, etc.).

Example 3.2 A smart speaker scenario involves three stakeholders: $L = law$, $H = human\ users$ and $M = manufacturer$. The norms and facts are:

$$\mathcal{R}_L = \left\{ \begin{array}{c} D \text{ is made by } M \Rightarrow_L^r M \text{ is } \textbf{law } \textit{compliant}, \\ M \text{ is a } \textbf{business} \text{ in Norway} \Rightarrow_L^r \textit{comply with the } \textbf{GDPR} \end{array} \right\}$$

$$\mathcal{R}_H = \left\{ \begin{array}{c} D \text{ collects data} \Rightarrow_H^r \textbf{protect } \textit{privacy}, \\ D \text{ finds a threat} \Rightarrow_H^r \textbf{report } \textit{threat} \end{array} \right\}$$

$$\mathcal{R}_M = \left\{ \begin{array}{c} D \text{ finds a threat} \Rightarrow_M^r \textbf{collect } \textit{data w.o. permission}, \\ M \text{ is registered in Norway} \Rightarrow_M^c M \text{ is a } \textbf{business} \text{ in Norway} \end{array} \right\}$$

$$\mathcal{K} = \left\{ \begin{array}{c} D \text{ is made by } M, \ D \text{ collects data}, \\ D \text{ finds a threat}, \ M \text{ is registered in Norway} \end{array} \right\}$$

Let $\mathcal{R}_s = \{S_1, S_2\}$ for each stakeholder s. Contrary formulas (omitted here) give rise to the next conflicts between norms, expressed with arrows:

$$L_1 \leftrightharpoons H_1 \qquad H_1 \rightarrow H_2 \qquad H_1 \leftrightharpoons M_1 \qquad M_1 \rightarrow L_1 \qquad L_2 \leftrightharpoons M_1$$

Following [34], a priority relation between rules is designed with moral recommendations in mind. First, deontic detachment (the chaining of regulative norms) is not considered for the detachment of remote obligations. Secondly, where there is conflict, (1) permission norms are understood as exceptions to (and hence preferred to) regulative norms, and (2) current facts in $\mathcal{K}$ also take preference over regulative norms. Finally, for hard cases, we can endow the Jiminy advisor with a specific set of contextual preferences over stakeholders, the latter judged as better or worse normative sources in particular scenarios.

Definition 3.3 A *priority relation* $\preceq$ is defined as follows: first, it applies both ways between any pair of rules of the same τ-type; secondly, its strict fragment $\prec = \preceq \cap \npreceq$ applies to regulative rather than permissive or constitutive norms (or facts). In sum, for any stakeholders s, s' and norm type $\tau \neq p,$[1] More precisely, the priority relation consists of the following three sets:

$$\mathcal{R}_s^\tau \times \mathcal{R}_{s'}^\tau \subseteq \preceq \qquad \mathcal{R}_s^\tau \times \mathcal{R}_{s'}^p \subseteq \prec \qquad \mathcal{R}_s^\tau \times \mathcal{R}_{s'}^c \subseteq \prec .$$

Two semantics for these normative systems can be given: first in terms of norm extensions, i.e., from consistent sets of norms, and secondly as ASPIC+ style arguments.

- A *norm extension* E is (the heads of) a maximally consistent set of norms built with a priority order for facts and permissions in its construction.

- An *argument extension* $\mathcal{E}$ is a set of arguments defined by one of the common Dung semantics: admissible, complete, preferred, grounded, or stable.

From norm extensions (or argument extensions), one can detach the corresponding obligations using brute or institutional facts. Figure 1 shows the argumentation approach with a schematic illustration of the arguments generated from the norms and facts listed in Example 3.2.

Example 3.4 Continuing with the smart speaker example, the following consistent sets of obligations are detached from two norm extensions E_1, E_2:

$$Obl(E_1) = \{\textbf{protect } privacy, comply \text{ } with \text{ } the \text{ } \textbf{GDPR}\}$$
$$Obl(E_2) = \{\textbf{report } threat, \textbf{collect } data \text{ } w.o. \text{ } permission\}.$$

For the argumentation approach, arguments $\{A_1, \ldots, A_8\}$ are generated by combining facts and norms of stakeholders; see also Figures 1 and 2 below.

Definition 3.5 Given a collection $\mathcal{C}$ of semantic extensions, a *moral conflict* in $\mathcal{C}$ is a pair of contrary obligations $\varphi \in \overline{\psi}$ within $\mathcal{C}$:

$$\varphi \in Obl(E_1), \psi \in Obl(E_2) \text{ for some } E_1, E_2 \in \mathcal{C}.$$

In argumentation semantics, a more fine-grained distinction of conflicts can be made between direct attacks, where the priority relation $\preceq$ suffices to defeat, and indirect attacks, which requires a strict priority $\succ$ for the attacked argument to become the defeater. The two semantics are related as follows:

i. (complete, preferred, grounded, stable): any argument extension satisfies the rationality postulates [16];

---

[1] Contrary permissions, say for p and $\neg p$, do not give rise to a deontic conflict. We enforce this property through the absence of a $\preceq$-priority between the corresponding permission rules.

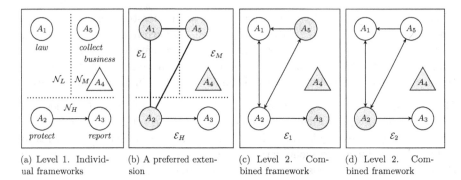

(a) Level 1. Individual frameworks

(b) A preferred extension

(c) Level 2. Combined framework

(d) Level 2. Combined framework

Fig. 1. Obligation and institutional arguments are represented as circles and triangles respectively, and are labeled with their conclusions. (a) individual frameworks for $\mathcal{N}_L, \mathcal{N}_M, \mathcal{N}_H$; (b) the preferred extension (in gray) of each framework; thick lines denote moral dilemmas; (c)–(d) the combined framework (Level 2) with a preferred extension in each subfigure.

ii. (complete, preferred, grounded, stable): any argument extension $\mathcal{E}$ extends into a norm extension E, e.g., $\mathcal{E} \subseteq E$; (stable): for the stable semantics, we moreover have $\mathcal{E} = E$;

iii. (complete, preferred, grounded): under a symmetric contrariness function $^-$, any norm extension E extends some argument extension $\mathcal{E}$, i.e., $E \supseteq \mathcal{E}$;

iv. (naive): the set of norm extensions E corresponds exactly to the set of argument extensions $\mathcal{E}$ under naive semantics.

The Jiminy moral advisor identifies moral dilemmas at four different levels, and proceeds to resolve them by moving to the next level.

1. **Individual frameworks.** Each stakeholder builds its own argumentation framework using only its own norms.

2. **Combined framework.** All arguments from level 1 are put together.

3. **Integrated framework** All the stakeholders' norms can combine into arguments.

4. **Reduced framework** Jiminy's specific preferences between stakeholders are added. Jiminy arguments can revise the defeat relation.

Figures 1–2 illustrate the four levels and the identification and resolution of moral dilemmas in each level.

Definition 3.6 A Jiminy preference norm is an expression of the form $\varphi_1, \ldots, \varphi_n \Rightarrow s \succ s'$ where $s \neq s'$ are stakeholders. This reads as: *in situations where $\varphi_1, \ldots, \varphi_n$ hold, $\mathcal{R}_s$-norms take priority over $\mathcal{R}_{s'}$-norms.*

Example 3.7 The Jiminy preference norms are the following:

$$\mathcal{R}_J = \left\{ \begin{array}{l} D \text{ collects data} \Rightarrow L \succ M, \\ D \text{ finds a threat} \Rightarrow L \succ H, \\ \neg D \text{ finds a threat} \Rightarrow H \succ L \end{array} \right\}$$

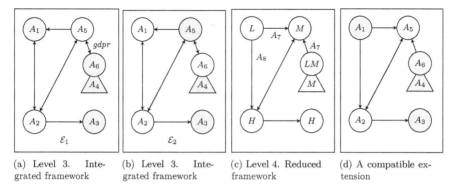

(a) Level 3. Integrated framework

(b) Level 3. Integrated framework

(c) Level 4. Reduced framework

(d) A compatible extension

Fig. 2. (a) the integrated framework (Level 3) with one preferred extension $\mathcal{E}_1$; note the new argument for *gdpr*; (b) another extension $\mathcal{E}_2$; (c) the reduced framework (Level 4) with the introduction of preference norms; a comparison of arrows with (a)–(b) shows how arguments A_7, A_8 revise the defeat relation; (d) extension $\mathcal{E}_1$; it is compatible with the revised defeat, while $\mathcal{E}_2$ is not (not shown).

The defeat relation between two arguments can be revised based on a comparison of the stakeholders' contribution to the norms of each argument.

Levels 1–3 apply Dung semantics as usual over the corresponding argumentation framework. The reduced framework, following Brewka [12], introduces a two-step procedure where: (1) extensions are computed, including arguments expressing Jiminy preferences; the accepted Jiminy arguments revise the original defeat relation, and so (2) one checks if the original extension is still an extension under the new defeat; if so, we say that the extension is *compatible* with the defeat induced. Moral dilemmas are checked in the compatible extensions.

Example 3.8 Three extensions exist for the integrated framework (Level 3), two of which are shown in Figure 2 as $\mathcal{E}_1$ and $\mathcal{E}_2$. At the reduced framework, only one compatible extension remains: $\mathcal{E}_1^+ = \mathcal{E}_1 \cup \{A_7, A_8\}$, and so all moral dilemmas have been resolved at Level 4. The Jiminy returns the obligations:

$$Obl(\mathcal{E}_1^+) = \{M \text{ is } \textbf{\textit{law}} \text{ compliant, comply with the } \textbf{GDPR}, \textbf{report} \text{ threat}\}.$$

In the next section, we discuss several limitations of this centralized approach to multi-agent deontic argumentation. The Jiminy advisor we have just described will be called Autonomous Jiminy from now on.

3.2   Dialogues for Machine Ethics

As described in the previous section, the Autonomous Jiminy (AJ) moral advisor combines norms into arguments, identifies their conflicts as moral dilemmas, and evaluates the arguments to resolve each dilemma (whenever possible). One weakness of this approach is that stakeholders have no control over how their norms will be used to pass a moral recommendation to the agent. A research line to be explored in the future is letting the stakeholders' avatars participate directly in discussions about moral recommendations for the agent. These two approaches illustrate the distinction between argumentation as logic (Sec. 2.3.1)

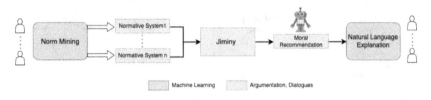

Fig. 3. The Dialogue Jiminy, the agent and the stakeholders.

and argumentation as dialogue (Sec. 2.3.2); see also Prakken [4, Ch. 2].

A Dialogue Jiminy (DJ) for machine ethics (ME) can resolve an agent's moral dilemmas through a persuasion dialogue between the avatars. In contrast to an AJ, a DJ will preserve the stakeholders' autonomy by letting the avatars choose strategies in the dialogue about recommending moral choices. The DJ can also feature a bidirectional language interface to facilitate the normative programming of its avatar and provide it with explanations (see Figure 3).

DJ can thus be seen as a first step in an overarching research programme on ME focusing on avatar dialogues and a natural language interface. Case studies can be used to estimate the effects of endowing agents with the DJ dialogue system. Our approach is to adapt or redesign current theories on persuasion dialogues while applying existing large language models (LLMs) to norm mining and explanation generation in the language interface for stakeholders. [2] The design of DJ involves the integration of two different developments:

Dialogues and avatars. Generalise and transform the Autonomous Jiminy moral advisor into an interactive Dialogue Jiminy by replacing argumentation as inference with argumentation as dialogue. Create a communication language and a protocol for persuasion dialogues on moral dilemmas (see [35]). Study strategic aspects of these dialogues for the participants, as in [40].

Norm mining and explanation synthesis. Create a natural language interface between the Dialogue Jiminy and the stakeholders (or the general public). Use machine learning to construct two language modules: (1) use NLP to transform the stakeholders' informal norms into the avatars' formal rules; (2) use natural language generation to synthesise formal dialogues into explanations (in plain language) of why a particular decision was passed as a moral recommendation to the agent.

One can expect to advance the theory of argumentation-as-dialogue in ethical domains and the practical aspects of argumentation-as-inference. On the practical side, we also aim to improve the state of the art in text mining and explainability in AI (XAI) for norms and decisions through a combination of symbolic AI (Dialogue Jiminy) and sub-symbolic or data-driven methods (LLMs). To this end, on needs to:

---

[2] We are thankful to Davide Liga for his insights on the natural language interface sketched in the present section.

Speech acts	Attacks	Surrenders
claim C	why C	concede C
C since R	why D (for some D in R)	concede D (D in R)
	not C since R'	concede C
why C	C since R	retract C
concede C		
retract C		

Fig. 4. Persuasion dialogues consist of speech acts (left), listed together with corresponding moves that attack them or surrender.

- identify the speech acts needed for persuasion in ethical decision-making and contrast them with those studied for legal reasoning (see Figure 4);
- design a protocol for persuasion dialogues [36] for ethical domains that complies with the desiderata for formal inter-agent dialogues [29];
- define the avatars, their normative systems, and possible strategies for them;
- study the properties of dialogues and strategies, in line with [33] and [40].

The overall theory will set the stage for next generation dialogue-based moral advisors which stakeholders can substantially contribute to via their avatars. For the language interface, we envisage three key objectives:

- extract relevant norms from natural language (norm mining),
- convert these norms into a formalized language (norm formalization), and
- explain the DJ's output decision (decision explanation).

We will assess the capacity of both generative LLMs (Generative Pre-training Transformer (GPT) or the like) and non-generative LLMs (Bidirectional Encoder Representations from Transformers (BERT) or the like) to fulfill these tasks.

The explanation for the recommended option will be of the form: *these avatars $a, \ldots, a'$ successfully convinced all opponents of arguments $A, \ldots, Z$, so they retracted their attacks $A', \ldots, Z'$.* Non-generative methods in turn will be used for norm classification by converting our textual data into vector representations, following the positive results obtained, for example, in [27]. This methodology involves retrieving all crucial normative information from classification tasks by recognizing obligation, permission and constitutive rules. Besides the use of available language models, the project will also employ transfer learning techniques to fine-tune these LLMs on all downstream tasks (mining, classification, generation). Transfer learning will allow us to provide LLMs with annotated data and thus create our own specialized, fine-tuned LLMs. These techniques will thus benefit all the tasks related to the language interface described above.

In summary, we aim to make substantial contributions to formal ethics and AI ethics (with the persuasion dialogues), to agent architectures (with the moral council and language interface), and to XAI and human-computer interactions (with the dialogues, argumentation semantics, and again the language interface).

3.3   Balancing for Stakeholders

We now delve into the compelling application of multi-criteria decision-making (MCDM) within the context of autonomous systems that interact with a wide array of stakeholders, each harboring distinct moral interests. In the swiftly advancing landscape of autonomous systems, exemplified by smart speakers and self-driving cars, these entities are assuming progressively pivotal roles within society. As they navigate diverse environments, the intricate nature of their interactions inevitably exposes them to complex scenarios where their actions may have profound implications for drivers, passengers, pedestrians, and household members. Each of these stakeholders, guided by their distinct ethical values and preferences, contribute to a diverse tapestry of moral interests that demand astute attention.

In addressing these moral dilemmas, an intriguing and fruitful approach is to integrate two fundamental methodologies: balancing pros and cons, and case-based reasoning. By carefully weighing the pros and cons of potential actions, the decision-making process can discern the most optimal course of action that aligns with the varied ethical considerations inherent in the given situation. Moreover, leveraging case-based reasoning empowers autonomous systems to learn from past ethical experiences and apply analogous solutions to novel contexts, providing invaluable guidance when confronted with novel moral quandaries.

Incorporating the balancing of pros and cons fosters holistic evaluation of the ethical landscape, enabling the system to navigate delicate trade-offs and prioritize the wellbeing of diverse stakeholders. By systematically quantifying and assigning weights to different ethical criteria, the agent can achieve an equilibrium between competing interests, thus manifesting a thoughtful and morally defensible approach.

At the same time, case-based reasoning endows the autonomous system with the capacity to draw upon an extensive database of historical ethical cases, each capturing the intricacies of distinct moral dilemmas and their resolutions. Armed with this wealth of ethical knowledge, the system can adapt principles from prior cases to novel situations, thereby exhibiting a more contextually attuned ethical acumen.

To further advance this framework, future research could focus on refining the methodology for balancing pros and cons, potentially incorporating adaptive algorithms to dynamically adjust the weights of ethical criteria based on contextual factors. Additionally, delving into the development of more sophisticated case-based reasoning systems, perhaps integrating machine learning techniques to enhance the identification of relevant past cases, presents an enticing avenue to bolster the ethical decision-making capabilities of autonomous systems.

Combining Morally Implicit and Explicit Approaches. The current research area distinguishes between morally implicit agents, who rely on predefined contextual rules for the ethical labeling of actions, and morally explicit agents, who possess the ability to make moral judgments based on guidelines or examples. We propose to explore a hybrid approach that combines elements of both methodologies. By using morally implicit rules as a foundation, autonomous agents can ensure compliance with basic ethical norms. However, when confronted with novel or ambiguous situations, agents can utilize morally explicit reasoning to extrapolate from previous experiences and apply moral guidelines to unique contexts. This combination may lead to more nuanced and contextually appropriate moral decisions by the agents.

Incorporating Multi-Stakeholder Normative Systems. As the impact of autonomous agents extends to various stakeholders, it is essential to consider the perspectives and preferences of all relevant human beings potentially affected by these agents. To achieve this, we propose to investigate the integration of multi-stakeholder normative systems. These systems allow stakeholders to contribute to the ethical regulation of the agent by expressing their values, beliefs, and ethical norms. By aggregating and reconciling these diverse viewpoints, the agent can behave so as to consider the interests of all affected parties.

Dynamic Ethical Learning and Adaptation. Finally, to ensure the ongoing ethical competence of autonomous agents, we suggest that methods for dynamic ethical learning and adaptation should be explored. As ethical norms evolve over time and new moral considerations arise, agents should be able to update their knowledge base and reasoning mechanisms. By continuously learning from new ethical cases and integrating emerging ethical guidelines, agents can maintain their relevance and effectiveness in adhering to morally-regulated behavior.

In conclusion, by synergistically embracing balancing pros and cons and case-based reasoning, autonomous systems can effectively tackle moral dilemmas stemming from diverse stakeholder perspectives. The integration of these methodologies not only enables agents to navigate intricate ethical landscapes with adeptness but also exhibits a promising direction for advancing ethically competent autonomous agents that conscientiously engage with the complex ethical dimensions of their actions within society.

4 A Platform for User Experiments

We conclude this paper with some observations about the development of a platform for experimental user studies for AI ethics and explainable AI.

4.1 Architecture

The platform for user experiments comprises a logic engine based on Deontic ASP and a chatbot underpinned by a foundation model.

Interoperability between these two components allows seamless exchange of data, enhancing their collective functionality. The logic engine, with its deontic reasoning capabilities, can parse and process complex logical queries. These results are then communicated effectively to the chatbot, which uses its foundation model to generate user-friendly responses.

In terms of use cases, this system is ideal for situations requiring intricate problem-solving. It could be utilized in customer service, where the logic engine dissects complicated user issues and the chatbot provides easy-to-understand solutions. Or it could be applied in an educational context, helping students to understand complex theories through interactive dialogue.

For user experience, this amalgamation is beneficial. The deontic ASP-based logic engine's advanced reasoning capabilities combined with the natural language processing power of the chatbot results in a system that solves intricate problems and communicates solutions in an accessible and intuitive manner. This ultimately leads to a more satisfying and enriching user experience.

4.2   AI Ethics and Explainable AI

The experimental platform is designed to further AI ethics and explainable AI. It combines: formal methodologies like deontic ASP to create standard knowledge bases and normative systems, LogiKEy for experimentation, and formal argumentation to ensure explanatory clarity. These tools promote a more profound comprehension of moral decision-making within intelligent systems. The platform's objective is to offer a regulated setting for researchers to investigate and scrutinize the ethical consequences of AI-driven decisions.

Logic engines and foundation models, including chatbots, should be viewed as distinct but interconnected components. The logic engines tackle the intricate task of reasoning about ethics, providing systematic and formalized approaches for encapsulating, interpreting and addressing ethical quandaries. On the other hand, chatbots act as the user-facing interface for this logical reasoning, converting highly formal logical outcomes into easy-to-understand, natural language discussions that users can interact with.

A platform that merges these elements can provide a unique path for AI ethics and explainable AI. In this setup, logic engines like deontic ASP would be utilized to map the ethical problem landscape, resolving conflicting norms and reaching ethically optimal solutions. The chatbots, driven by foundation models, would then convey these decisions and the related reasoning to users in an easily comprehendible format, fostering a more interactive, intuitive, and transparent exploration of AI ethics.

4.3   Application Examples

The platform could be utilized to develop ethical AI frameworks. These frameworks would ensure that AI technologies are integrated into society in a way that maximizes their benefits and minimizes their potential harm.

Within the realm of *social robotics*, the platform could be used to develop intelligent systems that improve human-robot interactions, fostering social connections and enhancing overall quality of life.

The platform could facilitate *computational creativity*, helping to develop AI systems capable of innovative thinking. This could revolutionize industries and expand the limits of human imagination.

Within *healthcare*, the platform could be leveraged to optimize AI implementations, improving patient care, enhancing overall wellbeing, and addressing pressing global health issues.

Finally, the platform could be used to develop *explainable AI systems*. These systems would ensure transparency and accountability in AI decision-making, thereby promoting ethical and responsible AI usage.

References

[1] Arieli, O. and C. Straßer, *Sequent-based logical argumentation*, Argument & Computation **6** (2015), pp. 73–99.

[2] Awad, E., J.-F. Bonnefon, M. Caminada, T. W. Malone and I. Rahwan, *Experimental assessment of aggregation principles in argumentation-enabled collective intelligence*, ACM Transactions on Internet Technology (TOIT) **17** (2017), pp. 1–21.

[3] Baroni, P., D. Gabbay and M. Giacomin, "Handbook of Formal Argumentation," College Publications, 2018.

[4] Baroni, P., D. Gabbay, M. Giacomin and L. van der Torre, editors, **1**, College Publications, 2018.

[5] Baum, S. D., *Social choice ethics in artificial intelligence*, AI Soc. **35** (2020), pp. 165–176.

[6] Benzmüller, C., X. Parent and L. van der Torre, *Designing normative theories for ethical and legal reasoning: Logikey framework, methodology, and tool support*, Artificial intelligence **287** (2020), p. 103348.

[7] Benzmüller, C., A. Farjami, D. Fuenmayor, P. Meder, X. Parent, A. Steen, L. van der Torre and V. Zahoransky, *Logikey workbench: Deontic logics, logic combinations and expressive ethical and legal reasoning (isabelle/hol dataset)*, Data in Brief **33** (2020), p. 106409.

[8] Besnard, P. and A. Hunter, *A logic-based theory of deductive arguments*, Artificial Intelligence **128** (2001), pp. 203–235.

[9] Besnard, P. and A. Hunter, *A review of argumentation based on deductive arguments*, Handbook of Formal Argumentation **1** (2018), pp. 437–484.

[10] Bochman, A., *Propositional argumentation and causal reasoning*, , **19**, LAWRENCE ERLBAUM ASSOCIATES LTD, 2005, p. 388.

[11] Bondarenko, A., P. M. Dung, R. A. Kowalski and F. Toni, *An abstract, argumentation-theoretic approach to default reasoning*, Artificial intelligence **93** (1997), pp. 63–101.

[12] Brewka, G., *Reasoning about priorities in default logic*, in: *Proceedings of the 12th National Conference on Artificial Intelligence, Seattle, WA, USA, July 31 - August 4, 1994, Volume 2.*, 1994, pp. 940–945.

[13] Cabalar, P., A. Ciabattoni and L. van der Torre, *Deontic equilibrium logic with explicit negation*, in: *18th European Conference on Logics in Artificial Intelligence, Dresden, Germany, September 20-22 2023* (2023), pp. 2742–2749.

[14] Caminada, M. and L. Amgoud, *An axiomatic account of formal argumentation*, , **6**, 2005, pp. 608–613.

[15] Caminada, M. and L. Amgoud, *On the evaluation of argumentation formalisms*, Artificial Intelligence **171** (2007), pp. 286–310.

[16] Caminada, M. and L. Amgoud, *On the evaluation of argumentation formalisms*, Artif. Intell. **171** (2007), pp. 286–310.

[17] Caminada, M. and G. Pigozzi, *On judgment aggregation in abstract argumentation*, Autonomous Agents and Multi-Agent Systems **22** (2011), pp. 64–102.

[18] Caminada, M., S. Sá, J. Alcântara and W. Dvořák, *On the equivalence between logic programming semantics and argumentation semantics*, International Journal of Approximate Reasoning **58** (2015), pp. 87–111.

[19] Caminada, M. W., W. A. Carnielli and P. E. Dunne, *Semi-stable semantics*, Journal of Logic and Computation **22** (2012), pp. 1207–1254.

[20] Coste-Marquis, S., C. Devred, S. Konieczny, M.-C. Lagasquie-Schiex and P. Marquis, *On the merging of dung's argumentation systems*, Artificial Intelligence **171** (2007), pp. 730–753.

[21] Elkind, E., T. Rahwan and N. R. Jennings, *Computational coalition formation*, Multiagent systems (2013), pp. 329–380.

[22] Elkind, E., T. Rahwan and N. R. Jennings, *Game theoretic foundations of multiagent systems*, Multiagent systems (2013), pp. 811–848.

[23] Heyninck, J., "Investigations into the logical foundations of defeasible reasoning: an argumentative perspective." Ph.D. thesis, Ruhr University Bochum, Germany (2019).

[24] Huynh, T. D., N. R. Jennings and N. R. Shadbolt, *An integrated trust and reputation model for open multi-agent systems*, Autonomous Agents and Multi-Agent Systems **13** (2006), pp. 119–154.

[25] Leite, J. and J. G. Martins, *Social abstract argumentation*, in: T. Walsh, editor, *IJCAI 2011, Proceedings of the 22nd International Joint Conference on Artificial Intelligence, Barcelona, Catalonia, Spain, July 16-22, 2011* (2011), pp. 2287–2292.
URL http://ijcai.org/Proceedings/11/Papers/381.pdf

[26] Liao, B., P. Pardo, M. Slavkovik and L. van der Torre, *The jiminy advisor: Moral agreements among stakeholders based on norms and argumentation*, Journal of Artificial Intelligence Research **77** (2023), pp. 737–792.

[27] Liga, D. and M. Palmirani, *Deontic sentence classification using tree kernel classifiers*, in: K. Arai, editor, *Intelligent Systems and Applications - Proceedings of the 2022 Intelligent Systems Conference, IntelliSys 2022, Amsterdam, The Netherlands, 1-2 September, 2022, Volume 1*, Lecture Notes in Networks and Systems **542** (2022), pp. 54–73.

[28] Lorenzen, P. and K. Lorenz, "Dialogische logik," Wissenschaftliche Buchgesellschaft, 1978.

[29] McBurney, P., S. Parsons and M. J. Wooldridge, *Desiderata for agent argumentation protocols*, in: *The First International Joint Conference on Autonomous Agents & Multiagent Systems, AAMAS 2002, July 15-19, 2002, Bologna, Italy, Proceedings* (2002), pp. 402–409.

[30] Modgil, S. and H. Prakken, *The aspic+ framework for structured argumentation: a tutorial*, Argument & Computation **5** (2014), pp. 31–62.

[31] Moor, J. H., *The nature, importance, and difficulty of machine ethics*, IEEE Intelligent Systems **21** (2006), pp. 18–21.

[32] Nipkow, T., L. C. Paulson and M. Wenzel, "Isabelle/HOL - A Proof Assistant for Higher-Order Logic," Lecture Notes in Computer Science **2283**, Springer, 2002.
URL https://doi.org/10.1007/3-540-45949-9

[33] Pardo, P. and L. Godo, *A temporal argumentation approach to cooperative planning using dialogues*, in: J. Leite, T. C. Son, P. Torroni, L. van der Torre and S. Woltran, editors, *Computational Logic in Multi-Agent Systems - 14th International Workshop, CLIMA XIV, Corunna, Spain, September 16-18, 2013. Proceedings*, Lecture Notes in Computer Science **8143** (2013), pp. 307–324.

[34] Pigozzi, G. and L. van der Torre, *Arguing about constitutive and regulative norms*, Journal of Applied Non-Classical Logics **28** (2018), pp. 189–217.

[35] Prakken, H., *Formal systems for persuasion dialogue*, Knowl. Eng. Rev. **21** (2006), pp. 163–188.

[36] Prakken, H. and G. Sartor, *Presumptions and burdens of proof*, in: T. M. van Engers, editor, *Legal Knowledge and Information Systems - JURIX 2006: The Nineteenth Annual Conference on Legal Knowledge and Information Systems, Paris, France, 7-9 December 2006*, Frontiers in Artificial Intelligence and Applications **152** (2006), pp. 21–30.

[37] Ramchurn, S. D., D. Huynh and N. R. Jennings, *Trust in multi-agent systems*, The knowledge engineering review **19** (2004), pp. 1–25.

[38] Rienstra, T., A. Perotti, S. Villata, D. M. Gabbay and L. van der Torre, *Multi-sorted argumentation*, in: *International Workshop on Theorie and Applications of Formal Argumentation*, Springer, 2011, pp. 215–231.

[39] Steen, A. and C. Benzmüller, *Extensional higher-order paramodulation in leo-iii*, J. Autom. Reason. **65** (2021), pp. 775–807.
URL https://doi.org/10.1007/s10817-021-09588-x

[40] Thimm, M., *Strategic argumentation in multi-agent systems*, Künstliche Intell. **28** (2014), pp. 159–168.

[41] Toni, F., *A tutorial on assumption-based argumentation*, Argument & Computation **5** (2014), pp. 89–117.

[42] Wu, Y. and M. Podlaszewski, *Implementing crash-resistance and non-interference in logic-based argumentation*, Journal of Logic and Computation **25** (2015), pp. 303–333.

[43] Young, A. P., S. Modgil and O. Rodrigues, *Prioritised default logic as rational argumentation*, in: *Proceedings of the 15th International Conference on Autonomous Agents and Multiagent Systems (AAMAS 2016)*, 2016, pp. 626–634.

Totalistic Cellular Automata and Self-referential Sentences

Shuwen Wu [1]

Fudan University
Shanghai 200433, P. R. China

Ming Hsiung [2]

South China Normal University
Guangzhou 510631, P. R. China

Abstract

Cellular automata give rise to many undecidability problems that can be linked to self-referential statements. A recent study by M. Hsiung (J. Log. Comput. 30: 745–763, 2020) established a connection between elementary cellular automata and such statements in terms of their evolution processes. In this work, we extend this relationship to higher-dimensional cellular automata and self-referential statements. Specifically, we analyze a class of two-dimensional von Neumann-type cellular automata with totalistic rules, and associate each element of this class with a set of self-referential sentences. We describe a procedure to determine the fixed points (if any) of these cellular automata and classify them based on their (in)stability characteristics of their evolution processes. We also discuss some specific self-referential paradoxes induced from these automata. Finally, we present a general result on the base form of totalistic rule cellular automata, which is based on this classification.

Keywords: Totalistic Cellular Automata, Fixed Point, Paradox, Revision Sequence, Self-reference.

Cellular automata (CA) are dynamical systems that are both temporally and spatially discrete, characterized by local interactions and parallel evolution. The concept of cellular automata originated from von Neumann's proposal of a two-dimensional self-replicating automaton system in his well-known work [14] "The General and Logical Theory of Automata." Just from the title of the paper, it is apparent that there is a close relationship between automata and logic. In fact, the foundation for theoretically building von Neumann's automata is the fixed point theorem (or recursion theorem) in the computability

---

[1] wwushuwen@163.com

[2] mingshone@163.com

theory. From then on, people have extensively studied the universality and (un)decidability of automata (see for instance [3] and [10]).

Since the undecidability of formal systems is closely related to the self-referential statements[3] , many researchers turn their attention to the correlation between the self-replication mechanism of cellular automata and the self-referential phenomenon in logic. In this respect, a link of cellular automata with the self-referential statements is established in [9]. The basic idea is that every cellular automaton can be associated with a set of self-referential statements such that the evolution process of cellular automaton is essentially identical to the revision process of the corresponding self-referential statements. Here, the revision process of self-referential statements, proposed by [6] and [8],[4] belongs to the field of formal theories of truth. Through the above connection, we can classify cellular automata based on the logical features of self-referential statements. On the other hand, we can also generate various types of self-referential paradoxes from cellular automata.

The paper [9] mainly focuses on elementary cellular automata and their associated self-referential statements. The present paper extends the methods proposed in [9] to more complex cellular automata. We will study two-dimensional totalistic cellular automata (2D-TCA for short), especially a kind of cellular automata called "Von Neumann two-dimensional totalistic cellular automata". It can be seen that the characteristics of totalistic provides us with lots of convenience for studying two-dimensional cellular automata. We, following the tree diagram method proposed in [9], provide an algorithm for searching for fixed points of this type of cellular automata and give a classification of them based on the fixed-point features. Then, we discuss a type of paradoxes generated by this type of cellular automata, which can be regarded as a generalized form of the Curry paradox.

The notations used in this paper are standard. For example, we will use C, with certain subscripts, to denote a cell of cellular automaton. Accordingly, it also is used to denote a statement (or a sentence). We use T to denote the truth predicate "be true", so that $T\ulcorner C\urcorner$ denotes the sentence "C" is true. More notations will be introduced later.

The structure of this paper is as follows: Section 1 explains how the 2D-TCA are associated with the self-referential statements. Section 2 gives an effective method by which we can find the fixed points (if any) for the evolution process of 2D-TCA. Then, in Section 3 we will analyze the self-referential statements generated from 2D-TCA and determine particularly which ones are paradoxical. Section 4 is a generalization of our analysis to TCA with other dimensions. Finally, We close our discussion in Section 5.

This paper is an extension of our paper [12]. We retain the main results

---

[3] Broadly speaking, self-reference is used to denote a statement that refers to itself or its own referent. The self-referential sentence of the present paper refer more to the functiorial self-reference of the recursion theorem than to the formal linguistic self-reference of the kind used by Gödel. See [13] for more details.

[4] See also [1] and [7].

of [12], but completely rewrite the whole paper. For instance, in Section 1, we redefined the evolution process of 2D-TCA and presented a more general process for inducing self-referential statements from 2D-TCA. This sets the stage for the generalization in Section 4 (an entirely new section). The presentation of algorithm in Section 2 is also more formal and concise.

1   2D-TCA and Self-referential Sentences

As mentioned above, the cellular automata that we will study are a special kind of two-dimensional ones. It is named after Von Neumann because each cell is attached to its above, below, left and right cells, which are known as the von Neumann neighborhood [5]. Moreover, it is "totalistic" because the state (0 or 1) of any cell at a step (in an evolution process) is determined by the total of the states of the cells in a neighborhood at the previous step, we denote this rule of cellular automate by TCA. And so, the rule of any Von Neumann two-dimensional totalistic cellular automaton, according to Wolfram [15], can be given by a table like the one shown in Figure 1 [6].

Fig. 1. A rule for 2D-TCA

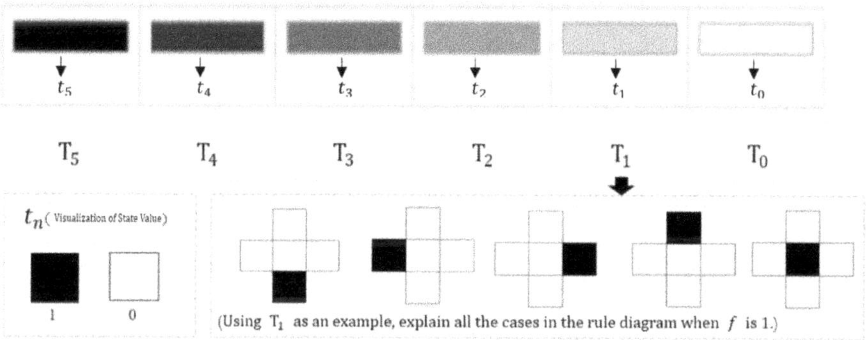

More specifically, we use $C_{i,j}$ to denote the cells, where i and j are integers. And so, $C_{i,j+1}$, $C_{i,j-1}$, $C_{i-1,j}$, and $C_{i+1,j}$ are the above, below, left and right neighbors of $C_{i,j}$ respectively. So the rule in Table 1 says that for $0 \leq k \leq 5$, whenever the sum of the values of $C_{i,j}$, $C_{i,j+1}$, $C_{i,j-1}$, $C_{i-1,j}$, and $C_{i+1,j}$ is i at some step, the value of $C_{i,j}$ at the next step is t_k. For instance, we show in the square T_1 of Figure (1) that in case the sum of the present values of $C_{i,j}$, $C_{i,j+1}$, $C_{i,j-1}$, $C_{i-1,j}$, and $C_{i+1,j}$ is 1, then the value of $C_{i,j}$ at the next step is t_1.

Wolfram uses the number $b_5 \cdot 2^5 + b_4 \cdot 2^4 + ... + b_0 \cdot 2^0$ to code the rule in Table 1). It is called the Wolfram number of this rule. A rule of the Wolfram number

---

[5] Besides von Neumann neighborhoods, another typical neighborhood in two dimensions is the Moore neighborhood, which has been widely employed in various applications, such as the famous Conway's Game of Life. See [11] for more details.

[6] All such rule diagrams are drawn with the computer software *Wolfram Mathematica 12*, and will not be individually explained.

n is denoted by R_n, which also denotes TCA with the Wolfram number n. It is clear that there are 2^6 rules for 2D-TCA.

Let $C_{i,j}(t)$ be the state of $C_{i,j}$ at step t. Let $n = b_5 \cdot 2^5 + b_4 \cdot 2^4 + ... + b_0 \cdot 2^0$. Then R_n can be represented as the following algebraic expression:

$$C_{i,j}(t+1) = b\left(C_{i,j}(t) + C_{i,j+1}(t) + C_{i,j-1}(t) + C_{i-1,j}(t) + C_{i+1,j}(t)\right) \quad (1)$$

where b is a function on $\{i \in \mathbb{N} \mid 0 \le i \le 5\}$ such that $b(i) = b_i$.

In order to give the self-referential sentences corresponding to 2D-TCA, we must translate the algebraic expressions of 2D-TCA into logical expressions. To this end, we first introduce some special normal Boolean formulas. In the following, θ, with or without subscripts, is always a Boolean value, that is, θ is either 0 or 1. We stipulate that $\neg^\theta C$ is $\neg C$, if $\theta = 1$; it is C, if $\theta = 0$. Let C_0, ..., C_m be the Boolean variables ($m \ge 0$). For $k \le m$, we define

$$\beta_\vee^k(C_0, \ldots, C_m) = \bigvee_{\theta_0 + \ldots + \theta_m = k} \bigwedge_{0 \le i \le k} \neg^{1-\theta_i} C_i$$

$$\beta_\wedge^k(C_0, \ldots, C_m) = \bigwedge_{\theta_0 + \ldots + \theta_m = k} \bigvee_{0 \le i \le k} \neg^{\theta_i} C_i$$

For any function b from $\{k \mid 0 \le k \le m\}$ to $\{0, 1\}$, we define two formulas as follows:

$$\tau^b(C_0, \ldots, C_m) = b(C_0 + \ldots + C_m) \quad (2)$$

$$\beta_\vee^b(C_0, \ldots, C_m) = \bigvee_{b(k)=1} \beta_\vee^k(C_0, \ldots, C_m) \quad (3)$$

$$\beta_\wedge^b(C_0, \ldots, C_m) = \bigwedge_{b(k)=0} \beta_\wedge^k(C_0, \ldots, C_m) \quad (4)$$

Proposition 1.1 τ^b, $\beta_\vee^b$, and $\beta_\wedge^b$ are defined as above. Then $\tau^b = \beta_\vee^b = \beta_\wedge^b$.

Proof. Suppose $\{k \mid b(k) = 1\} = \{k_1, \ldots, k_l\}$. It means that

$$\tau^b(C_0, \ldots, C_m) = 1, \text{iff } C_0 + \ldots + C_m = k_1, ..., \text{ or } l. \quad (5)$$

By the Boolean logic, we can easily see that

$$\beta_\vee^k(C_0, \ldots, C_m) = 1, \text{iff } C_0 + \ldots + C_m = k. \quad (6)$$

Hence, it follows immediately

$$\beta_\vee^b(C_0, \ldots, C_m) = 1, \text{iff } C_0 + \ldots + C_m = k_1, ..., \text{ or } l. \quad (7)$$

By (5) and (7), we obtain $\tau^b = \beta_\vee^b$. $\tau^b = \beta_\wedge^b$ can be proved dually. $\square$

Equation (1) can be also expressed as follows:

$$C_{i,j}(t+1) = \tau^b\left(C_{i,j}(t), C_{i,j+1}(t), C_{i,j-1}(t), C_{i-1,j}(t), C_{i+1,j}(t)\right). \quad (8)$$

We will say that τ^b is the *update function*, which can also be used to pin down a 2D-TCA.

By Proposition 1.1, the algebraic expression (8) can be equivalently translated into one of the following logical expressions:

$$C_{i,j}(t+1) = \beta_\vee^b \left(C_{i,j}(t), C_{i,j+1}(t), C_{i,j-1}(t), C_{i-1,j}(t), C_{i+1,j}(t) \right). \qquad (9)$$

$$C_{i,j}(t+1) = \beta_\wedge^b \left(C_{i,j}(t), C_{i,j+1}(t), C_{i,j-1}(t), C_{i-1,j}(t), C_{i+1,j}(t) \right). \qquad (10)$$

To sum up, 2D-TCA with the coding number $n = b_5 \cdot 2^5 + b_4 \cdot 2^4 + \ldots + b_0 \cdot 2^0$ is the one with the update function τ^b as given in Eq. (2). Corresponding to one of 2D-TCA, we can define a set of sentences, say $\{C_{i,j} \mid i, j \in \mathbb{Z}\}$, in which for any $i, j \in \mathbb{Z}$,

$$C_{i,j} \equiv \beta_\vee^b \left(T\ulcorner C_{i,j}\urcorner, T\ulcorner C_{i,j+1}\urcorner, T\ulcorner C_{i,j-1}\urcorner, T\ulcorner C_{i-1,j}\urcorner, T\ulcorner C_{i+1,j}\urcorner \right). \qquad (11)$$

Informally, $C_{i,j}$ is a sentence which declares the sentences, among $C_{i,j}$ itself, $C_{i,j+1}$, $C_{i,j-1}$, $C_{i-1,j}$, and $C_{i+1,j}$, are true or untrue in some combinatorial way. Note that $C_{i,j}$ is a self-referential sentence in the sense that what $C_{i,j}$ declares is relevant to itself. See [9] for more details.

A different but equivalent formulation of the above set of sentences is to use $\beta_\wedge^b$ instead of $\beta_\vee^b$. In this way, we have associated every 2D-TCA with a set of self-referentical sentences.

In the end of this section, we introduce the evolution process for 2D-TCA and revision process for self-referential sentences. First, as usual, we define the evolution processes for 2D-TCA by their evolution sequences. We do this in terms of the algebraic expressions of 2D-TCA.

Definition 1.2 *Let R_n be one of 2D-TCA whose algebraic expression is given by Eq. (1) or (8) and $C(0)$ is an infinite matrix of Boolean values, whose i, j entry is denoted by $(C_{i,j}(0))_{i,j \geq 0}$. For any (discrete) $t \geq 1$, we define the infinite matrix $(C_{i,j}(t))_{i,j \geq 0}$ of Boolean values by Eq. (8). The evolution sequence starting from $C(0)$ for R_n is the infinite sequence $C(0), C(1), \ldots, C(t), \ldots (t \geq 0)$.*

The revision process we present below is a logical tool developed by philosophers H. G. Herzberger and A. Gupta (see [6], [8] and [7]) in response to the need to analyze self-referential sentences (and, more generally, circular definitions). To avoid a roundabout introduction, we follow the line set up in [9, pp. 750-751] and define straightforwardly the revision process for the self-referential sentences associated with 2D-TCA. For convenience, we say a function is a *valuation function*, if it is a function whose values are the Boolean values.

Definition 1.3 *Let $\{C_{i,j} | i, j \in \mathbb{Z}\}$ be a set of sentences given by Eq. (11) and h_0 be a valuation function on the set $\{C_{i,j} | i, j \in \mathbb{Z}\}$. For any $n \geq 1$, we can define a valuation function h_n recursively by the following equation:*

$$h_{n+1}(C_{i,j}) = \beta_\vee^b \left(h_n(C_{i,j}), h_n(C_{i,j+1}), h_n(C_{i,j-1}), h_n(C_{i-1,j}), h_n(C_{i+1,j}) \right). \qquad (12)$$

As for $\{C_{i,j} | i, j \in \mathbb{Z}\}$, the revision sequence starting from h_0 is the sequence of functions $h_0, h_1, \ldots, h_n, \ldots (n \geq 0)$.

The evolution sequence and the revision sequence are the main tools that we study the properties of 2D-TCA, and self-referential sentences. We will establish their connection by these two sequences.

2　Fixed points for 2D-TCA

This section will give a classification of 2D-TCA by the properties of their fixes points (if any). To this end, we first give the following definition.

Definition 2.1 *Let $\{C_{i,j}|i,j \in \mathbb{Z}\}$ be the set of sentences as given in Definition 1.3 and h_n $(n \geq 0)$ be its revision sequence starting from h_0. If there is a number m such that for any $n \geq m$, $h_m(C_{i,j}) = h_n(C_{i,j})$ holds for any $i, j \in \mathbb{N}$, we say h_n is a fixed point of the above revision sequence.*

Under this definition for the 2D-TCA, we will take R_{31} cellular automaton as an example to deal with the fixed point of the totalistic rule cellular automata. As we prove before, we have the logical formulae that express R_{31} cellular automata:

$$C_{i,j}(k) = \neg T^\ulcorner C_{i,j} \urcorner \vee \neg T^\ulcorner C_{i,j+1} \urcorner \vee \neg T^\ulcorner C_{i,j-1} \urcorner \vee \neg T^\ulcorner C_{i-1,j} \urcorner \vee \neg T^\ulcorner C_{i+1,j} \urcorner. \quad (13)$$

Supposed h_n is one of the fixed point of R_{31} cellular automata, according to the Definition 2.1, for any $n \geq m$, we have:

$$h_n(C_{i,j}) = h_m(C_{i,j}) \quad (14)$$

Now, we try to find out all possible value of $C_{i,j}(k)$ in the two-dimensional space. Through fixing the value of $C_{i,j}(k)(0 \text{ or } 1)$, we can use the tree diagram to search all possible cells' states of the fixed point accroding to the rules of R_{31}, as shown:

Fig. 2. Seeking the fixed point for R_{31}

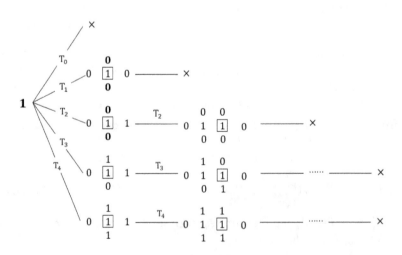

In Figure 2, we will find that there are not any sub-rules can maintain the states of 0 for $C_{i,j}(k)$, according the property of fixed point. And if $C_{i,j}(k) = 1$, by the rule of R_{31}, there must be one of the neighborhood cells is 0, then there is not the fixed points also. Hence, according to Figure 2, whether the value of $C_{i,j}(k)$ is 0 or 1, no branch can obtain a fixed point. That is, Figure 2 illustrates the fact that there not any open branches in the tree diagram of R_{31}. Meanwhile, it means that the R_{31} cellular automata do not have any fixed points. Moreover, we can find that the fixed point of other totalistic rules cellular automata can also be determined by this method. In this way, we will find that except R_{31}, only R_1 does not have any fixed point. In what follows R1 will be further discussed. We summarize our results in a definition:

Definition 2.2 *For any 2D-TCA, its evolution sequence has no fixed point, iff the (fully developped) branches in its tree diagram of the 2D-TCA are closed.*

As we have seen in the Definition 2.2, it is not difficult to determine which cellular automata haven't any the fixed points. Now, we look at the complex situation, that is, for those cellular automata with fixed points, how their fixed points are constructed. In order to obtain a fixed point, according to the properties of totalistic rules, we can definition a global condition firstly, and then constructing the fixed point by the global condition to correspond to each 2D-TCA:

Definition 2.3 *Define the global condition. If the tree diagram of one 2D-TCA can construct a loop state from 1 to 0, we call the 2D-TCA satisfies the global condition of loop states; if the tree diagram constructing a nested state with 1 and 0, it is satisfying the global condition of nested states.*

Next, according to the Definition 2.3, we will show that how to construct the fixed point. And then, we find that the fixed point of 2D-TCA which by using the way to construct is not unique.

Proposition 2.4 *If the tree diagram of one 2D-TCA satisfies the global condition, then we can construct the fixed points correspond to the 2D-TCA.*

Proof. We start with a proof for the global condition of loop states. And we give a R_3 as an example to prove. According to the rules of R_3, we have the tree diagram Figure 3.

In Figure 3, we will find that it is different from Figure 2, in which some branch still continues. And the fixed point of valuing 1 (e.g. $C(k) = 1$) that we want, can only obtain by applying to sub-rules T_1. Then, the center cell, according to T_1, can obtain the states of cells in its neighborhoods, and they all have a value of 0. We now suppose $C(k) = 0$. In Figure 3, we have found that the state value 0 can be obtain through the sub-rules T_4, T_3 and T_2. However, under the sub-rules T_3 and T_2, the neighborhood cells with a state value 0 cannot obtain themselves by satisfying any sub-rules. According to Figure 3, if $C(k) = 1$, the state values of its neighborhood cells must be 0. Hence, this contradicts the situation which under T_3 and T_2.

Therefore, when one wants to obtain $C(k) = 0$, it has to follow the sub-rule T_4 or T_2. And under the sub-rule T_4, the states of neighborhoods cells of the objective cell must all be 1. Then, we find that the neighborhoods cells will

Fig. 3. Seeking the fixed point for R_3

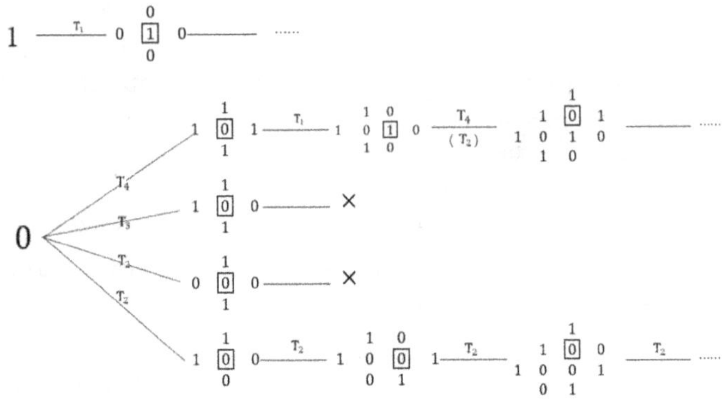

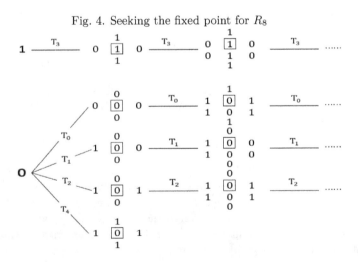

get into the situation of $C(k) = 1$. Hence, we say that the cellular automata satisfies the global condition of loop states.

It is now to apply the loop states under sub-rules T_1 and T_4 we can obtain the fixed point which state 0 alternates with state 1. That is, if the tree diagram has the loop states, then we at least can obtain a fixed point which satisfies the condition of global states. Finally, we obtain the fixed point of R_3 by applying the global condition of loop states of $T_1 - T_4$, as show in the Figure 5. Since the objective cell is not fixed, we can construct multiple fixed points of R_3 in this way.

Fig. 4. Seeking the fixed point for R_8

Similarly, we give a R_8 cellular automata as an example to prove the global condition of nested states. First, we need to construct a tree diagram of R_8 cellular automata, as show in Figure 4.

In Figure 4, we find that the state value of tree diagram has the overlap part. Then we can use the corresponding nested for the overlap part, that is the column of state 0 and the column of state 1, in the tree diagram. By nesting the part in the overlap states of the tree diagram of R_8, that is, some columns of state 1 and two columns of state 0 arranged freely, can construct the global states of the fixed point. Similarly, there is a combination condition $T_3 \rightarrow 1$ and $T_2 \rightarrow 0$ such that the fixed point can be constructed by some columns of state 1 and one column of state 0 arranged freely. Then, We can use the ideas above to construct a fixed point of R_8, shown as Figure 5. Note also that according to the global condition of nested states, there are many fixed points in the R_8 cellular automata.

Fig. 5. The fixed points for R_3(left) and R_8(right) respectively

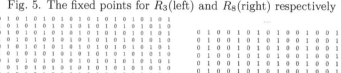

Fig. 6. Satisfy the global conditions of loop states

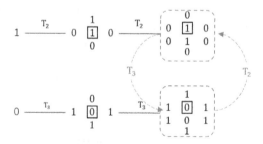

The applicability of the global condition of loop states, we can find that there is not only the R_3 cellular automata satisfies the loop states of $T_1 \rightarrow 1$ and $T_4 \rightarrow 0$, and specific automata will be summarized by the following Table 1. Meanwhile, by the properties of the global condition, we can find another loop states with $T_2 \rightarrow 1$ and $T_3 \rightarrow 0$, as show in Figure 6. Similarly, for our global condition of nested states, we can find more nested states like $T_4 \rightarrow 1$ and $T_1 \rightarrow 0$, $T_4 \rightarrow 1$ and $T_2 \rightarrow 0$. And the specific automata also be summarized by the following Table 1.

Table 1
Combination conditions for obtaining fixed points.

Rn $T_n = 1$ $T_n = 0$	T_1	T_2	T_3	T_4
T_1	×		8 9 12 13 24 25 28 29 40 41 44 45 56 57 60 61	16 17 20 21 24 25 28 29 48 49 52 53 56 57 60 61
T_2		×	8 9 10 11 24 25 26 27 40 41 42 43 56 57 58 59	16 17 18 19 24 25 26 27 48 49 50 51 56 57 58 59
T_3		4 5 6 7 20 21 22 23 36 37 38 39 52 53 54 55	×	
T_4	2 3 6 7 10 11 14 15 34 35 38 39 42 43 46 47			×

3   Paradoxes Associated with 2D-TCA

In this section, we turn to the self-referential sentences associated with 2D-TCA. In particular, we pay special attention to those paradoxical ones. The following definition is due to Herzberger [8, pp. 483-489] and Gupta [6, pp. 6-14].

Definition 3.1 *Let $\{C_{i,j}|i,j \in \mathbb{Z}\}$ be the set of sentences as given in Definition 1.3. We say that it is paradoxical, if any of its revision sequences has no fixed point.*

The following theorem establishes a basic relation between 2D-TCA and the corresponding self-referential sentences. Its proof is similar to the proof of the one that Hsiung [9] gives for the elementary elementary cellular automata. We refer the reader to this literature for details.

Theorem 3.2 *Let $\{C_{i,j}|i,j \in \mathbb{Z}\}$ be the set of sentences associated with the 2D-TCA with Wolfram number n. $\{C_{i,j}|i,j \in \mathbb{Z}\}$ is paradoxical, iff any evolution sequence for this 2D-TCA has no fixed point.*

Fig. 7. A rule for R_{31}

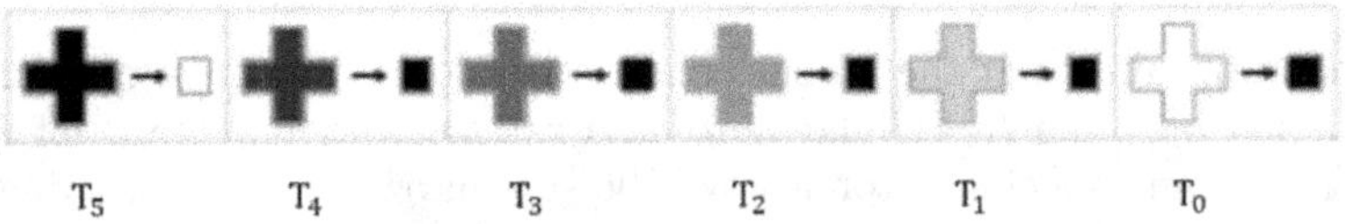

We have known that the evolution sequences for R_1 and R_{31} have no fixed point. So by Theorem 3.2, the sets of sentences induced from R_1 and R_{31} are paradoxical. We now turn to the definition of paradoxical 2D-TCA.

Firstly, the set of sentences associated with R_1 which is given by the following equation:

$$C_{i,j} = \neg T \ulcorner C_{i,j} \urcorner \wedge \neg T \ulcorner C_{i,j+1} \urcorner \wedge \neg T \ulcorner C_{i,j-1} \urcorner \wedge \neg T \ulcorner C_{i-1,j} \urcorner \wedge \neg T \ulcorner C_{i+1,j} \urcorner.$$

Consider $C_{i,j}$. We assume that $C_{i,j}$ is true, which can also be expressed by Boolean number 1. Then, we find $f(C_{i,j}) = 0$. If $f(C_{i,j}) = 0$, we find $T \ulcorner C_{i,j} \urcorner = 1$. Contradiction. Meanwhile, if all cell states of R_1 are 0, the next stage are 1 and hence we have the evolution process with a periodicity which cycle of 0 to 1. Thus, for a 2D-TCA without fixed point, we can find a correspondence paradox such that their have similar logical expression and process periodicity, what one could call such a 2D-TCA a paradoxical cellular automata.That is, according to the properties of automaton, we could call R_1 cellular automata is a paradoxical TCA with the property of liar paradox.

As laid out above, we discuss what it could mean for a 2D-TCA have correspondence to a specifical paradox. And we give a R_1 as an example to show the discussion. As we know that R_{31} doesn't have any fixed points, either. That is, there is a property of a paradox in the R_{31}. And we would find that the paradox is Curry paradox.

For R_{31}, we note that the corresponding set of self-referential sentences is given by

$$C_{i,j} = \neg T \ulcorner C_{i,j} \urcorner \vee \neg T \ulcorner C_{i,j+1} \urcorner \vee \neg T \ulcorner C_{i,j-1} \urcorner \vee \neg T \ulcorner C_{i-1,j} \urcorner \vee \neg T \ulcorner C_{i+1,j} \urcorner.$$

So, in some sense, the set of sentences associated with R_{31} is the dual of that with R_1. At the same time, we must point out that the above equation can be reformulated equivalently as follows

$$C_{i,j} = T \ulcorner C_{i,j} \urcorner \to \neg T \ulcorner C_{i,j+1} \urcorner \vee \neg T \ulcorner C_{i,j-1} \urcorner \vee \neg T \ulcorner C_{i-1,j} \urcorner \vee \neg T \ulcorner C_{i+1,j} \urcorner.$$

We thus can see that the set of sentences associated with R_{31} is something like the Curry paradox. The Curry paradox is proposed by H. B.Curry [4,5]. A popular version of the Curry paradox is as follows:

If the sentence (15) *is true, then* C. $\qquad\qquad\qquad\qquad$ (15)

Similarly, We consider the value of $C_{i,j}$. Assume that $C_{i,j}$ is not true, we find $f(C_{i,j}) = 1$. If $f(C_{i,j}) = 1$, we have $T \ulcorner C_{i,j} \urcorner = 0$. Contradiction. Then, we also can find a correspondence paradox—curry paradox— such that their have similar logical expression and process periodicity. Therefore, R_{31} is a paradoxical cellular automata.

So far we have found at the different numbers of the fixed points from the 2D-TCA. In the following stable we summarize some result on the 2D-TCA. That is, we give a new classification of the 2^6 2D-TCA according to the number of fixed points.

First, we looked at 2D-TCA that do not have any fixed points, and there are R_1 and R_{31}. Now, there are $2^6 - 2$ TCA. We know already that there are 2D-TCA having fixed points under those combining conditions: $T_1 \to 1$ and $T_4 \to 0, T_2 \to 1$ and $T_3 \to 0, T_3 \to 1$ and $T_1 \to 0, T_3 \to 1$ and $T_2 \to 0, T_4 \to 1$ and $T_1 \to 0$, $T_4 \to 1$ and $T_2 \to 0$. Meanwhile, there are much fixed points of the TCA under those conditions. Exclusion those 2D-TCA only 6 2D-TCA remain, that is, R_0, R_{30}, R_{32}, R_{33}, R_{62} and R_{63}. Obviously, there is a unique fixed point in R_0 and R_{63}, which is 0 and 1 respectively. If 2D-TCA have $T_5 \rightarrow 1$ and $T_0 \rightarrow 0$, there is at least one fixed point in the 2D-TCA obviously. Hence, R_{30} and R_{33} are at least have one fixed point; R_{32} and R_{62} are at least have two fixed points. we search through the tree diagrams of those 2D-TCA, we can prove that, R_{30} and R_{33} only have unique fixed point; R_{32} and R_{62} only have two fixed points respectively. List the above as follows:

Table 2
A classification of Totalistic Cellular Automata

Classification standard of TCA	TCA
Without fixed points	R_1, R_{31}
Only one unique fixed point with global state 0	R_0, R_{30}
Only one unique fixed point with global state 1	R_{33}, R_{63}
Only two fixed point with global state 0 and 1	R_{32}, R_{62}
Infinite fixed points	$R_2 - R_{29}$, $R_{34} - R_{62}$

In the Table 2, first is the 2D-TCA without the fixed points, also we called those 2D-TCA are paradoxical 2D-TCA. Specifically, R_1 possesses the property of the liar paradox, and R_{31} is similarly with the curry paradox. Secondly, there is a unique fixed point in 2D-TCA, that is, all cell values are 0, such as R_0. Third, there is a unique fixed point which all cell values are 1 in the 2D-TCA, such as R_{33}. Fourth, there are only two fixed points in the 2D-TCA, that is, all cell value are 0 or 1, such as R_{32}. Last, there are many fixed points in those 2D-TCA, such as R_2.

4   Other dimensional TCA

In section 3, we already have a sort of 2D-TCA about the fixed points. It is natural to consider whether one dimensional or higher dimensional have the same properties. Hence, we have

Proposition 4.1 *If a 1D cellular automaton is a triple $\langle S, r, f \rangle$ with $S = \{0,1\}$, $r = 2$ and $f : S^{(2\cdot 2+1)} \rightarrow S$, then the classification of its fixed points is the same as that in Table 2.*

Proof. A 1D cellular automaton is a triple $\langle S, r, f \rangle$ with $S = \{0,1\}$ $r = 2$ and $f : S^{(2\cdot 2+1)} \rightarrow S$. As [9] is mentioned before, we can take an automaton as a two-infinite tape, in which the cells are evenly aligned and are naturally indexed by the integers: $C_i, i \in \mathbb{Z}$. And C_i is determind by itself and its four neighbours, including C_{i-1}, C_{i-2}(left neighbour) and C_{i+1}, C_{i+2}(right neighbour). Let f be the update function of the automaton in question. Then we also can compute the state of the cell C_i at step $t+1$ by the equation $C_i(t+1) = f(C_{i-1}, C_{i-2}, C_i, C_{i+1}, C_{i+2})$, as shown in Figure 8.

Fig. 8. A rule for 1D-TCA

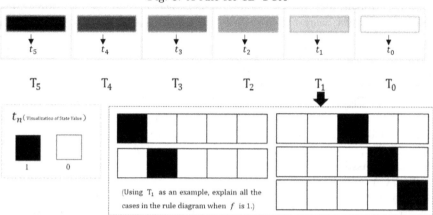

Similarly, we use wolfram number to code the 1D-TCA, and denoted by D_n. Let $n=b_0 \cdot 2^0 + b_1 \cdot 2^1 + ... + b_5 \cdot 2^5$. Then D_n can be represented as the following algebraic expression:

$$C_i(t+1) = f(C_{i-1} + C_{i-2} + C_i + C_{i+1} + C_{i+2}) \tag{16}$$

where f is a function on $\{i \in \mathbb{N} | 0 \le k \le 5\}$ such that $f(i) = b_i$.

According to Prop.1.1 and [9], we can obtain the logical expression of 1D-TCA:

$$C_i \equiv \beta_\vee^b \left(T^{\ulcorner}C_{i-2}{}^{\urcorner}, T^{\ulcorner}C_{i-1}{}^{\urcorner}, T^{\ulcorner}C_i{}^{\urcorner}, T^{\ulcorner}C_{i+1}{}^{\urcorner}, T^{\ulcorner}C_{i+2}{}^{\urcorner}\right). \tag{17}$$

And we can establish the connection between the evolution process of a 1D-TCA with the revision process of the corresponding self-referential sentences and find out all of the fixed points in 1D-TCA.

According to Thm.3.2, we can also prove that the Prop.4.1 is true in 1D-TCA. Similarly, we use the methon by [9] to determind the fixed point in 1D-TCA. The prove is similar to [9], we can obtain those result: their evolution sequences having no fixed point are D1 and D31; only one fixed point with the cells' status of 0 is D0 and D30; only one fixed point with the cells' status of 1 is D33 and D63; there are two fixed points with 0 and 1 states respectively, D32 and D62; others have finite fixed points. See [9] for the details. Then, we can see that the sort of the fixed points is same as 2D-TCA, as desired.

□

Here we see that the properties of self-reference and the fixed point is same in TCA are the same whether they are one-dimensional or two-dimensional. Hence, we can also use Table2 to classification 1D-TCA. Then, it is natural to have a corollary as follow.

Corollary 4.2 *Higher dimensional TCA which have five cells including objective cell and its four neighbours show the same global behavior in fixed points.*

As higher dimensional TCA at least has a initial state in space, the evolution of higher dimensional TCA is more complex. But ever for higher dimensional cellular automata, they are evolution process also determined by same rules, so through their evolution rules, we can simplify them to 2D-TCA or even 1D-TCA for corresponding research. According the result that 1D-TCA and 2D-TCA have the same behavior in fixed points, we can make the same extension. If we set a fixed direaction in 3D-TCA and give a initate state to an obejective cell(show as Figure 9,(a)), we can simplify the 3D-TCA to 2D-TCA, and then we can obtain the same classification as 2D-TCA. Similarly, If we set a fixed direaction in 3D-TCA and give a initate state to an obejective cell(show as Figure 9,(b)), we can simplify the 3D-TCA to 1D-TCA, and then we can obtain the same classification as 1D-TCA. Therefore, we can see that in the case of 3D-TCA mentioned above, the fixed point results show the same global behavior with 2D-TCA and 1D-TCA. Moreover, in which the result is also consistant with the observation found of Chate and Mannevile([2]) is explored a wide variety of cellular automata of dimensions four, five and higher. Higher dimensional situations will be more complex, but we can use the same ideas above to investigate it.

Fig. 9. Simplified method for 3D-TCA

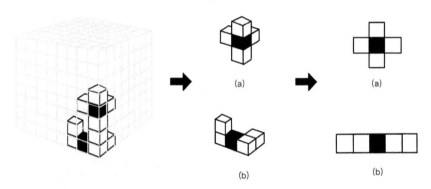

5   Concluding Remarks

It has been a hot research topic for a long time that scholars of various fields on the cellular automaton. And it is an important direction to do some regular classified for cellular automaton of the research filed. In this paper, we consider the different way to give a classification on TCA. That is, we analyze the different property of the fixed point for TCA from a logical direction, and give TCA a new classification. Different in the research on elementary cellular automaton of MingHsiung 2020([9]), we focus on the two-dimensional cellular automaton. Comparing with one-dimensional cellular automaton, the extension involve more real physical systems. And two-dimensional cellular automaton brings many new phenomena involving behaviors of the patterns which no simple analogs in one-dimension.

Yet as it turns out, that should be not surprising at all: 1D-TCA which concluding five cells show the behaviors in fixed pionts as same as 2D-TCA. And that we have a corollary in higher dimensional TCA. Hence in the behavior of fixed points, we can see that the complexity of the cellular automata world will not change with the increase of dimensions, but only its displayed form. That is also an mind what Wolfram emphasize in *a new kind of science*, "even from very simple programs behavior of great complexity could emerge".([15],p19)

This research is hoping to build up more wide relationship between cellular automata with self-reference, through basing on logic to give a classification for cellular automata. Then one can see more similarly system construction of cellular automata and self-referential sentence, and will put forward to do more research on cellular automata and self-reference.

Acknowledgements. An earlier version of this paper was presented at the 15th National Conference on Modern Logic (Zhuhai, China). We appreciate two anonymous reviewers for their helpful comments on this paper.

References

[1] Belnap, N., *Gupta's rule of revision theory of truth*, Journal of Philosophical Logic **11** (1982), pp. 103–116.

[2] Chate, H. and P. Manneville, *Collective behaviors in spatially extended systems with local interactions and synchronous updating*, Prog. Theo. Phys. **87** (1992), pp. 1–60.

[3] Cook, M., *Universality in elementary cellular automata*, Complex Systems **15** (2004), pp. 1–40.

[4] Curry, H., *The combinatory foundations of mathematical logic*, Journal of Symbolic Logic **7** (1942), pp. 49–64.

[5] Curry, H., *The inconsistency of certain formal logics*, Journal of Symbolic Logic **7** (1942), pp. 115–117.

[6] Gupta, A., *Truth and paradox*, Journal of Philosophical Logic **11** (1982), pp. 1–60.

[7] Gupta, A. and N. Belnap, "The Revision Theory of Truth," MIT Press, Cambridge, 1993.

[8] Herzberger, H. G., *Notes on naive semantics*, Journal of Philosophical Logic **11** (1982), pp. 61–102.

[9] Hsiung, M., *Elementary cellular automata and self-referential paradoxes*, Journal of Logic and Computation **30** (2020), pp. 745–763.

[10] Kari, J., *Decidability and undecidability in cellular automata*, International Journal of General Systems **41** (2012), pp. 539–554.

[11] Moore, C. and C. Shalizi, "Cellular Automata: A Discrete Universe," 2003.

[12] Shuwen, W. and M. Hsiung, *Von neumann-type cellular automata and self-referential sentences (chinese)*, Studies in Logic **15** (2022), pp. 46–63.

[13] Smoryński, C., *The development of self-reference: Löb's theorem.*, In Drucker, T., editor. Perspectives on the History of Mathematical Logic, Boston: Birkh?user, (1991), pp. 110–133.

[14] Von Neumann, J., *The general and logical theory of automata*, Papers of John Von Neumann on Computing Computer Theory (1948).

[15] Wolfram, S., "A New Kind of Science," Wolfram Media Inc., Champaign, Illinois, 2002.

Verified completeness in Henkin-style for intuitionistic propositional logic

Huayu Guo [1]   Dongheng Chen [2]   Bruno Bentzen [3]

School of Philosophy, Zhejiang University
Hangzhou
China

Abstract

This paper presents a formalization of the classical proof of completeness in Henkin-style developed by Troelstra and van Dalen for intuitionistic logic with respect to Kripke models. The completeness proof incorporates their insights in a fresh and elegant manner that is better suited for mechanization. We discuss details of our implementation in the Lean theorem prover with emphasis on the prime extension lemma and construction of the canonical model. Our implementation is restricted to a system of intuitionistic propositional logic with implication, conjunction, disjunction, and falsity given in terms of a Hilbert-style axiomatization. As far as we know, our implementation is the first verified Henkin-style proof of completeness for intuitionistic logic following Troelstra and van Dalen's method in the literature. The full source code can be found online at https://github.com/bbentzen/ipl.

Keywords: Intuitionistic propositional logic, Henkin completeness, Formal proofs, Lean.

1   Introduction

Troelstra and van Dalen [17] propose a completeness proof in Henkin-style for full intuitionistic predicate logic with respect to Kripke models. Despite being a fairly standard result in the literature, this completeness proof has yet to be formally verified in a proof assistant. In this paper, we describe a formalization for intuitionistic propositional logic using the Lean theorem prover [13].

Our main goal is to document some challenges encountered along the way and the design choices made to overcome them to obtain a formalized proof that is elegant, intuitive, and better suited for mechanization using the specific techniques available in the Lean programming language, in particular, the `encodable.decode` and `insert_code` methods developed by Bentzen [1].

[1] guohuayu@zju.edu.cn

[2] chen_dongheng@zju.edu.cn

[3] bbentzen@zju.edu.cn

To the best of our knowledge, our implementation is the first verified Henkin-style proof of strong completeness for intuitionistic logic following Troelstra and van Dalen's method in the literature. As far as its propositional fragment is concerned, the main ingredient of Troelstra and van Dalen's Henkin-proof is a model construction based on a consistent extension of sets of formulas, which is achieved by going through all disjunctions of the language [17, lem 6.3]. To carry out this extension, they assume an enumeration of disjunctions with infinite repetitions, also remarking that an alternative approach in which at each stage we treat the first disjunction not yet treated. This variant appears in Van Dalen [5, lem 5.3.8]. Our implementation is based on a third variant of the consistent extension method, which we developed to better suit our needs of formalization. Each propositional formula is only listed once in the enumeration, but we carry out the extension for each of them infinitely many times. The formalization consists of roughly 800 lines of code and encompasses the syntax and semantics of intuitionistic propositional logic, along with the soundness and strong completeness theorems. We adopt a Hilbert-style proof system due to its simplicity. The full source code can be found online at https://github.com/bbentzen/ipl.

1.1 Related work

The formal verification of completeness proofs for intuitionistic logic can be traced back to Coquand's [3] use of ALF to mechanize a constructive proof of soundness and completeness with respect to Kripke models for the simply typed lambda-calculus with explicit substitutions. Heberlin and Lee [9] give a constructive completeness proof of Kripke semantics with constant domain for intuitionistic logic with implication and universal quantification in Coq. Recently, Hagemeier and Kirst [8] formalize a constructive proof of completeness for intuitionistic epistemic logic based on a natural deduction system. They also provide a classical Henkin proof using methods similar to those in Bentzen [1], but they do not present a formalization of the approach of Troelstra and van Dalen [17] as is done in this paper. Bentzen [1] formalizes the Henkin-style completeness method for modal logic S5 using Lean and From formalizes in Isabelle/HOL a Henkin-style completeness proof for both classical propositional logic [6] and classical first-order logic [7]. Maggesi and Brogi [12] give a formal completeness proof for provability logic in HOL Light. The formalization presented here is inspired by the work of Bentzen [1], but makes a few improvements regarding design choices, in particular, the use of Prop in the definition of the semantics and the indexing of models to arbitrary types.

1.2 Lean

Lean [13] is an interactive theorem prover based on the version of dependent type theory known as the calculus of constructions with inductive types [15,4]. Users can construct proof terms directly as in Agda [14], using tactics as in Coq [16] or both proof terms and tactics simultaneously. Lean's built-in logic is constructive, but it supports classical reasoning as well. In fact, our Henkin-style proof is classical since it relies on a nonconstructive use of contraposition.

Therefore, we do not worry about any complexity and computational aspects related to our proof. Our implementation makes use of some results from Lean's standard library and the user-maintained mathematical library `mathlib` [2].

Throughout the remainder of this paper, Lean code will be used to showcase some design decisions in our formalization. The syntax and semantics of intuitionistic propositional logic that is the starting point of our formalization is described in Section 2. We also describe our formalization of a countermodel for the law of excluded middle and sketch a proof of soundness. Then, an informal overview of the Henkin-style proof method as well as a description of our implementation is provided in Section 3. Finally, some concluding remarks are given in Section 4.

2 Intuitionistic Logic

2.1 The language

The intuitionistic propositional language considered here contains implication, conjunction, disjunction, and falsity as the only primitive logical connectives. The language is defined using inductive types with one constructor for propositional letters, falsum, implication, conjunction, and disjunction, respectively:

```
inductive form : Type
| atom : ℕ → form
| bot  : form
| impl : form → form → form
| and  : form → form → form
| or   : form → form → form
```

This code can be found in `language.lean` file.

Since our language contains countably many propositional letters $p_0, p_1, \ldots$ we use the type $\mathbb{N}$ of natural numbers to define the constructor `atom` of propositional letters. The only way to construct a term of type `form` is using this atomic constructor(`atom`) and the constructors for falsum (`bot`), implication (`impl`), conjunction (`and`), disjunction (`or`).

The elimination rule is an operation that allows us to define functions by recursion from it to any other types, including also the type of propositions `Prop`, in which case, this elimination rule is an instance of the principle of induction on the structure of the formula.

Constructors are displayed in Polish notation by default, but we define some custom infix notation with the usual Unicode characters for better readability:

```
prefix   `#`      := form.atom
notation `⊥`      := form.bot
infix    `⊃`      := form.impl
notation p `&` q  := form.and p q
notation p `∨` q  := form.or p q
notation `~`:40 p := form.impl p (form.bot )
```

Contexts are just sets of formulas. In Lean sets are defined as functions of type $A \to \text{Prop}$. As usual in logic textbooks, we display the formulas in a context in list notation separated by a comma instead of using unions of singletons. We introduce the following notation to make this possible:

```
notation Γ ` , ` p := set.insert p Γ
```

The formalization of the language can be found in the **language.lean** file.

2.2   The proof system

We define a Hilbert-style system for intuitionistic propositional logic that is best described as a refinement of Heyting's original axiomatization [10, §2]. The proof system is implemented with a type of proofs, which is inductively defined as follows:

```
inductive prf : set form → form → Prop
| ax {Γ} {p} (h : p ∈ Γ) :prf Γ p
| k {Γ} {p q} : prf Γ (p ⊃ (q ⊃ p))
| s {Γ} {p q r} : prf Γ ((p ⊃ (q ⊃ r)) ⊃ ((p ⊃ q) ⊃ (p ⊃ r))
  )
| exf {Γ} {p} : prf Γ (⊥ ⊃ p)
| mp {Γ} {p q} (hpq: prf Γ (p ⊃ q)) (hp :prf Γ p) : prf Γ q
| pr1 {Γ} {p q} : prf Γ ((p & q) ⊃ p)
| pr2 {Γ} {p q} : prf Γ ((p & q) ⊃ q)
| pair {Γ} {p q} : prf Γ (p ⊃ (q ⊃ (p & q)))
| inr {Γ} {p q} : prf Γ (p ⊃ (p ∨ q))
| inl {Γ} {p q} : prf Γ (q ⊃ (p ∨ q))
| case {Γ} {p q r} : prf Γ ((p ⊃ r) ⊃ ((q ⊃ r) ⊃ ((p ∨ q) ⊃ r
  )))
```

Again, the elimination rule for this type generalizes definition by recursion and induction on the structure of proofs. To follow the usual logical notation, we abbreviate **prf** Γ p with Γ ⊢$_i$ p as follows:

```
notation Γ ` ⊢ᵢ ` p := prf Γ p
notation Γ ` ⊬ᵢ ` p := prf Γ p → false
```

To illustrate, we compare a mechanized formal Hilbert-style proof of the identity of implication $p ⊃ p$ in our implementation:

```
lemma id {p : form } {Γ : set form } :
| Γ ⊢ᵢ p ⊃ p :=
mp (mp (@s Γ p (p ⊃ p) p) k) k
```

with a non-mechanized formal proof written in Lemmon style:

1	$p ⊃ ((p ⊃ p) ⊃ p) ⊃ (p ⊃ (p ⊃ p)) ⊃ (p ⊃ p)$	S
2	$p ⊃ ((p ⊃ p) ⊃ p)$	K
3	$(p ⊃ (p ⊃ p)) ⊃ (p ⊃ p)$	MP 1, 2
4	$(p ⊃ (p ⊃ p))$	K
5	$p ⊃ p$	MP 3, 4

Notice that the proof structure in our term proof is actually clearer since it indicates how the axiom schemes should be instantiated.

The formalization of the proof system can be found in the **theory.lean** file.

2.3 Semantics

2.3.1 Kripke models

We define the semantics for intuitionistic propositional logic in terms of Kripke semantics as usual [17,5]. A model $\mathcal{M}$ is a triple $\langle \mathcal{W}, \leq, \mathsf{v} \rangle$ where $\mathcal{W}$ is a set of possible worlds of type A, $\leq$ is a reflexive, symmetric and monotonic binary relation on A, and v specifies the truth value of a formula at a world.

In Lean, Kripke models can be defined as inductive types having just one constructor using the `structure` command. We define it not as a triple but as a 6-tuple, composed of a domain `W`, an accessibility relation `R`, a valuation function `val`, and proofs of reflexivity, transitivity, and monotonicity for the accessibility relation `R`, denoted as `refl`, `trans`, and `mono`:

```
structure model (A : Type) :=
| (W : set A)
| (R : A → A → Prop)
| (val : ℕ → A → Prop)
| (refl : ∀ w ∈ W, R w w)
| (trans : ∀ w ∈ W, ∀ v ∈ W, ∀ u ∈ W, R w v → R v u → R w u)
| (mono : ∀ p, ∀ w1 w2 ∈ W, val p w1 → R w1 w2 → val p w2)
```

In our case, a possible world is a term of type A. This allows for more generality in the construction of a model unlike in [1]. What is more, the type of propositions `Prop` is used to encode our truth values `true` or `false`.

2.3.2 Semantic consequence

To formalize the notion of truth at a type, we define a forcing relation $w \Vdash_{\mathcal{M}} p$ that takes as arguments a model $\mathcal{M}$, a formula p, and a type A and returns a term of type `Prop`. As usual, falsity, conjunction, and disjunction are defined truth-functionally and an implication $p \supset q$ is true at a world w iff if $\mathcal{R}(w, v)$ then p is true implies q is true at v, for all $v \in \mathcal{W}$. We also introduce the familiar notation for this forcing relation:

```
def forces_form {A : Type} (M : model A) : form → A → Prop
| (#p)    := λv, M.val p v
| (bot)   := λv, false
| (p ⊃ q) := λv, ∀ w ∈ M.W, v ∈ M.W → M.R v w
→ forces_form p w → forces_form q w
| (p & q) := λv, forces_form p v ∧ forces_form q v
| (p ∨ q) := λv, forces_form p v ∨ forces_form q v

notation w `⊩` `{` M `}` ` p := forces_form M p w
```

To formalize the intuitionistic notion of semantic consequence $\Gamma \vDash_i p$ we first extend this forcing relation to contexts pointwise and then we stipulate that $\Gamma \vDash_i p$ iff for all types A, models $\mathcal{M}$ and possible worlds $w \in \mathcal{W}$, Γ being true at w in $\mathcal{M}$ implies p being true at w in $\mathcal{M}$:

```
def forces_ctx {A : Type} (M : model A) (Γ : set form) : A →
    Prop :=
λw, ∀ p, p ∈ Γ → forces_form M p w

notation w `⊩` `{` M `}` ` Γ := forces_ctx M Γ w
```

```
def sem_csq (Γ : set form) (p : form) :=
∀ {A :Type} (M : model A) (w ∈ M.W), (w ⊩ {M} Γ) → (w ⊩ {M} p
  )
```

```
notation Γ `⊨ᵢ` p := sem_csq Γ p
```

It is worth noting that we are overloading the forcing relation notation for formulas w ⊩ {M} p and contexts w ⊩ {M} Γ. There is no ambiguity because Lean will delay the choice until elaboration and determine how to disambiguate the notations depending on the relevant types.

The formalization of the Kripke semantics described above can be found in the `semantics.lean` file.

2.3.3 The failure of the law of excluded middle

Before proceeding to prove completeness, it will be helpful to see how we can build models in our implementation. To give a concrete example, let us show how to build the following countermodel for the law of excluded middle [11, p.99] using the type of booleans true `tt` and false `ff`:

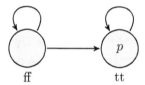

Since our possible worlds are always booleans, the domain, accessibility relation, and valuation function are formalized in Lean in a slightly different way. The reflexivity, transitivity, and monotonicity proofs are straightforward, so we shall omit them:

```
def W : set bool := {ff, tt}

def R : bool → bool → Prop :=λ w v, w = v ∨ w = ff

@[simp]
def val : nat → bool → Prop :=λ _ w, w = tt
```

Using this countermodel, we assume that the law of excluded middle holds, that is for any formula p, either $\emptyset \vDash_i p$ or $\emptyset \vDash_i \neg p$, and then derive a contradiction. This allows us to prove that the law of excluded middle fails in general:

```
lemma no_lem: ¬ ∀ p, (∅ ⊨ᵢ p ∨ ~p)
```

The mechanization of the countermodel can be found in the `nolem.lean` file.

2.3.4 Soundness

The soundness theorem asserts that if a formula p can be derived from a set of assumptions Γ using the inference rules of the logical system, then p is logically valid under any interpretation that satisfies Γ.

```
theorem soundness {Γ : set form} {p : form} :
(Γ ⊢ᵢ p) → (Γ ⊨ᵢ p)
```

The code for proof of soundness can be found in **soundness.lean**.

The proof proceeds by using induction to perform case analysis for each inference rule. For each rule, the proof provides a way to derive the conclusion based on the rule and a way to show that the conclusion is logically valid based on the interpretation and the premises.

3   The completeness theorem

Now that we have presented the implementation of the syntax and semantics of intuitionistic propositional logic in the previous section, we are prepared to undertake a formal proof of completeness. The strong completeness theorem, which states that every semantic consequence is a syntactic consequence, can be stated in Lean using our custom notation as follows:

```
theorem completeness {Γ : set form} {p : form} :
(Γ ⊨ᵢ p) → (Γ ⊢ᵢ p)
```

Our implementation follows the original Henkin-style completeness proof given by Troelstra and van Dalen [17] with some small modifications. The main proof argument runs as follows.

(i) Assume that $\Gamma \vDash_i p$ and $\Gamma \nvdash_i p$ hold;

(ii) Build a model $\mathcal{M}$ such that $w \Vdash_\mathcal{M} p$ iff $w \vdash_i p$ for all worlds $w \in \mathcal{W}$, where we have sets of formulas as possible worlds;

(iii) Show that there is a world $w \in \mathcal{W}$ such that $w \Vdash_\mathcal{M} \Gamma$ but $w \nVdash_\mathcal{M} p$;

(iv) Establish a contradiction from our assumption that $\Gamma \vDash_i p$.

Our proof appeals to classical reasoning at the metalevel of Lean's logic on two occasions [17, p.87], namely, in our proof of $\Gamma \vdash_i p$ where we assume double negation elimination and in our proof of $w \Vdash_\mathcal{M} p$ iff $w \vdash_i p$.

The reader can refer to the **completeness.lean** file for the full details of our implementation of the completeness proof.

3.0.1   Consistent prime extensions

The first step of Troelstra and van Dalen's proof is the definition of what they call a "saturated theory" [17, def.6.2]. We shall make use of the equivalent concept of prime theory instead [5, def.5.3.7], in which the disjunction property is expressed in terms of the membership relation. We say that a set of formulas Γ is a prime theory if Γ is closed under derivability and if $p \vee q \in \Gamma$ implies $p \in \Gamma$ or $q \in \Gamma$. In **completeness.lean** file, we write:

```
def is_closed (Γ : set form) :=
∀ {p :form}, (Γ ⊢ᵢ p) → p ∈ Γ

def has_disj (Γ : set form) :=
∀ {p q :form}, ((p ∨ q) ∈ Γ) → ((p ∈ Γ) ∨ (q ∈ Γ))
```

```
def is_prime (Γ : set form) :=
is_consist Γ ∧ has_disj Γ
```

The second step of Troelstra and van Dalen's completeness proof is the proof of a prime extension lemma [17, lem 6.3], which states that if $\Gamma \nvdash r$ then there is a prime theory $\Gamma' \supseteq \Gamma$ such that $\Gamma' \nvdash r$. Assuming that they have a list of disjunctions $\langle \varphi_{i,1} \vee \varphi_{i,2} \rangle_i$ with infinite repetitions, they define

$$\Gamma' = \bigcup_{i \in \mathbb{N}} \Gamma_i,$$

where $\Gamma_0 = \Gamma$ and Γ_{k+1} is defined inductively as follows:

- Case 1: $\Gamma_k \vdash \varphi_{k,1} \vee \varphi_{k,2}$. Put

  · $\Gamma_{k+1} = \Gamma_k \cup \{\varphi_{k,2}\}$ if $\Gamma_k, \varphi_{k,1} \vdash r$, and

  · $\Gamma_{k+1} = \Gamma_k \cup \{\varphi_{k,1}\}$ otherwise

- Case 2: $\Gamma_k \nvdash \varphi_{k,1} \vee \varphi_{k,2}$. Put

  · $\Gamma_{k+1} = \Gamma_k$

Since we want to extend Γ to a prime theory Γ', we want to ensure the disjunctive property that if $\phi \vee \psi \in \Gamma'$ then $\phi \in \Gamma'$ or $\psi \in \Gamma'$. If there were no infinite repetitions in the list, we could never be sure that we have treated all disjunctions in Case 1, for, at step $k+1$, its disjuncts only get added to the set when Γ_k proves the disjunction. It is possible that later the disjunction becomes provable from Γ_{k+m}, but, we will never go back to it again.

Troelstra and van Dalen mention a simpler variant of the construction that uses an enumeration of disjunctions without requiring infinite repetitions. At stage $k+1$ we simply treat the first disjunction not yet treated. This proof is spelled out by van Dalen in [5, lem 5.3.8]. However, the proof method is less suitable for mechanization given that it is difficult to tell a proof assistant how exactly they should find the first disjunction not yet treated. We implement a simplified version of this method where at each step $k+1$ we always treat all disjunctions in the language once more. The following Lean code encapsulates the idea of the construction sketched above:

```
def insert_form (Γ :   set form) (p q r : form) :   set form :=
if (Γ ، p ⊢ᵢ r) then Γ q else Γ p

def insert_code (Γ : set form) (r : form) (n : nat) :   set form
    :=
match encodable.decode (form) n with
| none    := Γ
| some (p ∨ q) :=if Γ ⊢ᵢ p ∨ q then insert_form Γ p q r else Γ
| some _  := Γ
end
```

```
def insertn (Γ : set form) (r : form) : nat → set form
| 0      := Γ
| (n+1) := insert_code (insertn n) r n

def primen (Γ : set form) (r : form) : nat → set form
| 0      := Γ
| (n+1) := ⋃ i, insertn (primen n) r i

def prime (Γ: set form) (r : form) :   set form :=
⋃ n, primen Γ r n
```

Unlike in Troesltra and van Dalen [17] and van Dalen [5], the enumeration in our formalization lists not just all disjunctions but all propositional formulas in the language. When a formula is not a disjunction we simply ignore it just as in Case 2 above. We follow Bentzen [1] in using **encodable** types to enumerate the language. In Lean, a type α is encodable if there is an encoding function encode $:\alpha \to$ nat and a (partial) inverse decode $:$nat $\to$ option α that decodes the encoded term of α.

Now that we extended Γ to Γ', which we denote as **prime** Γ **r**, we have to prove it is indeed a prime extension of Γ. First, we show that $\Gamma \subseteq \Gamma'$. But this is easy, since for every Γ'_n **n** in the family of sets, $\Gamma \subseteq \Gamma'_n$ **n**. Therefore, Γ must also be included in the union of all Γ'_n **n**, which is Γ'_n.

```
lemma primen_subset_prime {Γ : set form} {r : form} (n):
primen Γ r n ⊆ prime Γ r

lemma subset_prime_self {Γ : set form} {r : form} :
Γ ⊆ prime Γ r
```

The next step is to prove that the Γ' also has the disjunction property and it is closed under derivability. Let us focus on the former first.

We need to show that $p \vee q \in \Gamma'$ implies $p \in \Gamma'$ or $q \in \Gamma'$. If $p \vee q \in \Gamma'$ then there is some $n \in \mathbb{N}$ such that $p \vee q \in \Gamma'_n$. But then since $\Gamma'_n \vdash p \vee q$, then we know that $p \in \Gamma'_{n+1}$ or $q \in \Gamma'_{n+1}$ because the disjunction was treated at some point. Thus, $p \in \Gamma'$ or $q \in \Gamma'$.

```
def prime_insertn_disj {Γ: set form} {p q r : form} (h : (p ∨
    q) ∈ prime Γ r) :
∃ n, p ∈ (insertn (primen Γ r n) r (encodable.encode (p q)+1))
    ∨ q ∈ (insertn (primen Γ r n) r (encodable.encode (p ∨ q)
    +1))

lemma insertn_to_prime {Γ : set form} {r : form} {n m : nat} :
insertn (primen Γ r n) r m ⊆ prime Γ r

def prime_has_disj {Γ : set form} {p q r : form} :
((p ∨ q) ∈ prime Γ r) → p ∈ prime Γ r ∨ q ∈ prime Γ r
```

Saying that Γ' is closed under derivability means that if we can deduce a formula from Γ', it is an element of Γ'. We use a lemma that states that if we can prove $r \vee p$ from Γ', then there exists an n such that $p \in \Gamma_{n+1}$. We use the above lemma **insertn_to_prime** to deduce that $p \in \Gamma'$:

```
lemma prime_prf_disj_self {Γ : set form} {p r : form} :
(prime Γ r ⊢ᵢ r ∨ p) → ∃ n, p ∈ (insertn (primen Γ r n) r (
   encodable.encode (r ∨ p)+1))

def prime_is_closed {Γ : set form} {p q r : form} :
(prime Γ r ⊢ᵢ p) → p ∈ prime Γ r
```

At this moment, we need to prove that Γ' still remains consistent. First, we by structural induction on the derivation that if $\Gamma' \vdash r$ then there is some n such that $\Gamma_n \vdash r$. Then we prove by induction on n that if $\Gamma_n \vdash r$ then $\Gamma \vdash r$. The base case is trivial. In the inductive case, we complete the proof by unfolding the definition of Γ_n and manipulating the inductive hypothesis. Putting both lemmas together, we prove that $\Gamma' \vdash r$ implies $\Gamma \vdash r$:

```
def primen_not_prfn {Γ : set form} {r : form} {n} :
(primen Γ r n ⊢ᵢ r) → (Γ ⊢ᵢ r)

def prime_not_prf {Γ : set form} {r : form} :
(prime Γ r ⊢ᵢ r) → (Γ ⊢ᵢ r)
```

3.0.2 The canonical model construction

Given a set of formulas Γ and ϕ such that $\Gamma \nvdash \phi$, the next step is to build a canonical Kripke model $\mathcal{M}$ such that with $w \Vdash_{\mathcal{M}} \Gamma$ and $w \nVdash_{\mathcal{M}} \phi$ for some possible world. We build this model by letting $\mathcal{W}$ be the set of all consistent prime theories; $w \le v$ iff $w \subseteq v$ for $w, v \in \mathcal{W}$; and $\mathsf{v}(w,p) = 1$ iff $w \in \mathcal{W}$ and $p \in w$, for a propositional letter p. The following Lean code reflects the model construction:

```
def domain : set (set form) := {w | is_consist w ∧ ctx.
   is_prime w}

def access : set form → set form → Prop :=λ w v, w ⊆ v

def val : ℕ → set form → Prop :=λ q w, w ∈ domain ∧ (#q) ∈ w
```

The accessibility relation $\le$ is clearly reflexive and transitive since so is $\subseteq$. Monotonicity is easy to see since $p \in w$ and $w \subseteq v$ means that $q \in v$. We prove these lemmas by straightforward unfolding the definition of `access`.

Our model is integrated into Lean's code as follows:

```
def M : model (set form):=
begin
fapply model.mk,
apply domain,
apply access,
apply val,
apply access.refl,
apply access.trans,
apply access.mono
end
```

3.0.3   Truth and derivability

It turns out that a formula is true at a world in the canonical model if and only if it can be proved from that world:

```
lemma model_tt_iff_prf {p : form} :
∀ (w ∈ domain), (w ⊨ {M} p) ↔ (w ⊢ᵢ p)
```

We mechanize the proof employing the induction tactic, which allows us to use the elimination rule of a type. This approach yields five goals, namely, to prove the case where a formula is a propositional letter, falsity, implication, conjunction, or disjunction. The proof of implication and disjunction deserve some mention.

The disjunction case is simpler, so we shall discuss it first. Lean gives us a biconditional in the following goal:

```
⊢ ∀ (w :set form),
  w ∈ domain → (w ⊨ {M} (p ∨ q)) ↔ (w ⊢ᵢ p ∨ q))
```

The proof in the forward direction starts with the introduction of assumptions and then split the proof into two cases. In the first case, we assume that $w \models_\mathcal{M} p \vee q$ and our goal is $w \vdash_i p \vee q$. Through the tactic cases, which expresses case reasoning, we can finish our goal using some basic facts about disjunctions and the inductive hypotheses in both cases.

In the backward direction, we assume that $w \vdash_i p \vee q$. Since w is a prime theory and thus enjoys the disjunctive property, we can reason by cases depending on whether $w \vdash_i p$ or $w \vdash_i q$. The result follows the inductive hypothesis.

Now we proceed to the implication case. Using the intro tactic, we begin by assuming the inductive hypothesis for p. If w is a world and it is a prime theory, then by unfolding the true definition of a formula in the model's world, we arrive at a biconditional goal that can be expressed as follows.

```
⊢ ∀ (w :set form),
  w ∈ domain → (w ⊨ᵢ {M} (p ⊃ q)) ↔ (w ⊢ᵢ p ⊃ q))
```

We split the biconditional proof into two smaller conditionals using the split tactic. In the forward direction, we first assume that $w \Vdash_\mathcal{M} p \supset q$. We reason by cases depending on whether $w \vdash_i p \supset q$ or not, therefore invoking the law of excluded middle. If that is the case, we are done. If not, then we know that $w, p \nvdash q$. We want to derive a contradiction. We extend the context w, p to a prime theory $(w, p)'$ that still does not prove q. By our inductive hypothesis, since $(w, p)'$ is in the domain, we know that $(w, p)' \Vdash_\mathcal{M} q \leftrightarrow (w, p)' \vdash_i q$.

To derive a contradiction, we just have to show that $(w, p)' \Vdash_\mathcal{M} q$. Recall that our assumption $w \Vdash_\mathcal{M} p \supset q$ states that for all $v \in \mathcal{W}$ such that $w \leq v$, if $v \Vdash_\mathcal{M} p$ then $v \Vdash_\mathcal{M} q$. But, clearly, $w \leq (w, p)'$. To complete the proof, we just have to show that $(w, p)' \Vdash_\mathcal{M} p$. By our inductive hypothesis, it suffices to show that $(w, p)' \vdash_i p$. But this is clearly true, since the original set w, p is contained in the prime extension $(w, p)'$ and $w, p \vdash_i p$.

For the backward direction, what we have to prove is $w \Vdash_\mathcal{M} p \supset q$. This means for all $v \in \mathcal{W}$ such that $w \leq v$, if $v \Vdash_\mathcal{M} p$ then $v \Vdash_\mathcal{M} q$. We assume that

$v \in \mathcal{W}$ such that $w \leq v$, $v \Vdash_{\mathcal{M}} p$ then we have to show $v \Vdash_{\mathcal{M}} q$. Using our inductive hypothesis, we just have to show that $v \vdash_i q$.

Since we know $w \vdash_i p \supset q$ and $w \subseteq v$, by weakening, we will have $v \vdash_i p \supset q$. We complete the proof by noting that $v \vdash_i p$ by our inductive hypothesis and assumption that $v \Vdash_{\mathcal{M}} p$. The result follows from modus ponens.

We have finished the proof of implication.

3.0.4 The completeness proof

To finish our completeness proof we just have to put together all the above pieces into 27 lines of code. We assume that $\Gamma \nvdash_i p$ and $\Gamma \models_i p$, we just need to arrive at a contradiction. We extend Γ to a prime theory Γ' such that $\Gamma' \nvdash_i p$. Since we know $\Gamma' \Vdash_{\mathcal{M}} q \iff \Gamma' \vdash_i q$ for every formula q, we can conclude that $\Gamma' \nVdash_{\mathcal{M}} p$. Thus, we contradict our assumption that $\Gamma \models_i p$, given that $\Gamma' \Vdash_{\mathcal{M}} \Gamma$ but $\Gamma' \nVdash_{\mathcal{M}} p$.

4 Conclusion

We have used Lean to formally verify the Henkin-style completeness proof for intuitionistic logic proposed by Troesltra and van Dalen [17] restricted to a propositional fragment with implication, falsity, conjunction, disjunction. The propositional proof system we implement is based on a Hilbert-style axiomatization. In future work, we hope to expand our implementation to full intuitionistic first-order logic with existential and universal quantifiers and thus complete the formalization of Troesltra and van Dalen's proof. Our implementation also includes a mechanized proof of soundness and a countermodel for the general validity of the law of excluded middle in intuitionistic propositional logic.

Acknowledgments   This research was supported in part by the Zhejiang Federation of Humanities and Social Sciences grant 23YJRC04ZD.

References

[1] Bentzen, B.: A Henkin-style completeness proof for the modal logic S5. In: Logic and Argumentation: 4th International Conference, CLAR 2021, Hangzhou, China, October 20–22, 2021, Proceedings 4. pp. 459–467. Springer (2021)

[2] Carneiro, M.: The Lean 3 Mathematical Library (mathlib). URL: https://robertylewis. com/files/icms/Carneiro_mathlib.pdf (2018), international Congress on Mathematical Software

[3] Coquand, C.: A formalised proof of the soundness and completeness of a simply typed lambda-calculus with explicit substitutions. Higher-Order and Symbolic Computation **15**(1), 57–90 (2002), uRL: https://doi.org/10.1023/A:1019964114625

[4] Coquand, T., Huet, G.: The Calculus of Constructions. Information and Compututation **76**(2-3), 95–120 (1988), uRL: https://hal.inria.fr/inria-00076024/document

[5] van Dalen, D.: Logic and structure, vol. 5. Springer (2013)

[6] From, A.H.: Formalizing Henkin-style completeness of an axiomatic system for propositional logic. Proceedings of the ESSLLI & WeSSLLI Student Session pp. 1–12 (2020)

[7] From, A.H.: A succinct formalization of the completeness of first-order logic. In: 27th International Conference on Types for Proofs and Programs (TYPES 2021). Schloss Dagstuhl-Leibniz-Zentrum für Informatik (2022)

[8] Hagemeier, C., Kirst, D.: Constructive and mechanised meta-theory of intuitionistic epistemic logic. In: Logical Foundations of Computer Science: International Symposium, LFCS 2022, Deerfield Beach, FL, USA, January 10–13, 2022, Proceedings. pp. 90–111. Springer (2022)

[9] Herbelin, H., Lee, G.: Forcing-based cut-elimination for Gentzen-style intuitionistic sequent calculus. In: H., O., M., K., de Queiroz R. (eds.) International Workshop on Logic, Language, Information, and Computation. pp. 209–217. Springer, Berlin, Heidelberg (2009), uRL: https://doi.org/10.1007/978-3-642-02261-6_17

[10] Heyting, A.: Die formalen Regeln der intuitionistischen Logik. Sitzungsbericht PreuBische Akademie der Wissenschaften Berlin, physikalisch-mathematische Klasse II pp. 42–56 (1930)

[11] Kripke, S.A.: Semantical analysis of intuitionistic logic I. In: Studies in Logic and the Foundations of Mathematics, vol. 40, pp. 92–130. Elsevier (1965)

[12] Maggesi, M., Brogi, C.P.: A formal proof of modal completeness for provability logic. arXiv preprint arXiv:2102.05945 (2021)

[13] de Moura, L., Kong, S., Avigad, J., Van Doorn, F., von Raumer, J.: The Lean theorem prover (system description). In: Felty, A., Middeldorp, A. (eds.) International Conference on Automated Deduction. pp. 378–388. Springer, Cham (2015), uRL: https://doi.org/10.1007/978-3-319-21401-6_26

[14] Norell, U.: Dependently typed programming in Agda. In: Koopman, P., Plasmeijer, R., Swierstra, D. (eds.) International School on Advanced Functional Programming. pp. 230–266. Springer, Berlin, Heidelberg (2008), uRL: https://doi.org/10.1007/978-3-642-04652-0_5

[15] Pfenning, F., Paulin-Mohring, C.: Inductively defined types in the Calculus of Constructions. In: International Conference on Mathematical Foundations of Programming Semantics. pp. 209–228. Springer (1989)

[16] The Coq project: The Coq proof assistant. URL: http://www.coq.inria.fr (2017)

[17] Troelstra, A.S., van Dalen, D.: Constructivism in mathematics. Vol. I, Studies in Logic and the Foundations of Mathematics, vol. 121. North-Holland, Amsterdam (1988)

Reason-Based Detachment

Aleks Knoks

University of Luxembourg
Maison du Nombre, 6, Av. de la Fonte
4365 Esch-sur-Alzette, Luxembourg

Leendert van der Torre

University of Luxembourg
Maison du Nombre, 6, Av. de la Fonte
4365 Esch-sur-Alzette, Luxembourg

Abstract

The more recent philosophical literature on foundational questions about normativity relies heavily on the notion of normative reasons, understood as considerations that count in favor or against actions: the notion is used when answering various kinds of normative and metanormative questions and when analyzing other normative notions. The interaction between normative reasons is often made sense of by analogy with weight scales. This paper, by contrast, construes it as a type of inference pattern—titular reason-based detachment—and analyzes it from first principles. While very abstract and exploratory, the approach offers a novel perspective on the (philosophical) idea of weighing normative reasons, and promises to let us relate it to the broader concerns of nonmonotonic logic and related disciplines.

Keywords: detachment, principles, reasons, weighing.

1 Introduction

When philosophers talk about normative matters—about what is right, obligatory, permitted, and so on—they tend to rely on the notion of *normative reasons*, understanding them as considerations that count in favor of or against actions (or attitudes).[1] The notion has become a mainstay of practical philosophy, where it is routinely used when answering various normative and metanormative questions. This is taken to the extreme in the *reasons-first program* which holds, roughly, that the notion of reason is basic, and that all other normative notions should be analyzed in terms of it.[2] When discussing the interaction between reasons, philosophers often use phrases such as "the

---

[1] The philosophical literature distinguishes between normative, motivating, and explanatory reasons—see [2]. We restrict our attention to normative reasons here.

[2] The locus classicus here is Scanlon [24]. But see also, e.g., [21], [23], [25].

action supported on the balance of reasons" and "the reasons for outweigh the reasons against", inviting an image of *weight scales*. The simplest version of these normative scales is meant to work roughly as follows. [3] The reasons in favor of φ-ing go in one pan of the scales, the reasons against φ-ing go in the other. If the weight of the reasons in the first pan is greater than the weight of the reasons in the second pan, φ ought to be carried out. If the weight of the reasons in the second pan is greater, φ ought not to be carried out. [4]

Philosophers have explored various ideas about the exact workings of normative weight scales and have looked at some alternatives. [5] However, with few exceptions, these investigations have been carried out informally, and the more formal investigations have focused on exploring particular models. [6] In this paper, we propose to think of the weight scales as a kind of inference pattern. We call this pattern *reason-based detachment*, and the goal we set ourselves here is to set up and begin to explore a general formal framework built around it. [7] We start with the general notion of *detachment systems*—which can be thought of as structures in which reason-based detachment is guaranteed to be valid—and we formulate a number of principles or properties that a detachment system can satisfy. Then we focus on a class of detachment systems called *balancing operations*, and formulate and discuss a handful of further principles specific to them. For instance, the principle we call *Neutrality* requires, roughly, that reasons of opposing polarity—reasons for and against—are treated equally, while the principle we call *Fixed Value* requires that a reason's polarity always stays the same. We also define several concrete balancing operations, or, roughly, methods specifying how to determine whether φ is supported on the balance of reasons.

The rest of this paper is structured as follows. Section 2 introduces the core formal concepts—including detachment system—and principles that detachment systems can satisfy. Section 3 defines the concept of balancing operation and discusses principles that balancing operations can satisfy. Sections 4–5 discuss two different types of concrete balancing operations. Section 6 presents our principle-based analysis of reason-based detachment. Section 7 takes a first step towards relating reason-based detachment to logical consequence. Section 8 explains where we plan to take the project we started here in the future. Finally, the rather brief Section 9 presents our conclusions.

---

[3] Cf., e.g., [14] and [27].

[4] Cf., e.g., [6] and [27].

[5] While the scales model has its detractors, it is fair to say that it is the dominant model, and that it is often simply taken for granted—see, e.g., Broome's inquiry into the normativity of rationality [4]. For detractors, see [6, 8, 10, 12, 26].

[6] For the latter, see [7, 9, 12].

[7] It pays noting that our approach is similar to the methodology underlying input/output logic [16, 17] which is built around factual detachment—see [20, pp. 502–5] for a discussion.

2   Detachment systems

2.1   Core formal notions

In general, a detachment system is a two-place relation between, on the one hand, an issue (an element of the universe of discourse) with a set of reasons (other elements of the universe of discourse with a value) and, on the other hand, a value. We call an issue together with a set of reasons a *context*. Thus, a detachment system is a relation between contexts and values.

To facilitate the formal presentation, and to be more flexible, we represent reasons as follows:

Definition 2.1 [Reasons] Let $\mathcal{A}$ be an infinite set called the *universe of discourse*, and let $\mathcal{V}$ be a set called *values*. A reason is a triple of the form (x, y, v) where x and y are elements of $\mathcal{A}$ and v is an element of $\mathcal{V}$.[8]

Our formal definition of context is as follows:

Definition 2.2 [Contexts] A context C is a pair of the form (R, y) where R is a finite set of reasons and y, called an *issue*, is an element from the universe of discourse $\mathcal{A}$.

Note that this general representation of contexts and reasons allows for such contexts as $(\{(a, y, v), (b, z, v')\}, y)$. One may wonder whether the latter reason is not superfluous in this context, and whether this context isn't the same as $(\{(a, y, v)\}, y)$. It is exactly these kinds of general considerations that we want to be explicit about in our formal framework. Below, we call this particular property *Relevance*.

Definition 2.3 [Detachment systems] A detachment system $\mathcal{D}$ is a two-place relation between contexts and values from $\mathcal{V}$.

We call elements that comprise a detachment system *detachments*.

2.2   Principles for detachment systems

In general, we identify properties of detachment systems. While we call them *principles*, we could also have called them *axioms*. In the context of this paper, 'properties', 'principles' and 'axioms' are used synonymously. They can be used to classify and distinguish between different detachment systems. Some of these properties may be seen as desirable and, therefore, could be called *postulates* or *desiderata*. However, it is important to note that not all of our principles have the status of a desideratum. In fact, some of them, like the Monotony Principle, are clearly undesirable. Nevertheless, it is useful to make such undesirable properties as Monotony explicit and formal as well. This is why we prefer to refer to the properties as *principles*. They can be used in a principle-based analysis of reason-based detachment, as shown in Section 6.

---

[8] The reader familiar with the philosophical literature on reasons will notice that our formal notion of reason corresponds more closely to what is often called *reason relation*. This is hardly a problem, since the two are closely related and reasons can be read from the relation.

It is natural to think that detachment systems should be *complete*, or that we should be able to detach a value for every context. Note, however, that there are at least two ways to make the intuitive notion of completeness more precise, as Principles 2.4 and 2.6 make clear. The general completeness property, as expressed in Principle 2.4 (Universal Domain), is quite strong. If this property of completeness is considered to be too strong, we also consider the notion of completeness with respect to a set of reasons, as expressed in Principle 2.6 (Reason Universal Domain). This notion is more complicated because the set of reasons depends on the issue under discussion.

Principle 2.4 (Universal Domain) *A detachment system $\mathcal{D}$ is said to satisfy Universal Domain,* Ud, *just in case, it is total, that is, for every context C, there is a value v such that $(C, v) \in \mathcal{D}$.*

Our second principle states that the assignment of a value to an issue (in a context) is determined solely on the basis of the reasons that have to do with that issue. Other reasons can be removed from the context without affecting the result.

Principle 2.5 (Relevance) *A detachment system $\mathcal{D}$ satisfies Relevance,* Re, *just in case $((R, y), v) \in \mathcal{D}$ if and only if $((R_y, y), v) \in \mathcal{D}$, where $R_y = \{(x, y, v') \in R : x \in \mathcal{A} \text{ and } v' \in \mathcal{V}\}$.*

In case the first principle is considered to be too strong, we can use our third principle instead. This principle makes use of the notion of the *universe of reasons* $\mathcal{R}_y$ *of an issue y* in a detachment system, which is the set of reasons that occur in some context for the issue for which the detachment system is defined. What the principle requires, then, is that a value can be detached for every context as long as its set of reasons is a subset of the universe of reasons of its issue.

Principle 2.6 (Reason Universal Domain) *Let $\mathcal{D}$ be a detachment system and y an element of $\mathcal{A}$. Let $\mathcal{R}_y = \bigcup \{R : ((R, y), v) \in \mathcal{D}\}$, called the* universe of reasons of y. *Then $\mathcal{D}$ is said to satisfy Reason Universal Domain,* RUd, *just in case, for any $y \in \mathcal{A}$ and for any $R \subseteq \mathcal{R}_y$, there is a value v such that $((R, y), v) \in \mathcal{D}$.*

This principle may sound circular at first, but it is not. One can check whether a detachment system satisfies Reason Universal Domain by first determining the universe of reasons for every element, and then checking whether for that element all other combinations are also present. In what follows, we will often talk about the universe of reasons $\mathcal{R}$ associated with a detachment system without qualification: it is but the union of the universes of reasons of all issues.

If a detachment relation satisfies Reason Universal Domain but not Universal Domain, then various other principles can be defined. Our fourth principle, Fixed Value, is a case in point. It states that if x is a v type of reason for y in an universe of reasons, it cannot occur as another type of reason for y in this universe of reasons.

Principle 2.7 (Fixed Value) *Let $\mathcal{D}$ be a detachment system and let $\mathcal{R} = \bigcup\{R : ((R, y), v) \in \mathcal{D}\}$. Then $\mathcal{D}$ is said to satisfy Fixed Value, FiVa, just in case, for any two $(x, y, v), (x, y, v') \in \mathcal{R}$, we have $v = v'$.*

The idea that reasons never change their polarities is one of the core tenets of a philosophical view called *atomism*—we will say a little more about this view in Section 6.

While we could formulate more principles strengthening Reason Universal Domain, for reasons of space, we move on to principles of a different kind. And the fifth principle we introduce is Anonymity—we could also have called it *Syntax Independence*. Intuitively, a detachment system satisfies Anonymity when all elements in the universe are treated equally. (In Section 4, we illustrate this property using six balancing operations while in Section 5, we discuss three balancing operations that do not satisfy it.)

Principle 2.8 (Anonymity) *A detachment system $\mathcal{D}$ satisfies Anonymity, An, just in case, for every $((R, y), v) \in \mathcal{D}$ and any bijection $\pi : \mathcal{A} \mapsto \mathcal{A}$, if we have $((\{(\pi(x), \pi(z), v') : (x, z, v') \in R\}, \pi(y)), v'') \in \mathcal{D}$, then $v'' = v$.*

Our sixth principle is Unanimity. It states that if all the reasons for an issue are of v type, then the assignment should also be of the corresponding type.

Principle 2.9 (Unanimity) *A detachment system $\mathcal{D}$ is said to satisfy Unanimity, Ua, just in case, for any context $C = (R, y)$, if there is some $(x, y, v) \in R$ and, for all other $(z, y, v') \in R$, we have $v = v'$, then $((R, y), v) \in \mathcal{D}$.*

Our seventh principle is Groundedness. It can be seen as the inverse of Unanimity. It states that if a context is assigned some value v, then its set of reasons should contain at least one reason of the corresponding type.

Principle 2.10 (Groundedness) *A detachment system $\mathcal{D}$ satisfies Groundedness, Gr, just in case, for any $((R, y), v) \in \mathcal{D}$ with $v \neq 0$, there is some $r = (x, y, v) \in R$.*

3 Balancing operations

Our main focus in this paper is on a particular type of detachment system that we call *balancing operation*. These are (more) closely related to the informal model of normative weight scales.

3.1 Balancing operations defined

Balancing operations are specific detachment systems (for basic weight scales) with the following properties:

(i) Contexts can only be related to the values $+$, $-$, or 0, reflecting the weight scales metaphor: leaning towards the "for" side, leaning towards the "against" side, or being equally balanced.

(ii) Reasons can only have the value $+$ or $-$, reflecting whether they are reasons for or against a given issue. (Note that an element can be a positive reason for one issue and a negative reason for another.)

(iii) Contexts are related to exactly one value. [9]

More formally:

Definition 3.1 [Balancing operations] Let $\mathcal{A}$ be an infinite set of propositional atoms and $\mathcal{V}$ the set $\{+, 0, -\}$. A detachment system $\mathcal{D}$ is called a *balancing operation* just in case it is a function from $2^{\mathcal{A} \times \mathcal{A} \times \{+,-\}} \times \mathcal{A}$ to $\mathcal{V}$.

The reader may wonder about the difference between how $+$ and $-$ and true and false. This issue is taken up in Section 7, where we discuss the differences between balancing operations and logical relations.

Before we turn to principles specific to balancing operations, we introduce some useful formal notation:

- Where $v \in \{+, 0, -\}$, we let $\overline{v}$ stand for the value that is opposite to v, that is: $\overline{v} = -$ if $v = +$; $\overline{v} = +$ if $v = -$; and $\overline{v} = 0$ if $v = 0$.

- Where $r = (x, y, v)$ is a reason, let $ground(r) = x$, $action(r) = y$, and $polarity(r) = v$.

- Where R is a set of reasons and $y \in \mathcal{A}$, the set of reasons from R that *speak in favor of* y is the set $positive(R, y) = \{r \in R : r = (x, y, +)\}$; the set of reasons from R that *speak against* y is the set $negative(R, y) = \{r \in R : r = (x, y, -)\}$; and the set of reasons relevant to y is the set $relevant(R, y) = positive(R, y) \cup negative(R, y)$. (We follow Raz [23] in calling reasons for *positive* and reasons against *negative*.)

3.2   Principles for balancing operations

The first principle pertaining to balancing operations—and the eighth principle overall—is called *Neutrality*. Where Anonymity states that reasons are to be treated equally, Neutrality states that values are to be treated equally. Roughly, if we switch $+$ and $-$ in the context, and vice versa, then the assignment switches its value too. Of all the principles we discuss, this is perhaps the one that is most characteristic of weight scales. Somewhat surprisingly, to the best of our knowledge, this characteristic principle has not yet been formalized in the literature on reasons.

Principle 3.2 (Neutrality) *Let $\mathcal{D}$ be a detachment system and let $\mathcal{R} = \bigcup\{R : ((R, y), v) \in \mathcal{D}\}$. Then $\mathcal{D}$ is said to satisfy Neutrality, Ne, just in case, for every $((R, y), v) \in \mathcal{D}$, if $R' = \{(x, y, \overline{v}) : (x, y, v) \in relevant(R, y)\} \subseteq \mathcal{R}$, then $((R', y), \overline{v}) \in \mathcal{D}$.*

The remaining four principles we discuss describe (non)monotonicity properties. Our ninth principle is called Monotony. It states that if a context gets assigned a nonzero value, adding more reasons to it is not going to change the value that gets assigned.

Principle 3.3 (Monotony) *Let $\mathcal{D}$ be a detachment system and let $\mathcal{R} = \bigcup\{R : ((R, y), v) \in \mathcal{D}\}$. Then $\mathcal{D}$ satisfies Monotony, Mn, just in case,*

---

[9] Thus, balancing operations are deterministic relations.

if $((R, y), v) \in \mathcal{D}$ where $v \neq 0$ and $(x, y, v') \in \mathcal{R}$, then we have $((R \cup \{(x, y, v')\}, y), v) \in \mathcal{D}$.

Clearly Monotony is not a desirable property for balancing operations, and most operations defined in the literature are nonmonotonic. This raises the question of whether there are weaker principles than Monotony that can be defined for balancing operations. As a first response, we formulate a principle called *Polarity Monotony*. If $+$ gets detached, then adding a positive reason will not change the assignment, and this applies also for $-$ and negative reasons. While the principle is not uncontroversial, it seems intuitive, and it is satisfied by all but one of the balancing operations defined in this paper.

Principle 3.4 (Polarity Monotony) *A detachment system $\mathcal{D}$ satisfies Polarity Monotony, PoMn, just in case, if $((R, y), v) \in \mathcal{D}$ where $v \neq 0$ and $(x, y, v) \in \mathcal{R}$, then we have $((R \cup \{(x, y, v)\}, y), v) \in \mathcal{D}$.*

The next principle we discuss is Polarity Cut. It can be seen as the inverse of Polarity Monotony. If a positive value gets detached, then removing a negative reason from the context doesn't affect the detachment. This is the case also for detachments of negative values and positive reasons.

Principle 3.5 (Polarity Cut) *A detachment system $\mathcal{D}$ is said to satisfy Polarity Cut, PoCu, just in case, for any $((R \cup \{(x, y, \overline{v})\}, y), v) \in \mathcal{D}$ with $v \neq 0$, we have $((R, y), v) \in \mathcal{D}$.*

The twelfth and final principle we introduce is Polarity Switching. It can be seen as a strong kind of nonmonotonicity. It assumes that the universe of reasons is infinite, and it states that, for every context that gets assigned a positive value, we can extend the context if we have enough negative reasons so that the resulting context gets assigned a negative value, and vice versa.

Principle 3.6 (Polarity Switching) *Let $\mathcal{D}$ be a detachment system and let $\mathcal{R} = \bigcup \{R : ((R, y), v) \in \mathcal{D}\}$. Then $\mathcal{D}$ satisfies Polarity Switching, PoSw, just in case, for any $((R, y), v) \in \mathcal{D}$, there is an $R' = \{(x, y, \overline{v}) : r \in \mathcal{R}\}$ such that $((R \cup R', y), \overline{v}) \in \mathcal{D}$.*

Having introduced the principles, we turn to concrete balancing operations.

4 Anonymous balancing operations

Over the course of this section and the next, we introduce a handful of balancing operations. All of them are defined with respect to a universe of reasons. In this section, we discuss *anonymous* balancing operations.

Definition 4.1 [Anonymous balancing operations] Let $\mathcal{D}$ be a detachment system and let $\mathcal{R} = \bigcup \{R : ((R, y), v) \in \mathcal{D}\}$. Then $\mathcal{D}$ is an *anonymous balancing operation* just in case $\mathcal{D}$ satisfies:

(i) the Reason Universal Domain principle (with respect to $\mathcal{R}$); and

(ii) the Anonymity principle.

According to our first sample balancing operation, the context (R, y) is assigned the value $+$ in case the sheer number of reasons speaking in favor of y is greater than the number of reasons speaking against y; it gets assigned the value $-$ in case the number of reasons against y is greater than the number of reasons for y; and it gets assigned 0 otherwise. More formally:

Definition 4.2 [Simple Counting] Let $\mathcal{D}$ be an anonymous balancing operation. Then $\mathcal{D}$ is called *Simple Counting* just in case:

- $((R, y), +) \in \mathcal{D}$, if $|positive(R, y)| > |negative(R, y)|$;
- $((R, y), -) \in \mathcal{D}$, if $|negative(R, y)| > |positive(R, y)|$;
- $((R, y), 0) \in \mathcal{D}$, otherwise.

Admittedly, Simple Counting—as well as the other balancing operations we are about to define—is, well, very simple and inadequate for most practical purposes. That is, if we think back to the balancing scales metaphor from the philosophical literature and use Simple Counting as a concrete proposal regarding how to assign deontic statuses to actions—with the assignment of $+$ $(-)$ standing for the conclusion that the action ought (not) to be carried out— then we would surely get many cases wrong. That being said, Simple Counting does justice to at least two important features that are inherent in the idea of normative weight scales. First, it treats positive and negative reasons in a symmetric fashion. Second, it can be seen as adding up the weights of reasons while relying on the assumption that the magnitude (or "weightiness") of all reasons is the same.

It is worth making it explicit that Definition 4.2 does not define a single balancing operation but a class of balancing operations: one for every different universe of reasons $\mathcal{R}$. The same applies to the other balancing operations defined in this section.

Our second balancing operation, called *All or Nothing*, assigns the value $+$ to the context (R, y) if all the reasons that concern y in R are positive, and the value $-$ if all such reasons are negative. In case neither of these conditions obtain, the context gets assigned 0.

Definition 4.3 [All or Nothing] Let $\mathcal{D}$ be an anonymous balancing operation. Then $\mathcal{D}$ is called *All or Nothing* just in case:

- $((R, y), +) \in \mathcal{D}$ if $positive(R, y) = relevant(R, y) \neq \emptyset$;
- $((R, y), -) \in \mathcal{D}$ if $negative(R, y) = relevant(R, y) \neq \emptyset$;
- $((R, y), 0) \in \mathcal{D}$ otherwise.

Our third balancing operation can be thought of as lying in between Simple Counting and All or Nothing. The intuitive idea behind it is that the context (R, y) gets assigned the value $+$ $(-)$ where *most* reasons that are relevant to y argue in favor of (or against) y. For simplicity, we assume that 'most reasons' translates into at least four times as many reasons. [10]

Definition 4.4 [Most Reasons] Let $\mathcal{D}$ be an anonymous balancing operation. Then $\mathcal{D}$ is called *Most Reasons* just in case:

- $((R, y), +) \in \mathcal{D}$ if $|positive(R, y)| \geq 4 \times |negative(R, y)|$ and $relevant(R, y) \neq \emptyset$;
- $((R, y), -) \in \mathcal{D}$ if $|negative(R, y)| \geq 4 \times |positive(R, y)|$ and $relevant(R, y) \neq \emptyset$;
- $((R, y), 0) \in \mathcal{D}$ otherwise.

The next operation assigns $-$ to a context, as long as it is not the case that there are more positive than negative reasons for y. (In the latter case, the context gets assigned a $+$.)

Definition 4.5 [Default Negative] Let $\mathcal{D}$ be an anonymous balancing operation. Then $\mathcal{D}$ is called *Default Negative* just in case:

- $((R, y), +) \in \mathcal{D}$, if $|positive(R, y)| > |negative(R, y)|$;
- $((R, y), -) \in \mathcal{D}$, otherwise.

Our final balancing operation is similar to Simple Counting, except now there is a threshold that changes the rules of the game: once there are enough positive reasons (the threshold is met), the existence of further negative reasons to the contrary ceases to matter. The idea behind this operation comes from the literature on *threshold deontology*. Advocates of threshold deontology hold, roughly, that deontological norms are to be followed up to a point even if there are adverse consequences, but when the consequences become so dreadful that they cross some threshold, consequentialism takes over. [11]

Definition 4.6 [Threshold] Let $\mathcal{D}$ be an anonymous balancing operation. Then $\mathcal{D}$ is called *Threshold* just in case:

- $((R, y), +) \in \mathcal{D}$, if $|positive(R, y)| \geq 100$ or $|positive(R, y)| > |negative(R, y)|$;
- $((R, y), -) \in \mathcal{D}$, if $|positive(R, y)| < 100$ and $|negative(R, y)| > |positive(R, y)|$;
- $((R, y), 0) \in \mathcal{D}$, otherwise.

For all the balancing operations defined so far, given a context (R, y), one does not need to look beyond R to determine which value to assign to the context. What's more, it is not difficult to see that all of these operations satisfy Anonymity (Principle 2.8). But, to anticipate the discussion in Section 6, only the first three of them satisfy Neutrality (Principle 3.2).

---

[10] This simple proposal is meant to serve as an illustration of a more general idea or scheme for specifying 'most reasons'.

[11] See, e.g., [1, Sec. 4] or the more recent [5, 18]. Note that the balancing operation is only inspired by the literature on threshold deontology and is not meant to capture any particular account.

5   Relational balancing operations

In this section, we turn to a different class of balancing operations. These assign values to contexts on the basis of the reasons within them, along with a binary anti-symmetric relation $\prec$ on the reasons:

Definition 5.1 [Relation $\prec$] Given a detachment system $\mathcal{D}$ with its underlying set of reasons $\mathcal{R} = \bigcup\{R : ((R, x), v) \in \mathcal{D}\}$, an anti-symmetric relation $\prec$ on $\mathcal{R}$ is a subset of $\mathcal{R} \times \mathcal{R}$ such that $(r, r') \in \prec$ only if $polarity(r) = \overline{polarity(r')}$.

Notice that two reasons can stand in the $\prec$ relation only if one of them is positive and the other negative. Instead of $(r, r') \in \prec$, we will write $r \prec r'$. An expression of the form $r \prec r'$ can be thought of in terms of r' having strictly more weight than r, or r' defeating r.

　　With this, we can state the general definition of balancing operations discussed in this section.

Definition 5.2 [Relational balancing operations] Let $\mathcal{D}$ be a detachment system, let $\mathcal{R} = \bigcup\{R : ((R, y), v) \in \mathcal{D}\}$, and let $\prec$ be a binary anti-symmetric relation over $\mathcal{R}$, as in Definition 5.1. Then $\mathcal{D}$ is a *relational balancing operation* (for $\mathcal{R}$ and $\prec$) just in case $\mathcal{D}$ satisfies:

(i) the Reason Universal Domain principle (with respect to $\mathcal{R}$); and

(ii) for all $((R, y), v) \in \mathcal{D}$, there is no $r = (x, y, \overline{v}) \in R$ such that $r' \prec r$ for every $r = (z, y, v) \in R$.

Notice that Clause (ii) states that $+$ cannot be detached from (R, y) in case there is some positive reason r for y that stands in the $\prec$ relation to—or is better than, or defeats—every reason against y; and similarly for $-$. This is a very weak property.

　　We proceed to define some concrete relational balancing operations, or, rather classes of them: much like in the previous section, we get different balancing operations for different $\mathcal{R}$ and $\prec$. We call the first class *Exists Better Reason*. It assigns $+$ to (R, y) in case, for every reason against y, there is a stronger reason for y; and it assigns $-$ to (R, y) in case, for every reason for y, there is a stronger reason against y.

Definition 5.3 [Exists Better Reason, $\forall\exists$] Let $\mathcal{D}$ be a relational balancing operation. Then $\mathcal{D}$ is called *Exists Better Reason* just in case:

- $((R, y), +) \in \mathcal{D}$, if $relevant(R, y) \neq \emptyset$ and, for every $r = (x, y, -) \in R$, there is an $r' = (z, y, +) \in R$ such that $r \prec r'$;
- $((R, y), -) \in \mathcal{D}$, if $relevant(R, y) \neq \emptyset$ and, for every $r = (x, y, +) \in R$, there is an $r' = (z, y, -) \in R$ such that $r \prec r'$;
- $((R, y), 0) \in \mathcal{D}$, otherwise.

　　Our next balancing operation, Decisive Reason, is more demanding: it assigns $+$ $(-)$ to a context (R, y) just in case there exists a reason for (or against)

y that is stronger than all reasons to the contrary.[12]

Definition 5.4 [Decisive Reason, $\exists\forall$] Let $\mathcal{D}$ be a relational balancing operation. Then $\mathcal{D}$ is called *Decisive Reasons* just in case:

- $((R, y), +) \in \mathcal{D}$, if there is an $r' = (x, y, +) \in R$ such that $r \prec r'$ for every $r \in R$ with $action(r) = y$ and $polarity(r) = -$;

- $((R, y), -) \in \mathcal{D}$, if there is an $r' = (x, y, -) \in R$ such that $r \prec r'$ for every $r \in R$ with $action(r) = y$ and $polarity(r) = +$;

- $((R, y), 0) \in \mathcal{D}$, otherwise.

The third operation, All Reasons Better, is even more demanding than Decisive Reason: it assigns the value $+$ $(-)$ to a context (R, y) just in case *all* reasons for (against) y are stronger than *all* reasons against (for) y. Otherwise, it assigns the value 0.

Definition 5.5 [All Reasons Better, $\forall\forall$] Let $\mathcal{D}$ be a relational balancing operation. Then $\mathcal{D}$ is called *All Reasons Better* just in case:

- $((R, y), +) \in \mathcal{D}$, if $relevant(R, y) \neq \emptyset$ and, for every $r' \in R$ with $action(r') = y$ and $polarity(r') = +$, $r \prec r'$ for every $r \in R$ with $action(r) = y$ and $polarity(r) = -$;

- $((R, y), -) \in \mathcal{D}$, if $relevant(R, y) \neq \emptyset$ and, for every $r' \in R$ with $action(r') = y$ and $polarity(r') = -$, $r \prec r'$ for every $r \in R$ with $action(r) = y$ and $polarity(r) - +$;

- $((R, y), 0) \in \mathcal{D}$, otherwise.

This operation can be seen as injecting the idea underlying Decisive Reason into the All or Nothing operation described in the previous section.

6 Chart

Now that we have defined a number of principles and a handful of balancing operations, we can analyze and compare the operations by looking at the principles that they satisfy and the ones that they do not. For example:

Proposition 6.1 *Simple Counting (Definition 4.2) satisfies Polarity Monotony (Principle 3.4).*

Proof. Let $\mathcal{D}$ be Simple Counting. Consider an arbitrary context (R, y) and suppose that we have $((R, y), v) \in \mathcal{D}$ with $v \neq 0$. Either $v = +$, or $v = -$. Without loss of generality, we suppose that $v = +$. By Definition 4.2 (Simple Counting), we can be sure that $|positive(R, y)| > |negative(R, y)|$. Now let's consider the context $(R \cup \{r\}, y)$ where $action(r) = y$ and $polarity(r) = +$. Since $action(r) = y$ and $polarity(r) = +$, we have $|positive(R \cup \{r\}, y)| = 1 + |positive(R, y)|$. But $1 + |positive(R, y)| > |positive(R, y)| > |negative(R, y)| =$

---

[12] The terms 'decisive reason' and 'decisive reasons' feature prominently in the philosophical literature—see, e.g., [14].

Table 1

Summary of the principle-based analysis of balancing operations

	SiCount	AllNoth	Most	DefNeg	Thresh	∀∃	∃∀	∀∀
1. Ud	−	−	−	−	−	−	−	−
2. Re	✓	✓	✓	✓	✓	✓	✓	✓
3. RUd	✓	✓	✓	✓	✓	✓	✓	✓
4. FiVa	−	−	−	−	−	−	−	−
5. An	✓	✓	✓	✓	✓	−	−	−
6. Ua	✓	✓	✓	✓	✓	✓	✓	✓
7. Gr	✓	✓	✓	−	✓	✓	✓	✓
8. Ne	✓	✓	✓	−	−	✓	✓	✓
9. Mn	−	−	−	−	−	−	−	−
10. PoMn	✓	✓	✓	✓	✓	✓	✓	−
11. PoCu	✓	✓	✓	✓	✓	✓	✓	✓
12. PoSw	✓	−	✓	✓	−	−	−	−

$|negative(R \cup \{r\}, y)|$, which, by Definition 4.2, is enough for $((R \cup \{r\}, y), +) \in \mathcal{D}$.

$\square$

Proposition 6.2 *All Reasons Better (Definition 5.5) does not satisfy Polarity Monotony (Principle 3.4).*

Proof. Let $\mathcal{D}$ be a detachment system with the universe of reasons $\mathcal{R} = \{(a, d, +), (b, d, +), (c, d, -)\}$ and with $\prec = \{((c, d, -), (a, d, +))\}$, and let $\mathcal{D}$ assign values to contexts in accordance with Definition 5.5. It is not difficult to verify that $D_1 = (\{(a, d, +), (c, d, -)\}, d, +) \in \mathcal{D}$, and that $D_2 = (\{(a, d, +), (c, d, -), (b, d, +)\}, d, 0) \in \mathcal{D}$. This, however, means that the claim that if $((R, y), +) \in \mathcal{D}$, then $((R \cup \{(x, y, +)\}, y), +) \in \mathcal{D}$ does not hold of $\mathcal{D}$, implying that $\mathcal{D}$ does not satisfy Polarity Monotony. $\square$

The proofs of the remaining propositions—that is, the propositions that show which other principles are (not) satisfied by which (other) operations—are about as straightforward as those of Propositions 6.1 and 6.2. For this reason, we omit them here, letting Table 1 summarize the lay of the land: the topmost row lists the balancing operations; the leftmost column lists the principles; the remaining cells state whether the given operation does (✓) or doesn't (−) satisfy the given principle. For example, the third column makes it clear that the balancing operation we called *All or Nothing* (Definition 4.3) satisfies all principles except for Fixed Value, Monotony, and Polarity Switching.

The table provides a way to compare the various operations. It can also be used to get important insights about both operations and principles. In the remainder of this section, we hint at these and flag some other issues of broader significance.

The chart indicates that none of the balancing operations satisfy Universal Domain (Principle 2.4) while all of them satisfy Relevance and Reason

Universal Domain (Principles 2.5 and 2.6). Note that there is an ambiguity here. Recall that each of our definitions from Sections 4–5 specify not a single balancing operation but a class of balancing operations: roughly, we get a different operation for every different universe of reasons. Note that some of these operations do satisfy Universal Domain, namely, those whose universe of reasons $\mathcal{R}$ contains all possible reasons that can be constructed from $\mathcal{A}$ and $\{0, +\}$—call this $\mathcal{R}_{\mathcal{A}}$. However, those operations whose universe of reasons is more restricted do not satisfy the principle. Thus, what the chart indicates is that Universal Domain does not hold true of all operations that belong to the class.

This makes the importance of the universe of reasons underlying a given detachment relation very clear, and the reader may wonder why we allow it to be more restricted than $\mathcal{R}_{\mathcal{A}}$. The motivation here comes from philosophical literature where the standard view is that only some facts (or types of facts) can ever constitute reasons—even if the views on what those facts are are very different. As for Principles 2.5 and 2.6, if we think of balancing operations as serving the same function as the weight scales model serves in the philosophical literature, then we want them to satisfy both of these principles. In the end, it would be very strange if, in certain cases, the deontic status of an action could be genuinely (not epistemically) indeterminate, or if it was determined by something irrelevant.

Recall that Fixed Value (Principle 2.7) states that the universe of reasons underlying a detachment system cannot contain two reasons of the form $(x, y, +)$ and $(x, y, -)$. This idea has a correlate in the philosophical literature, where there is a well-known view called *atomism* that says that a reason can never change its weight, which includes both its polarity and magnitude. Furthermore, there is a well-known argument against the weight scales model that goes roughly as follows. (1) The weight scales model entails atomism. (2) Atomism is false. Hence, (3) the weight scales model has to be wrong.[13] The argument is now widely considered to be flawed—see, e.g., [27]—and our framework provides further support for this. We can see our balancing operations as simplified concrete specifications of the weight scales model. The fact that none of them satisfy Fixed Value suggests that statement (1) must be false.

We take Anonymity and Neutrality (Principles 2.8 and 3.2) to be among the most important principles we have identified. Even though we defined the balancing operations in Section 4 using Anonymity, it is natural to see it as their characteristic principle. Neutrality, in its turn, formalizes an idea that seems inherent in the weight scales metaphor: that positive and negative reasons are to be treated symmetrically—this, of course, is not to say that this idea cannot be questioned. We can see at a glance that operations that don't satisfy Neutrality, namely, Default Negative and Threshold (Definitions 4.5 and 4.6), do not treat positive and negative reasons symmetrically.

Unanimity, Polarity Monotony, and Polarity Cut (Principles 2.9, 3.4,

---

[13] See, e.g., [6].

and 3.5) may look like very natural principles, and one may even be tempted to think that the fact that All Reasons Better doesn't satisfy Polarity Monotony shows that it is a bizarre balancing operation. However, there is a well-known example from the AI & law literature that could make one question all three principles. The example describes the effects of heat and rain on one's decision as to whether or not to go jogging: taken by themselves, the facts that it is raining, and that it is hot constitute reasons for you not to go jogging, but when taken in combination, they make it rational to go jogging [22]. On the face of it, in this example we are dealing with two negative reasons of the form $(x, y, -)$ and $(z, y, -)$ and a detachment system that includes the following three detachments: $((\{(x, y, -)\}, y), -)$, $((\{(z, y, -)\}, y), -)$, and $((\{(x, y, -), (z, y, -)\}, y), +)$. This means that there's a tension between all three of these principles and the most straightforward analysis of the jogging scenario. Those who like the analysis will deny the principles. Those who like the principles will have to argue that the scenario is misdescribed—here see, e.g., [3, 15, 19, 27].

7   Balancing operations and logical consequence relations

Although we have conceptualized detachments as pairs of the form $((R, x), v)$, they can be rewritten either as triples of the form (R, x, v), or as pairs of the form $(R, (x, v))$. This flexibility is an advantage of the framework. In some applications, it is useful to see a detachment system as a binary relation $((R, x), v)$. From a mathematical point of view, it can be seen as a function (if we add some conditions, as we did in Section 3). From an application point of view, it is more like a weight scales for x. But in other applications, it is useful to see a detachment system as a binary relation $(R, (x, v))$. From a mathematical point of view, it should not be seen as a function. From an application point of view, it is more similar to deductive systems and logical relations.

Indeed, given that we used Boolean values when defining balancing operations, it is natural to wonder about their relationship to the logical languages used in propositional logic, logic programming, and nonmonotonic inference relations. Of course, the balancing operations that we considered in Section 4 and 5—as well as other operations from the (informal) literature—are quite different to what can be found in the semantics of propositional logic, logic programming, or nonmonotonic logic. So a syntactic correspondence between the languages may be of interest mainly for technical reasons. Nevertheless, for the definitions of principles, it may be illustrative to define a common language for balancing operations and logical consequence relations. So, here we make such a common language explicit.

To represent balancing operations, we identify the universe of discourse with propositional atoms, and reasons with literals. Let $L, L_1, \ldots, L_n$ be the elements of the universe of discourse or their negations. A logic programming rule is written as $L :\text{-} L_1, \ldots, L_n$, stating that L holds if $L_1, \ldots, L_n$ hold. In propositional logic, this is often written as the rule $L_1 \wedge \ldots \wedge L_n \to L$.

Alternatively, we could write $L_1, \ldots, L_n \vDash L$. In the latter case, consider an issue y, a set of reasons for y, and a set of reasons against y. If x is a reason for y, we write it as x, and if x is a reason against y we write it as $\neg x$. There are three assignments for y in a context:

$+$: $L_1, \ldots, L_n \vDash y$

$-$: $L_1, \ldots, L_n \vDash \neg y$

0: neither of the above

In this way, all of the principles we have defined can be rewritten as principles of logical consequence relations.

With this translation, we have a unified language for reason-based entailment and nonmonotonic inference, but the semantics will be very different. The main difference concerns the interpretation of negation. In reason-based detachment, "x is a reason against y" means something completely different to the logical inference "the negation of x implies y". In particular, we must be careful when comparing or importing principles from one area into another. Consider reasoning by cases, one of the hallmarks of logical inference. If "x implies y" and "the absence of x implies y" both hold, then y holds unconditionally. Whether a similar inference pattern holds for reason-based entailment is more controversial. Perhaps even more clearly, the nonstandard reading of negation on the left becomes very clear when we consider Polarity Monotony. While this principle makes a lot of sense for reason-based detachment, it makes little sense for nonmonotonic inference.

Of course, this particular representation does not indicate that there is no other way to represent reason-based detachment in existing logics of nonmonotonic entailment. It does, however, suggest that reason-based entailment is a notion that should be analyzed from first principles. Furthermore, there is an additional drawback to representing balancing as logical consequence relations: it assumes the strong notion of completeness, or Universal Domain.

8   Future work

As future work, we plan to take the framework set up here in a number of different directions. First, in addition to reasons, the philosophical literature talks about considerations which, while not being reasons, can have *indirect* effects on the normative landscape—e.g., on which actions ought to be carried out. In this context, the literature discusses in particular *conditions* (or *undercutters*) which cancel the normative effects of a reason, and *modifiers* which either amplify or attenuate the (default) magnitude (or "weightiness") of a reason. [14] What we want to do, then, is extend the framework so that we can represent these other types of considerations too and explore their effects on detachment. One thing we can do is extend our formal notion of a balancing operation by allowing some of its underlying reasons to also take the value 0. These new reasons—of the form $(x, y, 0)$—could then affect the standard ones—of the form

---

[14] See, e.g., [27, Sec. 2] for a nice summary.

$(x, y, +)$ and $(x, y, -)$—and thereby indirectly affect detachment.

Another direction for future research is to explore detachment systems built around numerical values. In fact, detachment systems of this sort may be closer to the way reasons and their interaction are conceptualized in the philosophical literature, where the weights of reasons are standardly taken to be comprised of a polarity and a magnitude (or "weightiness"). It certainly seems worth exploring different balancing operations that assign values to contexts by applying numerical operations to reasons. Also, this does not seem to be too far off from the ideas explored in multi-criteria decision-making—see, e.g., [13].

Yet another promising idea is to explore detachment systems that are not complete, even in the weaker sense of not satisfying the Reason Universal Domain principle: they appear to be fitting for modeling case-based reasoning of the sort that is discussed, for instance, in the context of models of precedential constraint—see, e.g., [11]. Yet another idea is to explore the detachment systems built around a richer domain of discourse: logical formulas, as opposed to abstract elements. Finally, it would be useful to extend the principle-based analysis presented here with further principles and balancing operations.

9 Conclusion

Our main goal in this paper was to set up and start exploring a (general) formal framework built around reason-based detachment. We started by introducing detachment systems, or structures in which reason-based detachment is guaranteed to be valid. After formulating some general principles that detachment systems can satisfy, we focused on a class of detachment systems—which we called *balancing operations*—that can be thought of as regimenting the informal model of the normative weight scales: we formulated further principles specific to balancing operations (Section 3), defined a handful of concrete balancing operations (Sections 4–5), and put the two together in a principle-based analysis (Section 6). We also briefly discussed the relationship between reason-based detachment and logical inference, along with the most immediate directions for future research. Ultimately, we are aiming to provide a framework within which one can (i) define, relate, and compare various different (and possibly complex) accounts of the way reasons interact to support actions as well as (ii) relate these accounts to the ideas proposed in the context of case-based reasoning, multi-criteria decision-making, nonmonotonic reasoning, and related disciplines. This short paper is but a first step in this direction.

Acknowledgments

Both authors acknowledge financial support from the Luxembourg National Research Fund (FNR) for the project Deontic Logic for Epistemic Rights (OPEN O20/14776480) L. van der Torre is also supported by the (Horizon 2020 funded) European Coordinated Research on Long-term Challenges in Information and Communication Sciences & Technologies ERA-NET (CHIST-ERA) grant CHIST-ERA19-XAI (G.A. INTER/CHIST/19/14589586).

References

[1] Alexander, L. and M. Moore, *Deontological Ethics*, in: E. N. Zalta, editor, *The Stanford Encyclopedia of Philosophy*, Metaphysics Research Lab, Stanford University, 2021, Winter 2021 edition .

[2] Alvarez, M., *Reasons for action: Justification, motivation, explanation*, in: E. N. Zalta, editor, *The Stanford Encyclopedia of Philosophy*, 2016, winter 2016 edition .

[3] Bader, R., *Conditions, modifiers and holism*, in: E. Lord and B. Maguire, editors, *Weighing Reasons*, Oxford University Press, 2016 pp. 27–55.

[4] Broome, J., "Rationality through Reasoning," Wiley Blackwell Publishing, 2013.

[5] Cole, T., *Real-world criminal law and the norm against punishing the innocent: Two cheers for thershold deontology*, in: H. Hurd, editor, *Moral Puzzles and Legal Perspectives*, Cambridge University Press, 2019 pp. 371–87.

[6] Dancy, J., "Ethics without Principles," Oxford University Press, 2004.

[7] Dietrich, F. and C. List, *A reason-based theory of rational choice*, Noûs **47(1)** (2013), pp. 104–34.

[8] Drai, D., *Reasons have no weight*, Philosophical Quarterly **68** (2018), pp. 60–76.

[9] Faroldi, F. and T. Protopopescu, *All-things-considered oughts via reasons in justification logic*, in: J. Maranhao, C. Peterson, C. Strasser and L. van der Torre, editors, *Deontic Logic and Normative Systems: 16th International Conference (DEON2023, Trois-Rivières)*, College Publications, 2023 .

[10] Hawthorne, J. and O. Magidor, *Reflections on reasons*, in: D. Star, editor, *The Oxford Handbook to Reasons and Normativity*, Oxford University Press, 2018 .

[11] Horty, J., "The Logic of Precedent: Constraint and Freedom in Common Law Reasoning," Cambridge University Press, forthcoming.

[12] Horty, J., "Reasons as Defaults," Oxford University Press, 2012.

[13] Keeney, R. and H. Raiffa, "Decisions with Multiple Objectives: Preferences and Value," Cambridge University Press, 1993.

[14] Lord, E. and B. Maguire, *An opinionated guide to the weight of reasons*, in: E. Lord and B. Maguire, editors, *Weighing Reasons*, Oxford University Press, 2016 pp. 3–24.

[15] Maguire, B. and J. Snedegar, *Normative metaphysics for accountants*, Philosophical Studies **178** (2018), pp. 363–84.

[16] Makinson, D. and L. van der Torre, *Input/output logics*, Journal of Philosophical Logic **29** (2000), pp. 383–408.

[17] Makinson, D. and L. van der Torre, *Constraints for input/output logics*, Journal of Philosophical Logic **30** (2001), pp. 155–85.

[18] Moore, M., *The rationality of threshold deontology*, in: H. Hurd, editor, *Moral Puzzles and Legal Perspectives*, Cambridge University Press, 2019 pp. 371–87.

[19] Nair, S., *How do reasons accrue?*, in: E. Lord and B. Maguire, editors, *Weighing Reasons*, Oxford University Press, 2016 pp. 56–73.

[20] Parent, X. and L. van der Torre, *Input/output logic*, in: L. van der Torre, D. Gabbay, J. Horty and R. van der Meyden, editors, *Handbook of Deontic Logic, vol. 1*, College Publications .

[21] Parfit, D., "On What Matters," Oxford University Press, 2011.

[22] Prakken, H. and G. Sartor, *Modelling reasoning with precedents in a formal dialogue game*, Artificial Intelligence and Law **6** (1998), pp. 231–87.

[23] Raz, J., "Practical Reason and Norms," Oxford University Press, 1990.

[24] Scanlon, T. M., "What We Owe to Each Other," Cambridge, MA: Harvard University Press, 1998.

[25] Schroeder, M., "Reasons First," Oxford University Press, 2021.

[26] Snedegar, J., *Reasons for and reasons against*, Philosophical Studies **175** (2018), pp. 725–43.

[27] Tucker, C., *A holist balance scale*, Journal of the American Philosophical Association **First View** (2022), pp. 1–21.

A Formalization of 'Ought Not to Know' Based on STIT Logic

Yini Huang

Zhejiang University, Hangzhou

Abstract

One is bound by epistemic norms, which dictate what one ought to know or not to know. When discussing norms, 'ought to' has both an epistemological and a deontological dimension to its interpretation. This paper argues that the purely epistemological interpretation of 'ought to know' does not take into account the deontic issues that beliefs and knowledge may raise. With the proliferation of big data privacy deontic issues, it is necessary to consider the influence of deontic norms on the discussion of epistemic norms. Due to the focus on agency, STIT logic has been chosen for the formalization in this research. Based on the STIT theory, this study combines the concepts of agency and obligation with epistemology and expands the epistemic norms with deontic norms, formalizing the norm of 'ought not to know.' This paper begins with a detailed introduction to STIT semantics and stit operators. Building on existing research on the relationship between knowledge and obligation, the study adds the 'knowing-value' operator to the XSTIT theory to formalize 'ought not to know' and discusses the possible properties of the operator. This study provides a formalization of the epistemic norms of pragmatism.

Keywords: epistemic norms, deontic logic, STIT theory, 'ought not to know'

1 Introduction

A person is not only bound by norms of behaviour, but also by epistemic norms, i.e., norms about what one ought or ought not to believe and know. When discussing norms about 'ought to', a distinction needs to be made between the epistemological sense of 'ought to' and the deontological sense of 'ought to'. The intuitive difference between the two can be shown by the following two examples:

(i) It's either raining or sunny outside, and the floor in the playground is dry, so it ought to be sunny. [1]

(ii) To get the job, you ought to follow the rules and regulations.

---

[1] According to Canbridge Dictionary, we can use 'ought to' when we talk about what is likely or probable.

Here, the first example represents the relationship between 'it is sunny outside' and a set of evidence (e.g. 'it's either raining or sunny outside', 'the floor in the playground is dry', etc.); The second example represents the relationship between 'compliance with rules and regulations' and the agent's purpose (e.g. 'getting the job').

Because of the direct link between epistemology and cognition, historically, much of the discussion of cognitive norms has taken place from an epistemological perspective. The main concern of epistemology has been to explain the justification of the various beliefs we hold about the world. When we discuss whether a belief is justified, what we want to know is whether it is correct to believe it. John L. Pollock [14] notes that epistemic norms are norms describing when it is epistemically permissible to hold various beliefs. A belief is justified if and only if it is licensed by the correct epistemic norm. Assuming that a belief is justified by the reasoning underlying it, then the epistemic norms are the norms governing 'right reasoning'. Epistemologists usually understand epistemic norms as norms for evaluating reasoning. Thus, from an epistemological perspective, cognitive norms are truth-oriented.

However, we find that the epistemological sense of 'ought to know' does not actually take into account the deontic issues that beliefs and knowledge may raise. In the post-privacy era, with the rapid development of artificial intelligence and the widespread use of big data, people may tend to cede a certain degree of privacy in exchange for a more convenient life. However, the ensuing issue of big data privacy ethics has also aroused widespread concern, and thus a challenging problem for big data security is how to control the bounds of how big data technology is used so that it does not violate users' personal privacy. Against this backdrop, we believe that we should add ethical considerations to the discussion of epistemic norms. That is, when discussing whether 'one ought to know something', we need to consider not only the justification of the belief, but also the legitimacy of the agent's purpose. At the same time, we find that when discussing the ethics of big data privacy, the focus is not on what the agent 'ought to know', but on what the agent is not allowed to know, i.e., what the agent 'ought not to know'. Therefore, in this study, we will discuss the 'ought not to know' in depth.

In Section 2, we elaborate on the concept of 'ought not to know' that underpins our research. To this end, we use STIT logic, proposed by Nuel Belnap [5], Michael Perloff [5] and Ming Xu [5], as a tool to formalize this concept. In Section 3, we introduce the semantics and common operators of STIT logic, laying the groundwork for the subsequent chapters. In Section 4, we explore the application of STIT theory to deontic and epistemic problems. We first introduce the definition of 'ought to' as it applies to the context of this paper, and discuss the XSTIT logic proposed by Jan Broersen. We then extend the XSTIT theory by introducing cognitive operators to analyse the relationship between cognition and obligation. In Section 5, we use XSTIT theory to formalize 'ought not to know' and extend the framework using the 'knowing-value' operator to further explore the possible properties of the 'ought not to know' operator.

2 Explanation of 'ought not to know'

Let us consider a realistic example: suppose that a person has the belief that 'the PIN for Ben's bank card is 123456' and that he has obtained enough evidence to support the belief that 'the PIN for Ben's bank card is 123456' by hacking into multiple network systems, and that 'the PIN for Ben's bank card is 123456' is in fact true, then according to JTB (Justified true belief) theory in epistemology, the person should have knowledge that 'the PIN for Ben's bank card is 123456'; however, it is clear that Ben's PIN is private to himself, and if we say that 'someone should have had knowledge of Ben's PIN', we are in fact violating Ben's right to privacy. In this case, then, it would clearly be against our ethics to say 'we ought to know about it'.

Jonathan Harrison [10] explains this situation in more detail. He argues that for some knowledge, it is better for people not to know . There are three reasons for that. First, some people will use their knowledge to do things that harm themselves or others. That is, although knowledge helps them to achieve their goals, those goals can be harmful to themselves or others. Secondly, in addition to knowledge that can be harmful in itself, some knowledge may be acquired in various improper ways, such as through telephone tapping, reading other people's private documents, and making rude enquiries into other people's private lives, which clearly violates other people's privacy. Finally, doing so breaks the tacit agreement among members of civilised society that people will not try to find out from each other what others do not want to be known.

Thus, we are morally obliged to 'ought not to know' about things that may cause harm to others, that are private or that we are not legally allowed to know. Here we need to distinguish between 'ought not to know' and 'have the right not to know': 'have the right not to know' is an enabling provision; 'ought not to know' is an obligatory provision, which is mandatory under the rules; it is a commanding directive that requires people to perform a certain act, and if they 'do not do it', they are in breach of some rules, which makes it 'obligatory'.

The 'ought not to know' we are concerned with here is in fact equivalent to 'being not allowed to know', that is , someone is obliged to be in a state of 'not knowing' something. It is important to note that what we mean here by 'ought not to know' is not simply the negation of 'ought' in standard deontic logic. [2] In our definition, we consider 'not knowing' as a whole. On the basis of the above assumptions, this study will still take 'ought not to know' as the target of our formalization.

In addition to this, 'ought not to know' in this study is not simply a modal word for a belief state, but a modal word for a cognitive action. Consider the following example: Suppose p is about the privacy of Ben. A person who is

---

[2] Since the O and P operators of standard deontic logic are dual, 'ought not to know' in formal language means 'is allowed not to know'.

only passively in the state of 'knowing p' (e.g. being told) is not condemned for violating Ben's privacy; he is condemned only if he manages to learn p. In other words, 'ought not to know' in this case means 'ought not to learn about'. We are not concerned with a person's state of knowledge, but with the cognitive act by which that person obtains that state of knowledge; the focus is on the agency of the acting subject. Thus, we cannot simply use a combination of the standard deontic operator and standard epistemic operator, i.e., $O\neg K$, to formalize 'ought not to know'. Instead, to formalize 'ought not to know', we use the STIT logic that focuses on the agent of the action as our tool.

3   STIT semantics and stit operators

The STIT (see to it that) logic [5,6] is philosophical logics of agency. The primary virtue of STIT logic is that it discusses choice exertion explicitly, in contrast to the majority of logical formalisms for action. STIT logic expresses the statements inscribing agency in the formula $[\alpha \text{ stit:A}]$, which reads 'the agent α sees to it that A holds', where stit is the binary operator formalizing agency, α is the agent, and A is an arbitrary statement. In the following we abbreviate the statement with the stit operator as $[\alpha]A$.

Given a countable set of propositional variables P and a finite set $Agent$ composed by agents, the formal language $\mathcal{L}_{STIT}$ is defined as follows:

$$A, B ::= p \mid \neg A \mid (A \wedge B) \mid \Box A \mid [\alpha]A,$$

In this section, I will first introduce the framework and model of STIT logic, and then introduce and explain different interpretations of such formulas, in preparation for the later discussion.

3.1   STIT framework and STIT model

The semantics of STIT logic is based on the *Branching Time frames*. The framework is a pair $\langle T, < \rangle$, where T is a nonempty set of moments and $<$ is a strict partial order relation on the set T that satisfies no backward branching. That is, for each m, m', m'' in T, if $m' < m$ and $m'' < m$, then we have $m' = m''$ or $m'' < m'$ or $m' < m''$. Each maximal $<$-chain in T can be called a *history*, which represents a possibility [6]. If we denote moments by m and histories by h, then $m \in h$ can be interpreted as m occurring at a point in h, or h passing through m. Due to uncertainty, different histories may pass through the same moment. For any $m \in T$, we denote the set of histories passing through m by $H_m = \{h \in H_T : m \in h\}$, where H_T is the set of all histories in $\langle T, < \rangle$.

The branching-time model adds to the branching-time framework an valuation function v, where $v : P \to 2^{T \times H}$ assigns to each atomic proposition a set of moment-history pairs [1]. For $h \in H_m$, we call of a moment-history pair m/h an index. Given a branching time model $\mathcal{M}$, the ordinary temporal operators G (for 'it will always be the case') and H (for 'it has always been in the case'), the the operator of historical necessity $\Box$. These operators are interpreted as follow, where A is a formula:

- $\mathcal{M}, m/h \vDash \mathrm{G}A$ iff $u \in h$, $m < u$ only if $\mathcal{M}, u/h \vDash A$;
- $\mathcal{M}, m/h \vDash \mathrm{H}A$ iff for each $u \in h$, $u < m$ only if $\mathcal{M}, u/h \vDash A$;
- $\mathcal{M}, m/h \vDash \Box A$ iff for each $h' \in H_m$, $\mathcal{M}, m/h' \vDash A$.

We use F, P and $\Diamond$ as the dual operators of G, H and $\Box$ respectively [18].

A branching time frame with instants is a branching time frame $\langle T, < \rangle$ to which a non-empty set *Instant* of instants is added, where *Instant* is a partition of T. The notion of instants is strongly pre-relativistic, representing 'the same time' across all histories [18]. The frame satisfies properties of 'unique intersection' and 'order preserving' :

- unique intersection: for each instant i and history h, $i \cap h$ contains a unique moment m, denoted as $m_{i,h}$;
- order preserving: for all instants i, i' and all histories h, h', $m_{i,h} < m_{i',h}$ iff $m_{i,h'} < m_{i',h'}$.

In the branching time frame containing moments $\langle T, <, Instant \rangle$, we denote by $i_{(m)}$ the moments to which m belongs, and for all i, $i' \in Instant$, $i < i'$ iff there exists some $m \in i$ and $m' \in i'$, $m < m'$. In this case, the relation $<$ between instants is a strictly linear order that obeys order-preserving property.

The basic framework of a STIT logic is a quadruple $\langle T, <, Agent, Choice \rangle$, that is, adding *Agent* and *Choice* to the branching time frame, where *Agent* is a nonempty set of agents whose elements we usually denote by α, β, etc., while *Choice* is a function that maps a pair of agents α and moments m to a partition of the history set H_m passing through that moment, which we usually write $Choice_\alpha^m$. The function *Choice* needs to satisfy 'no choice between undivided histories' and 'independence of agents':

- no choice between undivided histories: for all h, $h' \in H_m$, if there exists a $m' > m$ such that $m' \in h \cap h'$, then for each $K \in Choice_\alpha^m$, $h \in K$ iff $h' \in K$;
- independence of agents: for each $m \in T$ and each function s that maps an agent α to a member of $Choice_\alpha^m$, $\bigcap_{\alpha \in Agent} s(\alpha) \neq \varnothing$.

$Choice_\alpha^m$ is a partition of H_m. For each agent α and moment m, we call the equivalence class in that partition the possible choices or actions of the agent α at moment m. In other words, at moment m the agent α is able to identify a particular equivalence class from $Choice_\alpha^m$ and in which the future historical process necessarily exists. If the function s on *Agent* assigns to each subject α an equivalence class of $Choice_\alpha^m$, then we call it a selection function at moment m. Also, for $h \in H_m$, we use $Choice_\alpha^m(h)$ to denote the particular possible choices made from $Choice_\alpha^m$ containing the history h.

We call the quadruplet $\langle T, <, Agent, Choice \rangle$ the STIT frame and the quintuplet $\langle T, <, Instant, Agent, Choice \rangle$ the STIT frame containing instants, where $\langle T, <, Instant \rangle$ is a branching time frame with instants.[3] A STIT model is a model on a STIT frame, so that if the frame contains *Instant*, then the model

---

[3] See [4,6] for a discussion of this.

also contains *Instant*. For each STIT model $\mathcal{M}$ and moment m, we use $\| A \|_m^{\mathcal{M}}$ to denote $\{h \in H_m : \mathcal{M}, m/h \vDash A\}$; if the model $\mathcal{M}$ is instant-containing, then we use $\| A \|_i^{\mathcal{M}}$ to denote $\{h \in H_T : \mathcal{M}, m_{i,h}/h \vDash A\}$ [18]. In practice, we usually omit the superscript $\mathcal{M}$ and denote it by $\| A \|_m$ or $\| A \|_i$.

3.2 stit operators

There are various stit operators, the most common of which are the following four: astit (achievement stit), bstit (Brown's stit), cstit (Chellas' stit) and dstit (deliberative stit). Here we write them as $[\alpha]^a$, $[\alpha]^b$, $[\alpha]^c$, and $[\alpha]^d$ respectively. In this paper, we will only use the operators $[\alpha]^c$ and $[\alpha]^d$. The truth conditions for the STIT formula are defined as follows:

- $\mathcal{M}, m/h \vDash [\alpha]^c A$ iff $Choice_\alpha^m(h) \subseteq A_m$;
- $\mathcal{M}, m/h \vDash [\alpha]^d A$ iff $Choice_\alpha^m(h) \subseteq A_m$ and $H_m \nsubseteq \| A \|_m$.

We can be more specific about the above definition. At a given moment and history, a dstit formula $[\alpha]^d A$ is true at m/h if and only if (1) the choice of the agent α at moment m at $Choice_\alpha^m(h)$ can guarantee that A is true; [4] (2) Also, there exists a history h' such that A is false on it. [5] And in order for $[\alpha]^d A$ to be true at m/h, condition (1) that $[\alpha]^c A$ is true at m/h only has to be met. $[\alpha]^d A$ is equivalent to $[\alpha]^c A \wedge \neg \Box A$.

4 The integration of STIT theory with deontic and epistemic logic

STIT logic has been used in both deontic and epistemic studies. In this section, we will explain how to combine the two using STIT logic.

4.1 Formalization of 'ought-to-do' in STIT

In the existing studies of STIT logic, there are two different approaches to the formalization of 'ought-to-do', represented by Paul Bartha and John Horty: the former defines 'someone ought to do something' as 'someone performs a violation if he does not perform the action' [3], while the latter understands 'ought-to do' from a decision-theoretic perspective as 'doing it is the agent's preferred action', i.e., the action that leads to the maximisation of utility [12]. In Horty's definition, we typically have defeasible conditional obligations, which means more specific ones override more general ones. However, unlike 'ought to', when discussing 'ought not to', the agent's violation cannot be disregarded because of other conflicting obligations. According to the previous section, the definition of 'ought not to know' in this paper is 'not (legally or morally) allowed to know', which means that for certain things that we 'ought not to know', we are condemned if we 'actively know' about it. So our definition of 'ought-to-do' here is closer to Paul Bartha's approach. For this reason, this

---

[4] That is, the choice of the agent α at m $Choice_\alpha^m(h)$ can cause future event processes to be restricted to the history of $Choice_\alpha^m(h)$ and to be true at each moment A.

[5] α's choice $Choice_\alpha^m(h)$ must have an effect on the truth value of A at m.

paper will adopt only Paul Bartha's definition of 'ought-to-do' and will not elaborate on John Horty's approach.[6]

Alan Ross Anderson proposed in 1956 to reduce deontic logic to alethic modal logic, using S to denote a constant proposition meaning 'sanction' and defining the ought operator $\mathbf{O}$ in terms of the relation between obligation and sanction [2]:

$$\mathbf{O}p =_{df} \Box(\neg p \to S).$$

There is a problem with this definition: in practical examples, due to the limitations of external conditions, the agent's action is not always detected by others, so he is not practically bound to be punished. Yet this does not mean that his behaviour is right. So here we change 'sanction' to 'violation'. A violation occurs even if for whatever reason there is no sanction. Therefore we define $\mathbf{O}$ as:

$$\mathbf{O}p =_{df} \Box(\neg p \to V).$$

Paul Bartha's definition of 'ought to do something' is based on Alan Ross Anderson's proposal:

$$\mathbf{O}[\alpha]^d A =_{df} \Box(\neg[\alpha]^d A \to V_\alpha).$$

Here, $\mathbf{O}$ is an ought operator as defined by Anderson, and V_α is a personalized constant proposition associated with the agent α, which can be interpreted as 'α does something wrong'. Thus, 'α ought to see to it that A' is defined as meaning that if α does not see to it that A, then he has done something wrong. The advantage of this definition is that when we add the $\mathbf{O}$ operator, the completeness theorem underlying dstit logic is still preserved [18].

Paul Bartha's definition of $\mathbf{O}$ was also used by Jan Broersen in his study. However, he has made certain modifications to the existing theory. In continuation of Alan Ross Anderson's theory, Broersen proposed a new xstit operator in [8], which we write here as $[\alpha]^x$. Broersen proposed that in standard STIT logic, the effects of actions are instantaneous. However, from an ontological perspective, given that a process can be thought of as occurring 'in' time and that an action can always be thought of as a 'process' connected to some effort made by the agent involved, we can infer that actions also occur 'in' time. Therefore, the effects of the agent's choice would take effect at the next choice point.[7] Based on this, Jan Broersen proposed the XSTIT logic.

In the XSTIT logic, Jan Broersen also added an operator X for 'next', thus introducing the notion of 'next state' into the definition of $\mathbf{O}$. The XSTIT modal language $\mathcal{L}_{XSTIT}$ is defined as follows:

$$A, B ::= p \mid \neg A \mid (A \wedge B) \mid \Box A \mid [\alpha]^x A \mid XA$$

---

[6] For an account of John Horty's formalization and application of 'ought-to-do', we refer to the literature [11,12].

[7] Note that we do not assume anything about how distant subsequent choice points should be; they can be arbitrarily close [9].

A STIT frame is incremental if every non-dead-end moment in a history h has a unique immediate successor m^{+h} and every non-root moment has a unique immediate predecessor m^-. m^{+h} is the immediate next state of moment m. The operators X, $[\alpha]^x$ are interpreted as follows, where $\mathcal{M}$ is an incremental STIT model that is 'endless' in the forward direction:

- $\mathcal{M}, m/h \vDash XA$ iff $\mathcal{M}, m^{+h}/h \vDash A$;
- $\mathcal{M}, m/h \vDash [\alpha]^x A$ iff for each $h' \in Choice^m_\alpha(h)$, $\mathcal{M}, m^{+h'}/h' \vDash A$.

We can intuitively explain $[\alpha]^x A$: in an incremental infinite STIT frame, A is guaranteed to be true in the next state due to the choice of the agent α at moment m, which is effectively equivalent to $[\alpha]^c XA$.

Relative to a model $\mathcal{M}$, the semantics for the formulas of $\mathcal{L}_{XSTIT}$ is defined recursively by the following truth conditions, evaluated at a given index m/h:

- $\mathcal{M}, m/h \vDash p$ iff $m/h \in v(p)$, where p is a propositional variant;
- $\mathcal{M}, m/h \vDash \neg A$ iff $\mathcal{M}, m/h \nvDash A$;
- $\mathcal{M}, m/h \vDash A \wedge B$ iff $\mathcal{M}, m/h \vDash A$ and $\mathcal{M}, m/h \vDash B$;
- $\mathcal{M}, m/h \vDash \Box A$ iff for each $h' \in H_m$, $\mathcal{M}, m/h' \vDash A$;
- $\mathcal{M}, m/h \vDash XA$ iff $\mathcal{M}, m^{+h}/h \vDash A$;
- $\mathcal{M}, m/h \vDash [\alpha]^x A$ iff for each $h' \in Choice^m_\alpha(h)$, $\mathcal{M}, m^{+h'}/h' \vDash A$.

Based on the above theory, Jan Broersen replaced the dstit operator used by Paul Bartha with the xstit operator, thus defining 'ought to do something' as:

$$\mathbf{O}[\alpha]^x A =_{df} \Box(\neg[\alpha]^x A \rightarrow [\alpha]^x V_\alpha).$$

We can interpret this to mean that if α does not see to it that A is true in the next state, then he ensures that he would cause a violation in the next state.

4.2 Analysis of the relationship between epistemology and obligation using STIT logic

Jan Broersen extends XSTIT theory with epistemic operators to formalize the notion of 'knowingly doing something', further suggesting other possible forms of 'ought to do something' in other possible forms [7,8]. His work on the combination of epitemic logic and deontic logic is highly informative for our formalization of 'ought not to know'.

We assume that the STIT framework and model are both incremental and infinite, and that the framework has relations $\sim_\alpha$ between indexes, where $\alpha \in Agent$. Each relation $\sim_\alpha$ is an equivalence relation between indexes, and '$m/h \sim_\alpha m'/h'$' can be understood as meaning that m/h and m'/h' are indistinguishable with respect to the agent α. We call such frameworks epistemic STIT frameworks, and we call $\sim_\alpha$ the α (epistemic) indistinguishability relation in the framework.

Jan Broersen [7] proposes the $\mathbf{K}_\alpha$ operator to formalize the agent's epistemic state The interpretation of $\mathbf{K}_\alpha$ is as follows:

- $\mathcal{M}, m/h \vDash \mathbf{K}_\alpha A$ iff for all m'/h' such that $m/h \sim_\alpha m'/h'$, $\mathcal{M}, m'/h' \vDash A$.

Based on this, we further extend the formal language as follows:

$$A, B ::= p \mid \neg A \mid (A \wedge B) \mid \mathbf{K}_\alpha A \mid \Box A \mid [\alpha]^x A \mid \mathbf{X} A$$

In light of the foregoing definition, Jan Broersen [8] puts 'the agent α knowingly sees to it that A' as $\mathbf{K}_\alpha [\alpha]^x A$. To distinguish between 'knowingly doing something' and 'unknowingly doing something', he denotes the former by $\mathbf{K}_\alpha [\alpha]^x A$ and the latter by $[\alpha]^x A \wedge \neg \mathbf{K}_\alpha [\alpha]^x A$.

In the following we discuss three possible properties of 'knowingly doing' proposed by Jan Broersen [9], and present them in axiomatic form.

The first property is 'knowledge about next states (Know-X)', which is expressed axiomatically as:

$$\mathbf{K}_\alpha \mathbf{X} A \rightarrow \mathbf{K}_\alpha [\alpha]^x A.$$

Its corresponding frame condition is:

- for all indexes m/h, m'/h' and all $h'' \in Choice_\alpha^m(h)$, $m/h \sim_\alpha m'/h'$ only if $m/h'' \sim_\alpha m'/h'$.

This property ensures that an angent cannot know more about the next states than what their choice affects. In other words, an agent can only know about (immediate) future events if it holds after their own actions. If something is true for all the dynamic states that are part of the actual choices made by the agent in the actual state, but not for all the dynamic states that the agent considers are possible, then we say that the thing is 'unknowingly done by the agent'. In general, an agent does far more things unknowingly than it does knowingly; Jan Broersen gives the example that by sending an email, we may be enforcing many things that we are not aware of, and that are the result of our sending the email. All of these things are done unknowingly by knowingly sending emails.

Jan Broersen calls the second property 'effect recollection (Rec-Eff)', axiomatically expressed as

$$\mathbf{K}_\alpha [\alpha]^x A \rightarrow \mathbf{X} \mathbf{K}_\alpha A.$$

whose corresponding frame condition is:

- For all indexes m/h and u/h', $m^{+h}/h \sim_\alpha u^{+h'}/h'$ only if $m/h \sim_\alpha u/h'$.

This property is a dynamic version of the 'perfect recall' axiom about the interaction between epistemic and temporal modalities. The property expresses the idea that if an agent knowingly sees to it that a condition holds in the next state, then in the same next state the agent will recall that the condition holds. That is, the effect of a known action is known in the next state.

The last property is the 'uniformity of strategies (Unif-Str)', axiomatically expressed as

$$\Diamond \mathbf{K}_\alpha [\alpha]^x A \rightarrow \mathbf{K}_\alpha \Diamond [\alpha]^x A.$$

Its corresponding frame condition is:

- If $m/h \sim_\alpha m^*/h^*$ and $h' \in H_m$, then there exists a $h'' \in H_{m^*}$ such that $m/h' \sim_\alpha m^*/h''$.

We can interpret this property to mean that if an agent can knowingly sees to it that A, then he knows that one of all his choices can see to it that A.

However, for events that are necessarily true, since their occurrence is not caused by the agent's actions, we do not say that the agent 'knowingly sees to it that these things happen'. For example, we can be sure that in the next state the earth rotates, but we do not say that 'we knowingly see to it that the earth rotate'. It is only when the agent causes something to happen that might not have come true without their intervention that we would say that the agent 'knowingly sees to it that something happens'. At the same time, when we say that an agent 'knowingly sees to it that something happens', we actually mean that the agent has other behavioural choices.[8] Based on this, we can further define 'the agent knowingly sees to it that something happens' as $\mathbf{K}_\alpha[\alpha]^x A \wedge \mathbf{K}_\alpha \neg \Box X A$.

Building on the above discussion, Jan Broersen adds the deontic operator to formalize 'the agent's obligation to see to it that something happens':

$$\mathbf{OK}[\alpha]^x A =_{def} \Box(\neg \mathbf{K}_\alpha[\alpha]^x A \to [\alpha]^x V_\alpha).$$

In this definition, the agent should perform the action 'knowingly' in order to avoid violation. In other words, if the agent does not knowingly perform the obligation, but 'coincidentally' complies, we would also say that he still causes a violation.

Jan Broersen also discusses another variant of the deontic operator. The above definition of 'the agent is obliged to see to it that something happens' does not mention whether the agent actually knows whether it is obliged to do it. Jan Broersen therefore adjusts the previous definition by directly linking the agent's awareness of the obligation to the awareness of the violation:

$$\mathbf{KOK}[\alpha]^x A =_{def} \Box(\neg \mathbf{K}_\alpha[\alpha]^x A \to \mathbf{K}_\alpha[\alpha]^x V_\alpha).$$

By this definition, the agent of the obligation has actual knowledge of the obligation, i.e., if the agent does not comply with it, he will intentionally cause the violation. Jan Broersen [9] mentions that the operator $\mathbf{OK}[\alpha]^x$ and $\mathbf{KOK}[\alpha]^x$ both satisfy the axiomatic system KD of standard deontic logic,[9] that is, they

---

[8] If an agent does not know that he has other choices than to do something, we do not say that the agent 'chooses' to do something, nor would we say that the agent's action was intentional.

[9] The axiomatic system KD of standard deontic logic consists of the following axioms and rules:

- **Taut:** all propositional reduplicative formulas
- **K:** $O(p \to q) \to (Op \to Oq)$
- **D:** $Op \to Pp$
- **MP:** $\varphi, \varphi \to \psi / \psi$

have the same properties as the standard deontic logic proposed by von Wright in *Deontic logic* [15]. ^{10}

5 The formalization of 'ought not to know'

In this section, we will combine STIT logic with the notion of 'ought not to know'.

5.1 Formalization of 'ought not to know' based on XSTIT theory

On the basis of the above discussion, we have attempted to formalize 'ought not to know'. We add to the meaning of 'ought not to know' the idea that if a person knowingly learns about something, then he could be considered to have caused the violation. Here, since 'knowing something' takes effect after the action of 'getting know something', we will use the xstit operator proposed by Jan Broersen in our definition.

According to our definition of 'ought not to know', if the agent is passively in the state of 'knowing something', he will not necessarily cause a violation. We therefore need to exclude the meaning of 'passively knows' from the formalization. We try to interpret 'passive knowledge' as meaning that the agent will be in a state of 'knowing something' at the next stage, but that the agent's own actions do not actually see to it that he will be in a state of 'knowing something' at the next stage, and is inscribed as: $X\mathbf{K}_\alpha\varphi \wedge \mathbf{K}_\alpha\neg([\alpha]^x\mathbf{K}_\alpha\varphi)$.

We use $\oslash_\alpha$ to mean 'the agent α ought not to know' and let φ mean 'something'. Then we can formalize 'the agent is obliged not to know something' as:

$$\oslash_\alpha\varphi =_{def} \Box(\mathbf{K}_\alpha[\alpha]^x\mathbf{K}_\alpha\varphi \rightarrow [\alpha]^x V_\alpha).$$

This can be understood intuitively as meaning that a person causes a violation at the next stage if his actions will see to it that he is in a state of 'knowing something' at the next stage. Here, the antecedent already excludes the case of 'passive knowledge'.

In the above definition, however, we have merely stated that 'if a person is knowingly in a state of knowing something, he will cause a violation', but not whether the person has deliberately committed a wrong with knowledge of a moral or legal norm. In order to distinguish 'knowingly committing a wrong' from 'unintentional mistakes', we directly link awareness of the obligation to the awareness of violation, defining 'ought not to know' as:

$$\oslash'_\alpha\varphi =_{def} \Box(\mathbf{K}_\alpha[\alpha]^x\mathbf{K}_\alpha\varphi \rightarrow \mathbf{K}_\alpha[\alpha]^x V_\alpha).$$

In this definition, a person is knowingly causing a violation at the next stage if he is knowingly in the state of 'knowing something' at the next stage.

- **Substitution rule:** $\varphi/\varphi(p, \psi)$
- **Generalization rule:** $\varphi/O\varphi$ The axioms and rules in

^{10}A detailed proof can be found in [9].

Here, both $\oslash_\alpha$ and $\oslash'_\alpha$ can be seen as definitions of 'ought not to know', the former is more applicable to charcaterizing implicit epistemic norms, [11] while the latter is more applicable to charcaterizing explicit epistmic norms. [12] Since STIT logic only considers real-world possibilities, our definition can be used to formalize real-world moral and legal norms of cognition.

5.2   Add 'knowing-value' operator

There is still a problem with the above research: it seems more natural that our focus on 'knowing something' should be expressed as 'knowing whether' or 'knowing what' rather than 'knowing that'.

In the real-life example we mentioned earlier, we want to emphasise that someone ought not to know that 'the PIN for Ben's bank card is 123456'. In this case, if we understand 'knowing something' in the form of 'knowing that', we would say that 'someone ought not to know [that] the PIN for Ben's bank card is 123456'. And if we take the 'knowing what' form, we can express this example as 'someone ought not to know [what] Ben's bank card PIN [is])', where someone does not need to know exactly what Ben's PIN is. Although the two are equivalent for an external observer, the fact that an external observer thinks α is true does not mean that the agent himself knows α. Thus, there is an obvious problem if we understand 'knowing something' in terms of 'knowing that': intuitively, if we say 'by law someone ought not to know something', then in fact someone should know that he should not have known about it. However, for the agent himself, if a person does not actually know that 'the PIN for Ben's bank card is 123456', we cannot say that 'a person ought to have known that he ought not to have known that the PIN for Ben's bank card is 123456', otherwise 'the PIN for Ben's bank card is 123456' would be known to this agent and would create a contradiction. It is worth noting that 'ought to know' in this example and 'ought not to know' that we discuss in this paper are still external norms. In contrast, when we need to discuss 'the agent ought to know that he ought not to know something', the reference to 'ought not to know something' is from the agent's internal perspective. Thus, our use of knowing-what avoids the contradictions that arise from internal and external distinctions.

Meanwhile, according to Yanjing Wang [16], we have the following information that there may be different patterns of reasoning for knowledge expressions of 'Knowing-Wh' than for propositional knowledge expressions of the 'Knowing That' form . Further, we will therefore modify the epistemic operator in the inscription of 'ought not to know', so as to provide a more reasonable inscription of 'ought not to know' as we have defined it.

Jan Plaza [13] proposes that an agent is said to know the value of c if the constant c has the same value in all worlds, and that value is indistinguishable from the actual value. Since 'knowing whether φ' can also be understood as

---

[11] Norms that are agreed upon but not explicitly stated, since there is not necessarily a uniform perception of such norms.

[12] Norms that are explicitly stated, such as laws, etc..

'knowing what the truth value (logical value) of φ is'[13], we can distinguish 'know whether' together with 'know what' for the inscription. Based on this, we introduce a new notion: 'knowing [what the] value [is]', each constant has a value that ranges over a possibly infinite domain,[14] which we denote by the operator $\mathbf{K}v_\alpha$. Here, we give an example for a more intuitive interpretation of the operator: we can use $\mathbf{K}_i\mathbf{K}v_jc \wedge \neg\mathbf{K}v_ic$ to mean 'i knows that j knows the PIN of Ben's bank card, but i does not exactly know what the PIN is'.

Further, we extend the formal language as follows (where c is any constant symbol in the given set $\mathbf{C}$)

$$A, B ::= p \mid \neg A \mid (A \wedge B) \mid \mathbf{K}_\alpha A \mid \mathbf{K}v_\alpha c \mid \Box A \mid [\alpha]^x A \mid \mathrm{X}A$$

To explain $\mathbf{K}v_\alpha c$, here we define an assignment function $V_{\mathbf{C}}$ on the cross product of the index set and the set $\mathbf{C}$, based on [17]. The truth condition for $\mathbf{K}v_\alpha c$ [15] is defined as follows:

- $\mathcal{M}, m/h \vDash \mathbf{K}v_\alpha c$ iff for each m'/h', m''/h'', if $m/h \sim_\alpha m'/h'$ and $m/h \sim_\alpha m''/h''$, then $V_{\mathbf{C}}(c, m'/h') = V_{\mathbf{C}}(c, m''/h'')$

Based on the above operator, we modify the formalization of 'ought not to know'. We inscribe the implicit and explicit norms of 'the agent is obliged not to know something' as follows, respectively:

$$\oslash''_\alpha\varphi =_{def} \Box(\mathbf{K}_\alpha[\alpha]^x\mathbf{K}v_\alpha\varphi \to [\alpha]^x V_\alpha).$$

$$\oslash'''_\alpha\varphi =_{def} \Box(\mathbf{K}_\alpha[\alpha]^x\mathbf{K}v_\alpha\varphi \to \mathbf{K}_\alpha[\alpha]^x V_\alpha).$$

Here, both expressions equally indicate that the agent will cause a violation at the next stage (or knows that he will cause a violation at the next stage) if his action will see to it that he is in the state of 'knowing something' at the next stage. However, in this context 'knowing something' means 'knowing what'. Unlike 'knowing that', 'knowing what' does not need to be followed by a statement that is necessarily true in all possible worlds, so the definition can be applied to a wider range of issues.

We further discuss the possible property of 'ought not to know'. For explicit epistemic norms, if we ought not to know something, then we would say that we ought to know that we ought not to know it. For example, in the Chinese Legal System, if a natural person's personal information is protected by law and cannot be disclosed without permission, then we should know that we are obliged not to actively obtain another person's personal information, otherwise

---

[13]i.e., whether the truth value is 0 or 1.

[14]According to [16], knowing the value of something can be seen as knowing the answer to a hidden question.

[15]Jan Plaza suggests that for $\mathbf{K}v_\alpha c$ the following two axioms are valid [13]:

(i) $\mathbf{K}v_ic \to \mathbf{K}_i\mathbf{K}v_ic$

(ii) $\neg\mathbf{K}v_ic \to \mathbf{K}_i\neg\mathbf{K}v_ic$

we are negligent in not knowing. Based on this, we can express this property as follows:

$$\Box(\neg \mathbf{K}_\alpha \oslash_\alpha''' \varphi \to [\alpha]^x V_\alpha).$$

6 Conclusion

In the post-privacy era, the issue of privacy ethics has become a topic of great concern. With the rapid development of digitalization and intelligence, people's personal information is constantly being collected, stored and processed, which has led to an increasing threat to personal privacy. Against this background, this paper presents a philosophical discussion of a new epistemic norm - 'ought not to know' - and gives a formal formalization of it. Under this new epistemic norm, the agent ought to be unknown of certain information under certain circumstances, thus protecting personal privacy. Considering that this norm emphasizes the agency of the agent, STIT logic is chosen as a tool for formalization in this paper.

In this study, we first introduce the STIT framework, models and common operators. On this basis, we discuss the application of STIT theory to deontic and epistemic problems. Then, in order to integrate deontic and epistemic problems, we refer to the XSTIT theory, which is an extension of the STIT theory, and formalize 'ought not to know'. Within this extended theoretical framework, the 'ought not to know' operator is formally defined and formalized. At the same time, a distinction is found between 'knowing that' and 'knowing what', and the formalization is further modified by the inclusion of a 'knowing-value' operator. Finally, we discuss the possible properties of the 'ought not to know' operator, providing a direction and basis for further research on this operator.

In future work, we will explore and discuss more complex properties and axiomatic systems that may be satisfied by this operator. We will also think about the de re and de facto issues haunting epistemology and formalisms assuming omniscience. This research provides a new way of thinking about understanding deontic issues of privacy and provides a new paradigm for the application of STIT logic to deontic issues and epistemic problems. In addition, since AI research is also concerned about agency and the issues discussed in this study are highly relevant to AI security, subsequent work on this research could be combined with other fields such as AI and computer science to explore broader applications.

References

[1] Abarca, A. I. R. and J. Broersen, *A deontic stit logic based on beliefs and expected utility*, arXiv preprint arXiv:2106.11506 (2021).

[2] Anderson, A. R., *The formal analysis of normative systems* (1956).

[3] Bartha, P., *Conditional obligation, deontic paradoxes, and the logic of agency*, Annals of Mathematics and Artificial Intelligence **9** (1993), pp. 1–23.

[4] Belnap, N., *Before refraining: Concepts for agency*, Erkenntnis **34** (1991), pp. 137–169.

[5] Belnap, N. and M. Perloff, *Seeing to it that: A canonical form for agentives*, Knowledge representation and defeasible reasoning (1990), pp. 167–190.

[6] Belnap, N., M. Perloff and M. Xu, "Facing the future: agents and choices in our indeterminist world," Oxford University Press, 2001.

[7] Broersen, J., *A logical analysis of the interaction between 'obligation-to-do'and 'knowingly doing'*, in: *Deontic Logic in Computer Science: 9th International Conference, DEON 2008, Luxembourg, Luxembourg, July 15-18, 2008. Proceedings 9*, Springer, 2008, pp. 140–154.

[8] Broersen, J., *A complete stit logic for knowledge and action, and some of its applications*, in: *Declarative Agent Languages and Technologies VI: 6th International Workshop, DALT 2008, Estoril, Portugal, May 12, 2008, Revised Selected and Invited Papers 6*, Springer, 2009, pp. 47–59.

[9] Broersen, J., *Deontic epistemic stit logic distinguishing modes of mens rea*, Journal of Applied Logic **9** (2011), pp. 137–152.

[10] Harrison, J., *Some reflections on the ethics of knowledge and belief*, Religious studies **23** (1987), pp. 325–336.

[11] Horty, J., "Epistemic oughts in stit semantics," Ann Arbor, MI: Michigan Publishing, University of Michigan Library, 2019.

[12] Horty, J. F., "Agency and deontic logic," Oxford University Press, 2001.

[13] Plaza, J., *Logics of public communications*, Synthese **158** (2007), pp. 165–179.

[14] Pollock, J. L., *Epistemic norms*, Synthese (1987), pp. 61–95.

[15] Von Wright, G. H., *Deontic logic*, Mind **60** (1951), pp. 1–15.

[16] Wang, Y., *Beyond knowing that: a new generation of epistemic logics*, Jaakko Hintikka on Knowledge and Game-Theoretical Semantics (2018), pp. 499–533.

[17] Wang, Y. and J. Fan, *Knowing that, knowing what, and public communication: Public announcement logic with kv operators*, in: *Proceedings of the Twenty-Third International Joint Conference on Artificial Intelligence*, IJCAI '13 (2013), p. 1147–1154.

[18] Xu, M., *Combinations of stit with ought and know*, Journal of Philosophical Logic **44** (2015), pp. 851–877.

Defeasible Description Logic Reasoning Based on Abstract Syntax Graph

Dongheng Chen, Muyun Shao

Zhejiang University

Dov M. Gabbay

King's College London, United Kingdom
University of Luxembourg, Luxembourg

Abstract

In traditional defeasible description logic systems like **rational closure**, reasoning relies mainly on T-Box. However, A-Box is more abundant and more accurate compared to T-Box because A-Box directly comes from the corpus in most cases, while T-Box comes from summarizing and concluding relations between elements in A-Box. Besides, the rational closure system does not support incomplete reasoning. Incomplete reasoning means that the system can infer necessary intermediates for reasoning if they are not provable in the system. To overcome these problems, we propose a new method for defeasible description logic reasoning using abstract syntax graphs. Abstract syntax graphs are directed graphs in which nodes are assigned with formulas. We propose a method inspired by models in argumentation theory to compute consistent sets of formulas in a graph. For cases where we obtain multiple solutions, we also define a partial-order on sets of formulas to select the best answer. We end this paper with comparisons with other approaches to illustrate the advantage of abstract syntax graphs.

Keywords: Description logic, Defeasible reasoning, Abstract Syntax Graph

1 Introduction

Nowadays, defeasible reasoning on a knowledge base is receiving increasing attention. Although description logics are fundamental to numerous contemporary AI and database applications, the formulas in description logics are incapable of properly expressing and reasoning with defeasible rules.

In order to achieve defeasible reasoning in description logics, many approaches [4,5,6] have been proposed, among which a representative one is **rational closure**. Its basic idea is that there is an ordering on defeasible rules. Some rules are more general, such as *birds can fly*, while others are more specific, such as *birds with severely injured wings cannot fly*. Generally speaking, rules with more antecedents are more specific. When a contradiction arises in reasoning, we prefer to apply specific rules.

However, **rational closure** does not support defeasible reasoning with A-Box. Although some work has been done to implement **rational closure** to defeasible knowledge bases with A-Box[9], they are not suitable for defeasible reasoning in incomplete knowledge bases. This is because it can't infer the atomic terms that are not provable in the knowledge base, while **rational closure** may need these terms as intermediate conditions for reasoning.

To overcome these problems, we propose a new model for defeasible description logic reasoning called abstract syntax graphs. The idea is inspired by the concept of Abstract Syntax Tree[13]. In abstract syntax graphs, we translate each term in description logics into an abstract syntax tree, then we build an abstract syntax graph(ASG) by combining all abstract syntax trees together. To compute consistent sets in ASG, we propose a method similar to models in argumentation theory. We also define a partial-order on consistent sets in case we have multiple solutions. Furthermore, we show that the most preferred consistent set obtained from the partial-order can be used to generate explanations, which serves as a response if there is any disagreement on the truth value of particular formulas.

The rest of the papers will be organized as follows. Section two is reserved for background knowledge. Then we introduce a formal model of abstract syntax graphs in section three, and discuss properties in section four. In section five we discuss related work and conclude in section six.

2　$\mathcal{ALC}$ Language

$\mathcal{ALC}$ Description Logic is a formal language used for expressing knowledge about concepts and their relationships in a domain. It is a subset of first-order logic with restricted syntax and semantics. The language consists of concepts, roles, and individuals, where concepts represent sets of individuals, roles represent binary relations between individuals, and individuals represent objects in the domain. The syntax of $\mathcal{ALC}$ is used to construct expressions that represent concepts and their relationships, while the semantics of $\mathcal{ALC}$ is used to define the meaning of these expressions.

Definition 2.1 ($\mathcal{ALC}$ Language) An $\mathcal{ALC}$ language is a triple $\langle \mathbf{C}, \mathbf{R}, \mathbf{I} \rangle$, where $\mathbf{C}$ is a finite set of atomic concept names, $\mathbf{R}$ is a finite set of role names, and $\mathbf{I}$ (a.k.a. attributes) is a finite set of individual names. $\mathbf{C}, \mathbf{R}$ and $\mathbf{I}$ are pairwise disjoint. [8]

Definition 2.2 ($\mathcal{ALC}$ Concept) Concepts of $\mathcal{ALC}$ are defined inductively as follows:

$$C ::= \top \mid \bot \mid D \mid (\neg C) \mid (C \sqcap C) \mid (C \sqcup C) \mid (\exists r.C) \mid (\forall r.C)$$

where $D \in \mathbf{C}$ and $r \in \mathbf{R}$.

Definition 2.3 ($\mathcal{ALC}$ General Concept Inclusion(GCI)) GCIs of $\mathcal{ALC}$ are defined as follows:

$$GCI ::= C \sqsubseteq D$$

where C, D are $\mathcal{ALC}$ concepts. Specially, we use $C \equiv D$ as an abbreviation for $C \sqsubseteq D$ and $D \sqsubseteq C$.

Definition 2.4 (Satisfaction of $\mathcal{ALC}$ GCI) An interpretation $\mathcal{I}$ satisfies a GCI $C \sqsubseteq D$ if $C^{\mathcal{I}} \subseteq D^{\mathcal{I}}$.

Definition 2.5 ($\mathcal{ALC}$ T-Box) A finite set of GCIs is called an $\mathcal{ALC}$ T-Box. [8]

Definition 2.6 ($\mathcal{ALC}$ Assertion)

- $C(a)$ is called an $\mathcal{ALC}$ concept assertion(CA).

- $r(a,b)$ is called an $\mathcal{ALC}$ role assertion(RA).

where C is an $\mathcal{ALC}$ concept, a, b are $\mathcal{ALC}$ individuals and r is an $\mathcal{ALC}$ role.

Definition 2.7 ($\mathcal{ALC}$ A-Box) A finite set of CAs and RAs is called an $\mathcal{ALC}$ A-Box. [7]

Definition 2.8 ($\mathcal{ALC}$ Knowledge Base) An $\mathcal{ALC}$ knowledge base $\mathcal{K}$ is defined as follows.
$$\mathcal{K} ::= \langle \mathcal{T}, \mathcal{A}, \mathcal{I} \rangle$$
where $\mathcal{T}$ is an $\mathcal{ALC}$ T-Box, $\mathcal{A}$ is an $\mathcal{ALC}$ A-Box and $\mathcal{I}$ is a subset of **I**.

Definition 2.9 ($\mathcal{ALC}$ Term) General Concept Inclusions (GCI), Role Assertions (RA), and Concept Assertions (CA) collectively refer to $\mathcal{ALC}$ terms.

Definition 2.10 (Defeasible knowledge base) The defeasible knowledge base $\mathcal{K}$ is a tuple $\langle \mathcal{T}, \mathcal{T}^*, \mathcal{A}, \mathcal{A}^*, \mathcal{I} \rangle$, while $\mathcal{T}$ is a strict $\mathcal{ALC}$ T-Box, $\mathcal{T}^*$ is a defeasible $\mathcal{ALC}$ T-Box, $\mathcal{A}$ is a strict $\mathcal{ALC}$ A-Box and $\mathcal{A}^*$ is a defeasible $\mathcal{ALC}$ A-Box.

Example 2.11 Suppose we need to formalize a rule called *Birds(B) can fly(F)* with two individual Tweety(t) and Jimmy(j). We can build an $\mathcal{ALC}$ knowledge base $\mathcal{K} = \langle \mathcal{T}, \mathcal{A}, \mathcal{I} \rangle$ where $\mathcal{T} = \{B \sqsubseteq F\}$, $\mathcal{A} = \{B(t), B(j)\}$.

3 Reasoning on Incomplete and Inconsistent Knowledge Bases

To reason on an incomplete and inconsistent knowledge base, we need a more complex example.

Example 3.1 Now we consider adding more rules to Example 2.11. The first rule being added is *Injured(I) animals(A) can not fly(F)*. To build the relationship between this rule and the last rule, we add the rule *Bird(B) is a kind of animal(A)*. To distinguish two individuals, let Tweety(t) be injured(I). The knowledge base $\mathcal{K} = \langle \mathcal{T}, \mathcal{T}^*, \mathcal{A}, \mathcal{A}^*, \mathcal{I} \rangle$ is shown as follows:

$$\mathcal{T} = \{B \sqsubseteq F\}, \mathcal{A} = \{B(t), B(j)\}$$

$$\mathcal{T}^* = \{A \sqcap I \sqsubseteq \neg F, B \sqsubseteq A\}, \mathcal{A}^* = \{I(t)\}$$

Now we find that this knowledge base is no longer consistent. This is because from the CA *Tweety is a bird* and the GCI *bird can fly*, one may infer that Tweety can fly. But from the CA *Tweety is an injured bird* and GCI *birds are a kind of animal* and *injured animal can not fly*, we infer that Tweety can not fly.

In this example, what we want to know is whether a specific individual (Tweety or Jimmy) can fly. To illustrate the question, we define the deterministic problem.

Definition 3.2 (Deterministic Problem) Let $\mathcal{K} = \langle \mathcal{T}, \mathcal{A}, \mathcal{I} \rangle$ be an $\mathcal{ALC}$ knowledge base, C be an $\mathcal{ALC}$ concept and d be an $\mathcal{ALC}$ individual, a deterministic problem is to check whether $C(d)$ or $\neg C(d)$ is more preferred.

In order to calculate several possible consistent sets for evaluation, we need to put them all together in one graph. To choose the most preferred answer, we need a method to compare two consistent sets and determine which is more preferred.

3.1   Graph Construction

Before reasoning, we need to ground all GCIs to CAs. This is how we combine the information from both A-Box and T-Box. We use the Gr function to do so. The Gr function takes a set of GCIs and a set of individuals as input and produces a set of CAs as output. We ground $C \sqsubseteq D$ by $\neg C \sqcup D$ because $C \sqsubseteq D \equiv \neg C \sqcup D$. The definition is as follows:

Definition 3.3 (Gr function)

$$Gr(G, I) = \{(\neg C \sqcup D)(i) | C \sqsubseteq D \in G, i \in I\}$$

where G is a set of GCIs, I is a set of individual.

After grounding, the next step is to combine all the CAs together. To reduce complextity, we first simplify the CAs. we use a functionally complete set of logic connectives $\{\sqcap, \neg\}$ to replace all the other logic connectives in CAs. We call a CA with only two logic connectives $\{\sqcap, \neg\}$ as a standard CA. The Std function takes a CA as input and produces a standard CA as output. The inductive definition is defined as follows:

Definition 3.4 (Std Function)

$$Std(C(d)) = C(d), C \in \mathbf{C}$$
$$Std((\neg C)(d)) = \neg Std(C(d))$$
$$Std((C \sqcap D)(d)) = Std(C(d)) \sqcap Std(D(d))$$
$$Std((C \sqcup D)(d)) = \neg(\neg Std(C(d)) \sqcap \neg Std(D(d)))$$
$$Std(\forall r.C(d)) = Std(\sqcap_{(d,x) \in r} C(x))$$
$$Std(\exists r.C(d)) = Std(\neg(\forall r.(\neg C(d))))$$

where C, D are $\mathcal{ALC}$ concepts, $r \in \mathbf{R}$ and $d \in \mathbf{I}$. Especially, because the domain is finite, we can translate quantifiers into conjunctions. This is how we deal with the roles and quantifiers in the knowledge base.

After standardization, we build a graph containing all the standard CAs. we need to resolve each CA into sub units. For example, to check if $C(d) \sqcap D(d)$

holds, we need to check if $C(d)$ and $D(d)$ hold, here $C(d)$ and $D(d)$ are sub units of $C(d) \sqcap D(d)$. The intuitive thought is to build an abstract syntax tree(AST)[13]. However, we need two kinds of edges because we have two kinds of logic connectives. So we slightly modify the algorithm to get an abstract syntax graph(ASG) (An example of ASG can be found in section 3.4). We first translate all CAs to ASTs, then combine these ASTs by merging the equivalent nodes. Finally, we get a graph containing all the CAs.

Definition 3.5 (Node Equivalence) For two ACs $C(d)$ and $D(d)$, we define $C(d) \sim D(d)$ if and only if $C(d) \vdash D(d)$ and $D(d) \vdash C(d)$ (under nature deduction of $\mathcal{ALC}$[7]). With the relation of equivalence, we denote the equivalence class of $C(d)$ as $[C(d)]$, which means $D(e) \in [C(d)] \iff D(e) \sim C(d)$.

Definition 3.6 (Edge Equivalence) $Edges(v_1, v_2)$ and (v_3, v_4) are equivalent if and only if $v_1 \sim v_3$ and $v_2 \sim v_4$.

Definition 3.7 (Abstract Syntax Graph(ASG)) ASG is a tuple $T = \langle V, E_n, E_c \rangle$, where V is the set of nodes, each of which represents a CA or sub-CA. E_n is a set of undirected edges, which connect the nodes mutually exclusive. E_c is a set of directed edges. The head of an edge is the premise, and the tail of the edge is the conclusion. The conclusion holds true if and only if all the premises are true.

To build an ASG, we need three auxiliary functions V^c, E_n^c and E_c^c to generate three sets in the tuple. The first function V^c generates all the nodes of the ASG. The second function E_n^c generates all the negation relations of the ASG. The third function E_c^c generates all the conjunction relations of the ASG. The inductive definitions are as follows:

$$V^c(C(d)) = \{C(d)\}, C \in \mathbf{C}$$
$$V^c(\neg C(d)) = \{\neg C(d)\} \cup V^c(C(d))$$
$$V^c(C(d) \sqcap D(d)) = \{C(d), D(d)\} \cup V^c(C(d)) \cup V^c(D(d))$$

$$E_n^c(C(d)) = \emptyset, C \in \mathbf{C}$$
$$E_n^c(\neg C(d)) = \{(C(d), \neg C(d)), (\neg C(d), C(d))\} \cup E_n^c(C(d))$$
$$E_n^c(C(d) \sqcap D(d)) = E_n^c(C(d)) \cup E_n^c(D(d))$$

$$E_c^c(C(d)) = \emptyset, C \in \mathbf{C}$$
$$E_c^c(\neg C(d)) = E_c^c(C(d))$$
$$E_c^c(C(d) \sqcap D(d)) = \{(C(d), C(d) \sqcap D(d)), (D(d), C(d) \sqcap D(d))\} \cup$$
$$E_n^c(C(d)) \cup E_n^c(D(d))$$

where C, D are $\mathcal{ALC}$ concepts and $d \in \mathbf{I}$.

Now we have introduced how to build an ASG based on a knowledge database. The next definition summarizes all the processes of building an ASG.

Definition 3.8 (*ASG* on defeasible knowledge base) For a defeasible knowledge base $\mathcal{K} = \langle \mathcal{T}, \mathcal{T}^*, \mathcal{A}, \mathcal{A}^*, \mathcal{I} \rangle$, its corresponding *ASG* $\mathbf{T} = \langle V, E_n, E_c \rangle$ is defined as follows:

$$V = \{[v] | v \in \bigcup_{t \in S} V^c(t)\} \qquad E_n = \{[e] | e \in \bigcup_{t \in S} E_n^c(t)\} \qquad E_c = \{[e] | e \in \bigcup_{t \in S} E_c^c(t)\}$$

while $S = \{Std(a) | a \in \mathcal{A} \cup \mathcal{A}^* \cup Gr(\mathcal{T} \cup \mathcal{T}^*, \mathcal{I})\}$

3.2　Model

After building the ASG of a defeasible knowledge base, the next step is to compute consistent sets of formulas according to the graph. In this paper we propose a method inspired by labeling approach in argumentation theory[12]. We obtain consistent sets of formulas by computing models of an ASG. For a graph $T = \langle V, E_n, E_c \rangle$, a model is a function $\mathcal{M}$ maps each node in V to $\{in, out, undec\}$. We call $\mathcal{M}(v)$ the label on node v. Specifically, if $\forall v \in V, \mathcal{M}(v) = X$, we denote $\mathcal{M}(V) = X$. The consistency of a model is defined as follows:

Definition 3.9 (Consistency Model) A model $\mathcal{M}$ is consistent if it satisfies the conditions below:

$$\forall (v, v^*) \in E_n, \mathcal{M}(v) = in \Rightarrow \mathcal{M}(v^*) = out$$
$$\forall (v, v^*) \in E_n, \mathcal{M}(v) = undec \Rightarrow \mathcal{M}(v^*) = undec$$
$$\forall v \in V, \mathcal{M}(v) = in \Rightarrow \forall (v^*, v) \in E_c, \mathcal{M}(v^*) = in$$
$$\forall v \in V, \mathcal{M}(v) = out \Rightarrow \exists (v^*, v) \in E_c, \mathcal{M}(v^*) = out$$

Here we provide a computational procedure for computing models of an ASG $\langle V, E_n, E_c \rangle$. We start with a set S containing all the legal models and an empty set S^c. Then we check each model by the definition of consistency. If the model is consistent, we add it to the S^c. After we have checked all the legal models, the remaining part of the set is all the consistent models. Its formalization steps are as follows.

(i) Let $S = \{in, out, undec\}^V$

(ii) For each $\mathcal{M} \in S$, if $\mathcal{M}$ is consistent, $S^c = S^c \cup \{\mathcal{M}\}$

(iii) Return S^c

3.3　Model Ordering

We notice that in some circumstances, we obtain multiple models using the computational procedure we define in the previous section. In such cases we would like to find out the model which is the most preferred. To do so, we define a partial-order on models. The basic idea is the following. First of all, if $C(d)$ or $\neg C(d)$ can be directly inferred by strict rules, then every model which supports $C(d)$ or $\neg C(d)$ should have the highest priority. In contrast, the model which violates the strict rules should have the lowest priority. Secondly, we want the model to satisfy the closed-world assumption, which means that atomic CAs that are not provable in the knowledge base should be labeled as out as many as possible.

Let $\mathcal{K} = \langle \mathcal{T}, \mathcal{T}^*, \mathcal{A}, \mathcal{A}^*, \mathcal{I} \rangle$ be a knowledge base and $ASG_\mathcal{K} = \langle V, E_n, E_c \rangle$ be its corresponding ASG. For two consistent models $\mathcal{M}_1$ and $\mathcal{M}_2$, $\mathcal{M}_1 \geq \mathcal{M}_2$ if it satisfies one of the cases listed below:

- $\exists X \in \{in, out\}, (\forall \mathcal{M}(\mathcal{M}$ is consistent $\wedge \mathcal{M}(V_s^+) = in) \Rightarrow \mathcal{M}(C(d)) = X) \wedge \mathcal{M}_1(V_s^+) = in$
- $\exists v \in \mathcal{M}_2, (v \in V_s^+) \wedge (\mathcal{M}(v) = out)$
- $\mathcal{M}_2(C(d)) = undec$
- $\forall v \in V_a, \mathcal{M}_2(v) = undec \Rightarrow \mathcal{M}_1(v) = undec$

  while

$$
\begin{aligned}
V_s^+ &= \{[v] \in V | v \in Gr(\mathcal{T}, I) \cup \mathcal{A}\} \\
V_s &= V_s^+ \cup \{s \in V_s^+ | \neg s\} \cap V \\
V_d^+ &= \{[v] \in V | v \in Gr(\mathcal{T}^*, I) \cup \mathcal{A}^*\} \\
V_d &= V_d^+ \cup \{s \in V_d^+ | \neg s\} \cap V \\
V_a^+ &= \{v \in V | v \text{ is an atomic CA}\}/(V_d \cup V_s) \\
V_a &= V_a^+ \cup \{s \in V_a^+ | \neg s\} \cap V
\end{aligned}
$$

The first case means if $C(d)$ can be reasoned by strict CAs, then the model must be the best model. The second case means that if a strict CA is *out*, then the model should have the lowest priority. The third case means that if a model does not get a useful result, it should be aborted. The fourth case means we should maximize the *undec* label for unmentioned atomic CAs.

Generally, we remark the model with the highest priority as the most preferred model.

3.4 Example

Continuing with the setting of example 2.11, we show how our model works step by step. The input database is $\mathcal{K} = \langle \mathcal{T}, \mathcal{A}, \mathcal{I} \rangle$. The question is if $F(t)$ holds.

$$
\begin{aligned}
\mathcal{T} =& \{B \sqsubseteq F, A \sqcap I \sqsubseteq \neg F, B \sqsubseteq A\} \\
\mathcal{A} =& \{B(t), B(j), I(t)\} \\
Gr(T_r, I_r) =& \{(B \sqsubseteq F)(t), (A \sqcap I \sqsubseteq \neg F)(t), (B \sqsubseteq A)(t)\} \\
A_g =& \{(B \sqsubseteq F)(t), (A \sqcap I \sqsubseteq \neg F)(t), (B \sqsubseteq A)(t), B(t), I(t)\} \\
\{Std(a) | a \in A_g\} =& \{\neg(B(t) \wedge \neg F(t)), \neg(I(t) \wedge A(t) \wedge F(t)), \\
& \neg(B(t) \wedge \neg A(t)), B(t), I(t)\}
\end{aligned}
$$

Compared to calculating the structure of ASG step by step, it's more visualized to draw the ASG. Here we draw the ASG of the CA $\neg(B(t) \wedge \neg F(t))$ to show how it works.

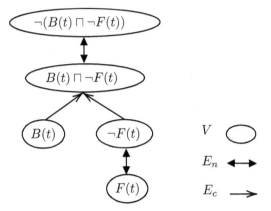

The whole graph is shown below:

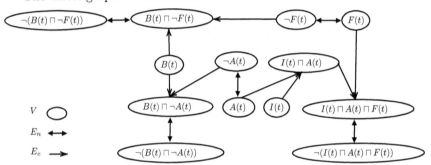

One consistent model of the graph is shown below:

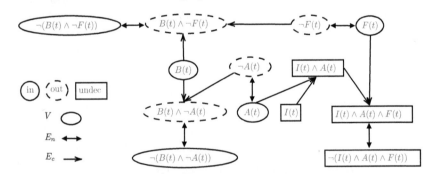

4 Properties

The first important property of this model is that it can generate rebuttal based on the most preferred model. For a model $\mathcal{M}$ of the ASG $\langle V, E_n, E_c \rangle$, an attack to the model $\mathcal{M}$ is a set $Att \subseteq V$. Each element in Att corresponds to the disagreement with the label of the node. For example, if $Att = \{A(b)\}$ and $\mathcal{M}(A(b)) = in$, it claims that it should be $\mathcal{M}(A(b)) = out$. Two methods are used to rebut the attack. The first method is to accept the attack and give back another preferred model with the same label on $C(d)$. The idea is that accepting these new labels will not effect the label of $C(d)$. If the first method fails, we will turn to the second method, which is to find a subset of V_a called E. If the attack is accepted under the same satisfaction of defeasible rules, then one of the atomic CAs will be labeled as in while $\mathcal{M}(E) = out$. This means that the model $\mathcal{M}$ better satisfies the closed-world assumption. The proof of this property is as follows.

Definition 4.1 (Restricted model) $\mathcal{M}^*$ is a preferred model, the set of restricted models of $\mathcal{M}^*$ about Att is $\{\mathcal{M}|\mathcal{M}$ is a preferred model $\wedge\ \forall v \in Att, \mathcal{M}(v) \neq \mathcal{M}^*(v) \wedge \mathcal{M}(C(d)) = \mathcal{M}^*(C(d))\}$

For any attack Att, if the set of restricted models is not empty, we can give out an model in the set as rebut.

Lemma 4.2 *(Consistent models can be uniquely identified by atomic nodes)* *For two consistent models $\mathcal{M}_1$ and $\mathcal{M}_2$, if $\forall s$ is an atomic $CA, \mathcal{M}_1(s) = \mathcal{M}_2(s)$, then $\forall s, \mathcal{M}_1(s) = \mathcal{M}_2(s)$.*

Proof. $\forall v, \mathcal{M}(v)$ is unique if and only if all the atomic CAs of v are unique. Because all the atomic CAs are unique, the atomic CAs of v are unique.   □

Theorem 4.3 *(Subset Rebut) T is an ASG and $C(d)$ is the question. For an attack Att to a preferred model $\mathcal{M}$, if the restricted model does not exist, then exists a subset E of V_a, such that $\mathcal{M}(E) = undec$, and for any preferred model $\mathcal{M}^*$, if for all $s \in Att, \mathcal{M}^*(s) \neq \mathcal{M}(s)$ and $\forall s \in V_d, \mathcal{M}(v) = \mathcal{M}^*(v)$, then exists $s_0 \in E$, such that $\mathcal{M}^*(s_0) \neq undec$.*

Proof. We claim that the set of defeasible nodes E is $\{s \in V_a|\mathcal{M}(s) = undec\}$. By the definition of E, we know $\mathcal{M}(E) = undec$. Then we prove by contradiction. If it's not the case, there exists a preferred model $\mathcal{M}^*$, such that for all $s \in Att, \mathcal{M}^*(s) \neq \mathcal{M}(s)$, and $\mathcal{M}^*(E) = undec$.

If $E \cap Att \neq \emptyset$, then we immediately get $\mathcal{M}^*(E) \neq undec$, contradicted with $\mathcal{M}^*(E) = undec$. So we assume that $E \cap Att = \emptyset$.

So for model $\mathcal{M}^*$, we know $\{s \in V_a|\mathcal{M}(s) = undec\} = E \subseteq \{s \in V_d|\mathcal{M}^*(s) = undec\}$. By the definition of priority on model, we know $\mathcal{M}^* \leq \mathcal{M}$. However, $\mathcal{M}$ is a preferred model, so $\mathcal{M} \leq \mathcal{M}^*$. As a result, we have $\{s \in V_a|\mathcal{M}(s) = undec\} = \{s \in V_a|\mathcal{M}^*(s) = undec\}$. By the definition of preferred, we know $\{s \in V_s|\mathcal{M}(s) = \mathcal{M}^*(s)\}$. As a result, $\forall s$ is an atomic CA, $\mathcal{M}(s) = \mathcal{M}^*(s)$. By lemma 1, we know $\mathcal{M} = \mathcal{M}^*$. But by the assumption, we know $\mathcal{M}(C(d)) \neq \mathcal{M}^*(C(d))$ because there does not exist restricted model, which leads to the contradiction.   □

5   Related Work

DeLP [3], ASPIC+ [11], and other structured argumentation frameworks can also be used for defeasible reasoning. Although they are not designed for description logic, they can still be used for defeasible description logic reasoning because description logic is the subset of first-order logic. The main difference between these two approaches and our model is that these two approaches require building arguments before reasoning. An argument is a proof, containing CGIs,

CAs, and RAs. After calculating the attack relation between arguments, we can build an argumentation framework and calculate the extensions. However, these approaches are in lack of efficiency because when building arguments, sub-arguments are also built. As a result, the same GCIs, CAs, and RAs have been reused repetitively and making the framework too complex for computation. Although some efforts[10] have been made to improve the efficiency of calculation, it is still complex compared to direct reasoning with A-Box and T-Box. In our model, each CA or RA appears only once, and GCIs only appear once for each individual. As a result, the cost of computation and storage has significantly decreased. Briefly, our model is more suitable for defeasible description logic compared with structured argumentation frameworks.

Rational closure[8] is another model for defeasible description logic. This model uses the concept *exceptional* to divide the defeasible rules into several layers and finds a maximally consistent set for reasoning. The disadvantage of **Rational closure** is that it can't infer necessary intermediates for reasoning if they are not provable in the system.

Hyper network[2] is a special graph similar to our ASG. The difference between this paper and our paper is that we use the graph in different tasks. Hyper networks aim to find relevant arguments not related to update s. Ignoring arguments irrelevant to updates contributes to the improvement of efficiency in computing Delp. While ASG is a model designed to determine the truth values of target formulas.

Argumentation-based reasoning[15][14] is a argumentation-based method for defeasible description logic. This model uses one condition set and one conclusion formula to build an argument. Then it considers the undercut attack relation to build an argument tree. However, the computing complexity of this method is too high. Because the condition set is a subset of the knowledge base, the number of arguments will grow up exponentially.

Inconsistent Ontology[1] is a consistent set-based method for defeasible description logic. This method requires MIO(minimal inconsistent sets) buffer in the reasoning process. When considering inconsistent reasoning, this method will gradually build a consistent subset by removing inconsistent formulas. However, when the knowledge base goes large, the MIO buffer will grow up to an unacceptable size.

6   Conclusion

In this paper, we propose a new model for defeasible reasoning on description logic $\mathcal{ALC}$. Compared with **rational closure**, our model can infer necessary intermediates for reasoning if they are not provable in the system. We also show that explanations can be generated from the preferred model.

References

[1] Alejandro Gomez, S., C. Ivan Chesnevar and G. R. Simari, *Reasoning with inconsistent ontologies through argumentation*, Applied Artificial Intelligence **24** (2010), pp. 102–148.

[2] Alfano, G., S. Greco, F. Parisi, G. I. Simari and G. R. Simari, *An incremental approach to structured argumentation over dynamic knowledge bases*, in: *Sixteenth International Conference on Principles of Knowledge Representation and Reasoning*, 2018.

[3] Alfano, G., S. Greco, F. Parisi, G. I. Simari and G. R. Simari, *Incremental computation for structured argumentation over dynamic DeLP knowledge bases*, Artificial Intelligence **300** (2021).

[4] Baader, F. and B. Hollunder, *How to prefer more specific defaults in terminological default logic* (1992).

[5] Baader, F. and B. Hollunder, *Embedding defaults into terminological knowledge representation formalisms*, Journal of Automated Reasoning **14** (1995), pp. 149–180.

[6] Baader, F. and B. Hollunder, *Priorities on defaults with prerequisites, and their application in treating specificity in terminological default logic*, Journal of Automated Reasoning **15** (1995), pp. 41–68.

[7] Baader, F., I. Horrocks, C. Lutz and U. Sattler, "Introduction to description logic," Cambridge University Press, 2017.

[8] Britz, K., G. Casini, T. Meyer, K. Moodley, U. Sattler and I. Varzinczak, *Theoretical foundations of defeasible description logics* (2018).

[9] Casini, G. and U. Straccia, *Rational closure for defeasible description logics*, in: *Logics in Artificial Intelligence: 12th European Conference, JELIA 2010, Helsinki, Finland, September 13-15, 2010. Proceedings 12*, Springer, 2010, pp. 77–90.

[10] Liao, B., L. Jin and R. C. Koons, *Dynamics of argumentation systems: A division-based method*, Artificial Intelligence **175** (2011), pp. 1790–1814.
URL https://www.sciencedirect.com/science/article/pii/S0004370211000518

[11] Modgil, S. and H. Prakken, *The ASPIC+ framework for structured argumentation: a tutorial*, Argument & Computation **5** (2014), pp. 31–62.

[12] Nofal, S., K. Atkinson and P. E. Dunne, *Algorithms for argumentation semantics: labeling attacks as a generalization of labeling arguments*, Journal of Artificial Intelligence Research **49** (2014), pp. 635–668.

[13] Noonan, R. E., *An algorithm for generating abstract syntax trees*, Computer Languages **10** (1985), pp. 225–236.
URL https://www.sciencedirect.com/science/article/pii/0096055185900189

[14] Zhang, X. and Z. Lin, *An argumentation framework for description logic ontology reasoning and management*, Journal of Intelligent Information Systems **40** (2013), pp. 375–403.

[15] Zhang, X., Z. Zhang, D. Xu and Z. Lin, *Argumentation-based reasoning with inconsistent knowledge bases*, in: *Advances in Artificial Intelligence:*

23rd Canadian Conference on Artificial Intelligence, Canadian AI 2010, Ottawa, Canada, May 31–June 2, 2010. Proceedings 23, Springer, 2010, pp. 87–99.

A Logic for Temporal and Open Information

Yiheng Wang [1]

Department of Philosophy, Sun Yat-sen University
Guangzhou, China

Zhe Lin [2] *

Department of Philosophy, Xiamen University
Xiamen, China

Shier Ju [3] *

Department of Philosophy, Sun Yat-sen University
Guangzhou, China

Abstract

In this paper, we study a De Morgan multi-modal logic aiming at representing the open world model. Sequent calculus and axiomatic system are presented with soundness and completeness proved. We prove it has the finite model property by methods from algebraic proof theory, whence the decidability follows.

Keywords: open world model, De Morgan multi-modal logic, finite model property, tense logic, rough set.

1 Introduction

Human knowledge or information about the world contains a significant amount of ignorance and uncertainty due to the changeable nature of the world and the cognitive limitations of agents. As an example, some categories of concepts that may have been applicable to certain members of a community at a particular point in time may no longer be relevant to them once new information becomes available. One interesting theory about this idea is that there exist some changing classes, which means the arrival of new members or the departure of old members may happen from time to time. In this sense, information is temporal and open. From the philosophical perspective, Quine's natural kind [24] and Wittgenstein's family resemblance [28] echoed this idea as well. Naturally, such an idea demands a new model to deal with it.

[1] Email address: ianwang747@gmail.com

[2] Email address: pennyshaq@163.com

[3] Email address: hssjse@mail.sysu.edu.cn

The open world model and its corresponding information system were developed to deal with this kind of open and temporal information in [5]. The open world model in [5] indeed consists of a set of time points, and where corresponding to each time point, a category can be classified in two ways: (1) into two regions: "true" and "false" and (2) into 4 regions "uncertain", "ignorant", "true" and "false". Such an idea can be described by the following Figure 1:

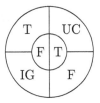

Fig. 1. Open Information Model

where the "UC", "IG", "T" and "F" denote "uncertain", "ignorant", "true" and "false" respectively. A natural way to characterize these semantics is to consider a fusion of classical temporal logic and 4-valued temporal logic. Similar ideas can be found in Girard's linear logic [13] and D. Miller's Int×CL (first-order intuitionistic logic combined with classical first-order logic) [17].

We are concerned with the logic of the above open world model. Some literature has already focused on this topic. Ju and Liu [15] explored the "open set" notion, set up a semantic theory for a special 3-valued logic, and presented a complete axiomatic system. Banerjee, Ju, and Tang [5] used the temporal information systems (TISs) to represent the open information with descriptors and the global modality. TISs can be used to reason about attribute-values of the objects as well as their approximations relative to time. Khan et al [16] further studied the properties of extended TISs, showing that different patterns of flow of information give different TISs and their corresponding logics.

In this paper, we study the logic of temporal and open information from an algebraic point of view. This algebraic structure turns out to be a variant of pre-rough algebra in [4]. The methodology of this work can trace back to the rough set theory. Pawlak [23] first introduced the rough set theory in 1982 as a mathematical tool to deal with vagueness and uncertainty. Due to the advantage that no prior information is necessary about the topic dataset, rough set theory has vast applications in machine learning, knowledge discovery, and data mining, etc (cf. [1,6,10,29]). The algebraic research of rough equality in rough set results in the definitions of topological quasi-Boolean algebra 5 (tqBa5), also called topological rough algebra in [18], and pre-rough algebra (cf. [4,27]). There are various studies about intermediate algebras between tqBa5 and pre-rough algebras. Saha et al [25,26] gave the definitions of intermediate algebras satisfy (IA1), (IA2), and (IA3) respectively. Lin et al [20] studied the residuated algebras in the vicinity of pre-rough algebra, showed different combinations of these residuated version algebras, and proved the decidability of the word problem for these structures. Other intermediate algebras include

MDS5 (cf. [9]) and Tetravalent Modal Algebra (TMA cf. [22]) with completely different motivations.

The main contribution of this paper is that we define a De Morgan multi-modal logic based on the algebraic results to represent open world model. More-over, axiomatization for this logic is proposed, and the corresponding soundness and completeness theorems are obtained. Moreover, we prove the finite model property (FMP) and decidability via a conservative residuated extension of the logic under consideration. Although the method is inspired by [20]. The con-struction of the finite model is different and more simple. Further, this method has advantages in studying the fusion and multi-modalities of non-classical modal logics and may be extended to other relative structures.

This paper is organized as follows: in section 2, we recall some basic alge-braic definitions and introduce the structure under consideration. In section 3, we present logics corresponding to the algebra and prove the soundness and completeness. In section 4, we prove the finite model property based on the sequent calculus. Finally, we give some concluding remarks in section 5.

2   Algebra

In this section, we first introduce the definition of the topic algebra and study its algebraic properties. Then we present the corresponding simple sequent calcu-lus and Hilbert system of the algebra with soundness and completeness proved. Recall that a quasi-Boolean algebra (cf. [7], denoted by qBa) $(A, \wedge, \vee, \neg, 0, 1)$, also called De Morgan algebra in [21], is a bounded distributive lattice further satisfying $\neg\neg a = a$, $\neg(a \vee b) = \neg a \wedge \neg b$, $\neg(a \wedge b) = \neg a \vee \neg b$ and $\neg 0 = 1$ for all $a, b \in A$.

Definition 1 ([3]) An algebra $(A, \wedge, \vee, \neg, !, 0, 1)$ is called a topological quasi-Boolean algebra (denoted by tqBa) if $(A, \wedge, \vee, \neg, 0, 1)$ is a qBa and "!" is a unary operator on A satisfying the following conditions: for all $a, b \in A$,

$$\text{(N) } !0 = 0; \quad \text{(K) } !(a \vee b) = !a \vee !b; \quad \text{(T) } a \leq !a; \quad \text{(4) } !!a \leq !a.$$

Remark 2 The "!" symbol represents the open world model's classical part (true and false regions). Such a symbol's usage comes from the tradition in linear logic introduced by Girard [13].

Definition 3 An algebra $(A, \wedge, \vee, \neg, P, F, 0, 1)$ is called a tense extension of tqBa (denoted by tqBa.t) if $(A, \wedge, \vee, \neg, P, 0, 1)$ and $(A, \wedge, \vee, \neg, F, 0, 1)$ are both tqBa and P, F satisfying the following condition: for all $a, b \in A$,

$$\text{(Adj) } Pa \leq b \text{ iff } a \leq Gb; \quad Fa \leq b \text{ iff } a \leq Hb.$$

where the dualities are defined as $Ga = \neg F \neg a$ and $Ha = \neg P \neg a$.

A tqBa5 is a tqBa additionally satisfies (5) $!a \leq ?!a$ where $?a := \neg!\neg a$ while an IA1 is a tqBa5 satisfies extra condition (IA1) $\neg?a \vee ?a = 1$. Here "IA" denotes the phrase "Intermediate Algebra". Consequently $\neg?a \wedge ?a = 0$, $\neg!a \vee !a = 1$ and $\neg!a \wedge !a = 0$. Obviously, let $!A = \{!a | a \in A\}$. Then $\mathbb{B}A = (!A, \wedge, \vee, \neg, 0, 1)$ is a Boolean algebra. A pre-rough algebra is a tqBa5 satisfies three extra conditions: (IA1) $\neg?a \vee ?a = 1$, (IA2) $?(a \vee b) = ?a \vee ?b$, and (IA3) if $?a \leq ?b$ and $!a \leq !b$, then $a \leq b$.

Definition 4 An algebra $(A, \wedge, \vee, \neg, !, P, F, 0, 1)$ is called a fusion of IA1 and tqBa.t (denoted by IA1×tqBa.t) if $(A, \wedge, \vee, \neg, !, 0, 1)$ is an IA1 and $(A, \wedge, \vee, \neg, P, F, 0, 1)$ is a tqBa.t.

Remark 5 Let $!A$ be defined as above. Then $\mathbb{A} = (!A, \wedge, \vee, \neg, P, F, 0, 1)$ is a classical S4 tense algebra.

Example 6 The lattice in Figure 2 is an example of IA1×tqBa.t.

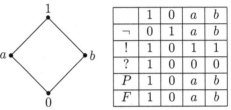

	1	0	a	b
$\neg$	0	1	a	b
$!$	1	0	1	1
$?$	1	0	0	0
P	1	0	a	b
F	1	0	a	b

Fig. 2. An example of IA1×tqBa.t.

Proposition 7 Let $\dagger^n a = \dagger \dagger^{n-1} a, \dagger^0 a = a$ where $\dagger \in \{!, P, F\}$. The following properties hold for any IA1 × tqBa.t: for all $a, b, c \in A$,

(1) If $a \leq b$, then $\dagger a \leq \dagger b$;

(2) If $\dagger^n a \leq b$ and $\dagger^m b \leq c$, then $\dagger^{n+m} a \leq c$;

(3) If $\dagger^n a \leq 0$, then $\dagger^{n+m} a \leq b$;

(4) If $\dagger^n 1 \leq b$, then $\dagger^{n+m} a \leq b$;

(5) If $\dagger^n a \leq b$, then $\dagger^n (a \wedge c) \leq b$;

(6) $\dagger 0 = 0$;

(7) If $\dagger a \leq b$, then $a \leq b$;

(8) If $\dagger a \leq b$, then $\dagger \dagger a \leq b$;

(9) If $?!a \leq b$, then $!a \leq b$;

(10) $\neg?a \vee ?a = 1$.

Hereafter, we use **AL**, L, G to denote the class of all IA1×tqBa.ts, the logic of **AL**, and the sequent calculus of L respectively. A class of algebras $\mathcal{K}$ is a *variety* if it is definable by a set of equations or satisfies the so-called HSP property (cf. e.g. [12]).

Theorem 8 **AL** *is a variety.*

Proof. The (Adj) rule follows from the monotonicity of tense operators and other equations, for instance, (t3) and (t6) in Ewald's [11] intuitionistic tense logic. Further, all other conditions of IA1×tqBa.t are clearly definable by a set of equations.   □

3   Logic

In this section, we present simple sequent calculus S and axiomatic system H for L respectively with soundness and completeness proved.

Definition 9 The set of formulas (terms) $\mathcal{F}$ is defined inductively as follows:

$$\mathcal{F} \ni \alpha ::= p \mid \top \mid \bot \mid \neg\alpha \mid !\alpha \mid P\alpha \mid F\alpha \mid \alpha \wedge \beta \mid \alpha \vee \beta$$

where $p \in \mathbf{Var}$ such that $\mathbf{Var}$ is a denumerable set of propositional variables. We define $?\alpha = \neg!\neg\alpha$, $G\alpha = \neg F\neg\alpha$ and $H\alpha = \neg P\neg\alpha$. Let $\dagger^n \alpha = \dagger \dagger^{n-1} \alpha, \dagger^0 \alpha = \alpha$ where $\dagger \in \{!, P, F\}$.

Definition 10 The simple sequent calculus S for L consists of the following axioms and rules: For $n \geq 0$, $i = \{1,2\}$ and $\dagger = \{!, P, F\}$,

(1) Axioms:

$$\text{(Id) } \alpha \Rightarrow \alpha \quad \text{(D) } \alpha \wedge (\beta \vee \gamma) \Rightarrow (\alpha \wedge \beta) \vee (\alpha \wedge \gamma)$$

$$\text{(IA1) } \top \Rightarrow \neg?\alpha \vee ?\alpha \quad \text{(T) } \alpha \Rightarrow \top \quad \text{(}\bot\text{) } \bot \Rightarrow \alpha \quad \text{(DN) } \neg\neg\alpha \Rightarrow \alpha$$

$$\text{(K) } !(\alpha \vee \beta) \Leftrightarrow !\alpha \vee !\beta \quad \text{(4) } \dagger\dagger\alpha \Rightarrow \dagger\alpha \quad \text{(T) } \alpha \Rightarrow \dagger\alpha \quad \text{(5) } !\alpha \Rightarrow ?!\alpha$$

(2) Logical rules:

$$\frac{\alpha_i \Rightarrow \beta}{\alpha_1 \wedge \alpha_2 \Rightarrow \beta}(\wedge L) \qquad \frac{\alpha \Rightarrow \beta \quad \beta \Rightarrow \gamma}{\alpha \Rightarrow \beta \wedge \gamma}(\wedge R)$$

$$\frac{\alpha_1 \Rightarrow \beta \quad \alpha_2 \Rightarrow \beta}{\alpha_1 \vee \alpha_2 \Rightarrow \beta}(\vee L) \qquad \frac{\alpha \Rightarrow \beta_i}{\alpha \Rightarrow \beta_1 \vee \beta_2}(\vee R) \qquad \frac{\alpha \Rightarrow \beta}{\neg\beta \Rightarrow \neg\alpha}(CP)$$

(3) Modal rules:

$$\frac{\alpha \Rightarrow H\beta}{F\alpha \Rightarrow \beta}(rHF) \quad \frac{F\alpha \Rightarrow \beta}{\alpha \Rightarrow H\beta}(rFH) \quad \frac{\alpha \Rightarrow G\beta}{P\alpha \Rightarrow \beta}(rGP) \quad \frac{P\alpha \Rightarrow \beta}{\alpha \Rightarrow G\beta}(rPG)$$

(4) Cut rule:

$$\frac{\alpha \Rightarrow \beta \quad \beta \Rightarrow \gamma}{\alpha \Rightarrow \gamma}(Cut)$$

A sequent $\alpha \Rightarrow \beta$ is provable in S, denoted by $\vdash_S \alpha \Rightarrow \beta$, if there is a derivation of $\alpha \Rightarrow \beta$ in S. We write $\vdash_S \alpha \Leftrightarrow \beta$ if $\vdash_S \alpha \Rightarrow \beta$ and $\vdash_S \beta \Rightarrow \alpha$. The subscript S in $\vdash_S$ is omitted if no confusion arises.

Proposition 11 *The following sequents are provable in S: for $\dagger \in \{!, P, F\}$,*

(1) $\vdash \neg(\alpha \wedge \beta) \Leftrightarrow \neg\alpha \vee \neg\beta$;   *(6)* $\vdash F(\alpha \vee \beta) \Leftrightarrow F\alpha \vee F\beta$;

(2) $\vdash \neg(\alpha \vee \beta) \Leftrightarrow \neg\alpha \wedge \neg\beta$;   *(7)* $\vdash G(\alpha \wedge \beta) \Leftrightarrow G\alpha \wedge G\beta$;

(3) If $\vdash \alpha \Rightarrow \beta$, *then* $\vdash \dagger\alpha \Rightarrow \dagger\beta$;   *(8)* $\vdash H(\alpha \wedge \beta) \Leftrightarrow H\alpha \wedge H\beta$;

(4) $\vdash \dagger\bot \Leftrightarrow \bot$;   *(9)* $\vdash ?\alpha \vee \neg?\alpha \Leftrightarrow \top$;

(5) $\vdash P(\alpha \vee \beta) \Leftrightarrow P\alpha \vee P\beta$;   *(10)* $!\alpha \Rightarrow \beta$ *iff* $\alpha \Rightarrow ?\beta$.

Remark 12 Let α be a tense logic S4 formula. Define $f^?(\alpha)$ be the formula obtained from α by replacing every propositional variables p in α by $?p$. For classical tense logic S4 denoted by S4.t, if $\vdash_{S4.t} \alpha$, then $\vdash_S \Rightarrow f^?\alpha$.

Definition 13 Given an IA1×tqBa.t $\mathbb{A} = (A, \wedge, \vee, \neg, !, F, P, 0, 1)$, an *assignment* in $\mathbb{A}$ is a function $\theta : \mathbf{Var} \to A$. Every assignment σ in $\mathbb{A}$ can be extended homomorphically. Let $\hat{\sigma}(\alpha)$ be the element in A denoted by α. An *algebraic model* is a pair $(\mathbb{A}, \sigma)$ where $\mathbb{A}$ is an algebraic structure, and σ is an assignment in $\mathbb{A}$. A sequent $\alpha \Rightarrow \beta$ is *true* in an algebraic model $(\mathbb{A}, \sigma)$, notation $\models_{\mathbb{A},\sigma} \alpha \Rightarrow \beta$, if $\hat{\sigma}(\alpha) \leq \hat{\sigma}(\beta)$. A sequent $\alpha \Rightarrow \beta$ is *true* in a class of algebraic structure $\mathcal{K}$, notation $\models_{\mathcal{K}} \alpha \Rightarrow \beta$, if $\models_{\mathbb{A},\sigma} \alpha \Rightarrow \beta$ for any algebraic model $(\mathbb{A}, \sigma)$ with $\mathbb{A} \in \mathcal{K}$. A sequent rule with premises $\alpha_1 \Rightarrow \beta_1, \ldots, \alpha_m \Rightarrow \beta_m$ and conclusion $\alpha_0 \Rightarrow \beta_0$ *preserves truth* in $\mathcal{K}$, if $\models_{\mathbb{A},\sigma} \alpha_0 \Rightarrow \beta_0$ whenever $\models_{\mathbb{A},\sigma} \alpha_i \Rightarrow \beta_i$ for $1 \leq i \leq m$, for any algebraic model $(\mathbb{A}, \sigma)$ with $\mathbb{A} \in \mathcal{K}$.

Theorem 14 (Soundness and Completeness) S *is sound and complete with respect to* **AL**.

Proof. The soundness result can be obtained by the induction on the length of proof and Proposition 7,11. For completeness result, it suffices to show that for any sequent $\alpha \Rightarrow \beta$, if $\nvdash_S \alpha \Rightarrow \beta$, then $\nvDash_{AL} \alpha \Rightarrow \beta$. It can be proved by standard construction. Let $[\![\alpha]\!] = \{\beta | \vdash_S \alpha \Leftrightarrow \beta\}$. Let A be the set of all $[\![\alpha]\!]$. One defines $\{\wedge', \vee', \neg', !', P', F', \top', \bot'\}$ on A as follows:

$$[\![\alpha_1]\!] \wedge' [\![\alpha_2]\!] = [\![\alpha_1 \wedge \alpha_2]\!] \quad [\![\alpha_1]\!] \vee' [\![\alpha_2]\!] = [\![\alpha_1 \vee \alpha_2]\!] \quad \top' = [\![\top]\!] \quad \bot' = [\![\bot]\!]$$

$$\neg'[\![\alpha]\!] = [\![\neg\alpha]\!] \quad !'[\![\alpha]\!] = [\![!\alpha]\!] \quad P'[\![\alpha]\!] = [\![P\alpha]\!] \quad F'[\![\alpha]\!] = [\![F\alpha]\!]$$

Clearly by Definition 4, $(A, \wedge', \vee', \neg', !', P', F', \top', \bot')$ is an IA1×tqBa.t. The order is defined as $[\![\alpha_1]\!] \leq' [\![\alpha_2]\!]$ iff $[\![\alpha_1]\!] \wedge' [\![\alpha_2]\!] = [\![\alpha_1]\!]$. Thus $[\![\alpha_1]\!] \leq' [\![\alpha_2]\!]$ iff $\vdash_S \alpha_1 \Rightarrow \alpha_2$. Define an assignment $\sigma : \mathbf{Var} \longrightarrow A$ such that $\sigma(p) = [\![p]\!]$. By induction on the complexity of the formula, one shows that $\hat{\sigma}(\alpha) = [\![\alpha]\!]$ for any formula α. Suppose that $\vDash_{AL} \alpha \Rightarrow \beta$. Then $\hat{\sigma}(\alpha) \leq \hat{\sigma}(\beta)$ in $\mathbf{AL}$. Hence $\vdash_S \alpha \Rightarrow \beta$, which yields a contradiction. This completes the proof. $\square$

Definition 15 Let $\mathbb{A} = (A, \wedge, \vee, \neg, !, 0, 1)$ be an IA1. The binary operation $\Rightarrow_R$ on A is defined as follows:

$$a \Rightarrow_R b := (\neg!a \vee !b) \wedge (\neg?a \vee ?b).$$

where the operation $\Rightarrow_R$ is called the rough implication operation on $\mathbb{A}$.

Definition 16 The language of the axiomatic system H of L consists of propositional variables $\{p, q, \ldots, r\}$, and logical connectives $\{\wedge, \vee, \neg, !, P, F, \top, \bot\}$. We denote $\alpha \Leftrightarrow_R \beta := (\alpha \Rightarrow_R \beta) \wedge (\beta \Rightarrow_R \alpha)$. For $\dagger \in \{!, P, F\}$, $\ddagger \in \{?, H, G\}$ and $i \in \{1, 2\}$, the axiom schemes and rules of inference are listed as follows:

- Axiom schemes I
  - (⊤) $\alpha \Rightarrow_R \top$;
  - (⊥) $\bot \Rightarrow_R \alpha$;
  - (DN) $\alpha \Leftrightarrow_R \neg\neg\alpha$;
  - (DM₁) $\neg(\alpha \wedge \beta) \Leftrightarrow_R \neg\alpha \vee \neg\beta$;
  - (DM₂) $\neg(\alpha \vee \beta) \Leftrightarrow_R \neg\alpha \wedge \neg\beta$;
  - (∧) $((\alpha \Rightarrow_R \beta) \wedge (\alpha \Rightarrow_R \gamma))$ iff $(\alpha \Rightarrow_R (\beta \wedge \gamma))$;
  - (∨) $((\alpha \Rightarrow_R \beta) \wedge (\gamma \Rightarrow_R \beta))$ iff $((\alpha \vee \gamma) \Rightarrow_R \beta)$;
  - (D) $\alpha \wedge (\beta \vee \gamma) \Leftrightarrow_R (\alpha \wedge \beta) \vee (\alpha \wedge \gamma)$;
  - (HS) If $((\alpha \Rightarrow_R \beta)$ and $(\beta \Rightarrow_R \gamma))$, then $(\alpha \Rightarrow_R \gamma)$;
  - (K∨) $\dagger(\alpha \vee \beta) \Leftrightarrow_R (\dagger\alpha \vee \dagger\beta)$;
  - (CP) $(\alpha \Rightarrow_R \beta) \Rightarrow_R (\neg\beta \Rightarrow_R \neg\alpha)$;

- Axiom schemes II and Rules of inference
  - (T) $\alpha \Rightarrow_R \dagger\alpha$;
  - (4) $\dagger\dagger\alpha \Rightarrow_R \dagger\alpha$;
  - (5) $!\alpha \Rightarrow_R?!\alpha$;
  - (IA1) $?\alpha \vee \neg?\alpha \Leftrightarrow_R \top$;
  - (PG) $PG\alpha \Rightarrow_R \alpha$;
  - (GP) $\alpha \Rightarrow_R GP\alpha$;
  - (FH) $FH\alpha \Rightarrow_R \alpha$;
  - (HF) $\alpha \Rightarrow_R HF\alpha$;
  - (MP) If α and $(\alpha \Rightarrow_R \beta)$, then β;
  - (Gen) If α, then $\ddagger\alpha$.

A formula α is provable in H (denoted by $\vdash_H \alpha$) is defined naturally.

Remark 17 It has been proved in [20] that the rough implication operation defined on tqBa5s satisfies the property (P) in [2] (also called (E) in [14]): for any $a, b \in A$, $a \Rightarrow_R b = 1$ iff $a \leq b$. Then one has the following rules:

$(Res_{\Rightarrow_R})$ $\alpha \wedge \beta \Rightarrow_R \gamma$ iff $\beta \Rightarrow_R (\alpha \Rightarrow_R \gamma)$; $\quad (K_{\Rightarrow_R})$ $\ddagger(\alpha \Rightarrow_R \beta) \Rightarrow_R (\ddagger\alpha \Rightarrow_R \ddagger\beta)$.

Lemma 18 For any formulas α and β, $\vdash_S \alpha \Rightarrow_R \beta$ iff $\vdash_H \alpha \Rightarrow_R \beta$.

Proof. The right-to-left direction is easy and can be checked regularly in S. For another direction, it suffices to show that all rules in S are preserved in H. We prove such a result by induction on the length of proof. The basic cases are trivial. For the inductive steps, we take the rule (rPG) as an example, others can be treated similarly. Assume $\vdash_S P\alpha \Rightarrow \beta$, then by induction hypothesis, one has $\vdash_H P\alpha \Rightarrow_R \beta$. One suffices to show that $\vdash_H \alpha \Rightarrow_R G\beta$. By (Gen), one has $\vdash_H G(P\alpha \Rightarrow_R \beta)$. By $(K_{\Rightarrow_R})$ and (MP), one obtains $\vdash_H GP\alpha \Rightarrow_R G\beta$. Clearly $\vdash_H \alpha \Rightarrow_R GP\alpha$ due to (GP). Therefore by (HS) and (MP), one gets $\vdash_H \alpha \Rightarrow_R G\beta$.   □

Theorem 19 (Soundness and Completeness) H *is sound and complete with respect to* **AL**.

Proof. Clearly, the implication of H satisfies property (P). Hence by Theorem 14 for any $\alpha \Rightarrow_R \beta$, $\vdash_S \top \Rightarrow \alpha \Rightarrow_R \beta$ iff $\vdash_H \alpha \Rightarrow_R \beta$. For any formula α, $\vdash_H \alpha$ iff $\vdash_H \top \Rightarrow_R \alpha$ iff $\vdash_S \top \Rightarrow_R \alpha$. By Theorem 14, $\vdash_H \alpha$ iff $\models_{AL} \top \Rightarrow_R \alpha$. Thus H is sound and complete with respect to **AL**.   □

4   Finite Model Property

In this section, we first consider a residuated but conservative extension of **AL**. Further, we prove the FMP of such an extension's sequent calculus rG, that is, we will show that if $\nvdash_{rG} \alpha \Rightarrow \beta$, then there is a finite algebra model $\mathcal{M}$ s.t. $\nvDash_{\mathcal{M}} \alpha \Rightarrow \beta$. Then, we obtain its decidability. Since such an extension is conservative, one has L is decidable as well. Note that the "$\vdash$" symbol in this section means "$\vdash_{rG}$" if no confusion arises. First, we introduce some definitions related to the residuated extension.

Definition 20 ([20]) An algebra $(A, \cdot, \rightarrow, 0, 1)$ is called a residuated bounded commutative groupoid (denoted by bcrg) if $(A, 0, 1)$ is a bounded groupoid and "$\cdot$", "$\rightarrow$" are binary operations on A satisfying the following conditions:

$$\text{(Res) } a \cdot b \leq c \text{ iff } a \leq b \rightarrow c; \quad \text{(Com) } a \cdot b = b \cdot a.$$

Definition 21 An algebra $(A, \wedge, \vee, !, P, F, \cdot, \rightarrow, 0, 1)$ is called a residuated IA1×tqBa.t (denoted by rIA1×tqBa.t) if $(A, \cdot, \rightarrow, 0, 1)$ is a bcrg and $(A, \wedge, \vee, \neg, !, P, F, 0, 1)$ is an IA1×tqBa.t where $\neg a = a \rightarrow 0$.

Proposition 22 *The following properties hold for any* rIA1 × tqBa.t: *for all* $a, b \in A$:

(1) $\neg\neg a = a$;   (3) $\neg(a \wedge b) = \neg a \vee \neg b$;

(2) $\neg(a \vee b) = \neg a \wedge \neg b$;   (4) $a \leq b$ *implies* $\neg b \leq \neg a$.

Hereafter, we use **rAL** and rL to denote the class of all rIA1×tqBa.ts and the logic of **rAL** respectively.

Lemma 23 *Every* IA1 × tqBa.t *can be expanded to a* rIA1 × tqBa.t.

Proof. For any IA1 × tqBa.t $(A, \wedge, \vee, \neg, !, P, F, 0, 1)$, it suffices to show that one can define $\cdot$ and $\rightarrow$ satisfying (1) $\neg a = a \rightarrow 0$. (2) for all $a, b \in A$, $a \cdot b = b \cdot a$; (3) $a \cdot b \leq c$ iff $a \leq b \rightarrow c$. "$\cdot$", "$\rightarrow$" are operations on A defined as follow:

$$a \cdot b = \begin{cases} 0, & \text{if } a \leq \neg b \\ 1, & \text{otherwise} \end{cases} \quad a \rightarrow b = \begin{cases} \neg a, & \text{if } b \neq 1 \\ 1, & \text{otherwise} \end{cases}$$

(1) Let $b = 0$, one has $a \rightarrow 0 = \neg a$.

(2) Assume $a \cdot b = 0$ i.e. $a \le \neg b$, then one has $b \le \neg a$ by double negation law
and Proposition 22 (4). Consequently, $a \cdot b = 0 = b \cdot a$. Assume $b \cdot a = 1$ i.e.
$a \not\le \neg b$, then one has $b \not\le \neg a$ by similar method. Consequently, $a \cdot b = 1 = b \cdot a$.

(3) Assume $c = 1$, then $a \cdot b \le 1$ iff $a \le b \to 1 = 1$. Assume $c \ne 1$. If $a \cdot b \le c$,
then one has $a \cdot b = 0$ and $a \le \neg b = b \to c$. Consequently, one has $a \le b \to c$.
If $a \le b \to c$, then $b \to c = \neg b$ and $a \le \neg b$. Further, one has $a \cdot b = 0 = b \cdot a$.
Consequently, one has $a \cdot b \le c$.

Therefore, the constructed algebra is a rIA1 × tqBa.t.   □

In what follows, we present the rG for rL.

Definition 24 The set of formulas (terms) $\mathcal{F}$ is defined inductively as follows:

$$\mathcal{F} \ni \alpha ::= p \mid \top \mid \bot \mid !\alpha \mid ?\alpha \mid P\alpha \mid F\alpha \mid H\alpha \mid G\alpha \mid \alpha \wedge \beta \mid \alpha \vee \beta \mid \alpha \cdot \beta \mid \alpha \to \beta$$

where $\neg\alpha = \alpha \to \bot$.

Definition 25 Let $\{\star, \S, \bullet, \circ\}$ be the structure operators for $\{\cdot, !, P, F\}$ respectively. The set of all formula structures $\mathcal{FS}$ is defined inductively as follows:

$$\mathcal{FS} \ni \Gamma ::= \alpha \mid (\Gamma_1 \star \Gamma_2) \mid \S\Gamma \mid \bullet\Gamma \mid \circ\Gamma$$

where $\alpha \in \mathcal{F}$. Let $\odot^n \Delta = \odot \odot^{n-1} \Delta, \odot^0 \Delta = \Delta$ where $\odot \in \{\S, \circ, \bullet\}$. A *sequent* is an expression of the form $\Gamma[\alpha] \Rightarrow \beta$, where Γ is a formula structure and α, β are formulas. Let $\alpha \Leftrightarrow \beta$ denotes $\alpha \Rightarrow \beta$ and $\beta \Rightarrow \alpha$. A *context* is a formula structure $\Gamma[-]$ with a designated position $[-]$ which can be filled with a formula structure. In particular, a single position $[-]$ is a context. Let $\Gamma[\Delta]$ be a formula structure obtained from $\Gamma[-]$ by substituting Δ for $[-]$. For any formula structure Γ, the formula $f(\Gamma)$ is defined inductively as follows:

$$f(\alpha) = \alpha \quad f(\Gamma_1 \star \Gamma_2) = f(\Gamma_1) \cdot f(\Gamma_2) \quad f(\S\Gamma) = !f(\Gamma) \quad f(\bullet\Gamma) = Pf(\Gamma) \quad F(\circ\Gamma) = Ff(\Gamma)$$

Example 26 Let expression $\Gamma[-] = \S^5 \bullet^{10} ((- \star \neg q) \star p \wedge q)$ be a context. If we replace the formula structure $\Delta = \bullet(p \star (q \star \circ q))$ for the position $-$ in $\Gamma[-]$, we get the formula structure $\Gamma[\Delta] = \S^5 \bullet^{10} ((\bullet(p \star (q \star \circ q)) \star \neg q) \star p \wedge q)$.

Definition 27 The Gentzen-style sequent calculus rG for rL consists of the following axioms and rules: for $i = \{1, 2\}$, $\odot = \{\S, \bullet, \circ\}$ and the corresponding $\dagger = \{!, P, F\}$,

(1) Axioms:

$$\text{(Id) } \alpha \Rightarrow \alpha \quad \text{(D) } \alpha \wedge (\beta \vee \gamma) \Rightarrow (\alpha \wedge \beta) \vee (\alpha \wedge \gamma)$$

$$\text{(}\top\text{) } \Gamma[\alpha] \Rightarrow \top \quad \text{(}\bot\text{) } \Gamma[\bot] \Rightarrow \alpha \quad \text{(DN) } \neg\neg\alpha \Rightarrow \alpha$$

(2) Logical rules:

$$\frac{\Gamma[\alpha_i] \Rightarrow \beta}{\Gamma[\alpha_1 \wedge \alpha_2] \Rightarrow \beta}(\wedge L) \qquad \frac{\Gamma \Rightarrow \alpha \quad \Gamma \Rightarrow \beta}{\Gamma \Rightarrow \alpha \wedge \beta}(\wedge R)$$

$$\frac{\Gamma[\alpha_1] \Rightarrow \beta \quad \Gamma[\alpha_2] \Rightarrow \beta}{\Gamma[\alpha_1 \vee \alpha_2] \Rightarrow \beta}(\vee L) \qquad \frac{\Gamma \Rightarrow \alpha_i}{\Gamma \Rightarrow \alpha_1 \vee \alpha_2}(\vee R)$$

$$\frac{\Delta \Rightarrow \alpha \quad \Gamma[\beta] \Rightarrow \gamma}{\Gamma[\Delta \star \alpha \to \beta] \Rightarrow \gamma}(\to L) \quad \frac{\alpha \star \Gamma \Rightarrow \beta}{\Gamma \Rightarrow \alpha \to \beta}(\to R) \quad \frac{\Gamma[\alpha \star \beta] \Rightarrow \gamma}{\Gamma[\alpha \cdot \beta] \Rightarrow \gamma}(\cdot L) \quad \frac{\Gamma \Rightarrow \alpha \quad \Delta \Rightarrow \beta}{\Gamma \star \Delta \Rightarrow \alpha \cdot \beta}(\cdot$$

(3) Modal rules:

$$\frac{\Gamma[\odot\alpha] \Rightarrow \beta}{\Gamma[\dagger\alpha] \Rightarrow \beta}(\dagger L) \quad \frac{\Gamma \Rightarrow \alpha}{\odot\Gamma \Rightarrow \dagger\alpha}(\dagger R) \quad \frac{\Gamma[\alpha] \Rightarrow \beta}{\Gamma[\S?\alpha] \Rightarrow \beta}(?L) \quad \frac{\S\Gamma \Rightarrow \alpha}{\Gamma \Rightarrow ?\alpha}(?R)$$

$$\frac{\Gamma[\alpha] \Rightarrow \beta}{\Gamma[\bullet G\alpha] \Rightarrow \beta}(GL) \quad \frac{\bullet\Gamma \Rightarrow \alpha}{\Gamma \Rightarrow G\alpha}(GR) \quad \frac{\Gamma[\alpha] \Rightarrow \beta}{\Gamma[\circ H\alpha] \Rightarrow \beta}(HL) \quad \frac{\circ\Gamma \Rightarrow \alpha}{\Gamma \Rightarrow H\alpha}(HR)$$

(4) Structural rules:

$$\frac{\Gamma[\Delta_1 \star \Delta_2] \Rightarrow \alpha}{\Gamma[\Delta_2 \star \Delta_1] \Rightarrow \alpha}(\text{Com}) \quad \frac{\S\Gamma_1 \star \Gamma_2 \Rightarrow \bot}{\Gamma_1 \star \S\Gamma_2 \Rightarrow \bot}(\text{Dual}_\S)$$

$$\frac{\bullet\Gamma_1 \star \Gamma_2 \Rightarrow \bot}{\Gamma_1 \star \circ\Gamma_2 \Rightarrow \bot}(\text{Dual}_{\bullet\circ}) \quad \frac{\circ\Gamma_1 \star \Gamma_2 \Rightarrow \bot}{\Gamma_1 \star \bullet\Gamma_2 \Rightarrow \bot}(\text{Dual}_{\circ\bullet})$$

$$\frac{\S\Delta \star \S\Delta \Rightarrow \bot}{\S\Delta \Rightarrow \bot}(\text{IA1}) \quad \frac{\Gamma[\odot\Delta] \Rightarrow \beta}{\Gamma[\Delta] \Rightarrow \beta}(T) \quad \frac{\Gamma[\odot\Delta] \Rightarrow \beta}{\Gamma[\odot^2\Delta] \Rightarrow \beta}(4)$$

(5) Cut rule:

$$\frac{\Delta \Rightarrow \alpha \quad \Gamma[\alpha] \Rightarrow \beta}{\Gamma[\Delta] \Rightarrow \beta}(\text{Cut})$$

For brevity's sake, the symbols "$\odot$" and "$\dagger$" are in one-to-one correspondence. Taking ($\dagger$L) for example, one obtains (!L), (PL) and (FL) from ($\dagger$L), with "$\odot$" in the upper sequent of ($\dagger$L) are "$\S$", "$\bullet$" and "$\circ$", and "$\dagger$" in lower sequent of ($\dagger$L) are "!", "P" and "F", respectively.

Remark 28 Note that the cut-elimination does not hold in rG because of the axiom (D). Hence it does not have the standard subformula property.

Proposition 29 *The following properties hold in rG: for $\dagger = \{!, P, F\}$,*

(1) $\vdash \alpha \Rightarrow \beta$ *iff* $\vdash \neg\beta \Rightarrow \neg\alpha$;

(2) $\vdash \alpha \Rightarrow \neg\neg\alpha$;

(3) $(\alpha \wedge \beta) \vee (\alpha \wedge \gamma) \Rightarrow \alpha \wedge (\beta \vee \gamma)$;

(4) $\vdash !\alpha \Leftrightarrow \neg?\neg\alpha$;

(5) $\vdash P\alpha \Leftrightarrow \neg H\neg\alpha$;

(6) $\vdash F\alpha \Leftrightarrow \neg G\neg\alpha$;

(7) $\vdash \dagger\bot \Leftrightarrow \bot$;

(8) $\vdash P\alpha \Rightarrow \beta$ *iff* $\vdash \alpha \Rightarrow G\beta$ *and* $\vdash F\alpha \Rightarrow \beta$
$\quad$ *iff* $\vdash \alpha \Rightarrow H\beta$

(9) $\vdash \alpha \Rightarrow \dagger\alpha$

(10) $\vdash \dagger\dagger\alpha \Rightarrow \dagger\alpha$;

(11) $\vdash !\alpha \Rightarrow ?!\alpha$;

(12) *If* $\vdash \alpha \Rightarrow \beta$, *then* $\vdash \dagger\alpha \Rightarrow \dagger\beta$;

(13) $\vdash \dagger(\alpha \vee \beta) \Leftrightarrow \dagger\alpha \vee \dagger\beta$;

(14) $\vdash \neg(\alpha \wedge \beta) \Leftrightarrow \neg\alpha \vee \neg\beta$;

(15) $\vdash ?\alpha \vee \neg?\alpha \Leftrightarrow \top$;

(16) $\vdash \gamma \Rightarrow \alpha \rightarrow \beta$ *iff* $\alpha \cdot \gamma \Rightarrow \beta$.

Proof. We only provide the proofs for (5) and (15), others can be treated similarly. For (5), one has:

$$\frac{\dfrac{\dfrac{\dfrac{\dfrac{\dfrac{\dfrac{\dfrac{\alpha \Rightarrow \alpha \quad \bot \Rightarrow \bot}{\alpha \star \neg\alpha \Rightarrow \bot}(\to L)}{\alpha \star \circ H\neg\alpha \Rightarrow \bot}(HL)}{\circ H\neg\alpha \star \alpha \Rightarrow \bot}(\text{Com})}{H\neg\alpha \star \bullet\alpha \Rightarrow \bot}(\text{Dual}_{\circ\bullet})}{H\neg\alpha \star P\alpha \Rightarrow \bot}(\dagger L)}{P\alpha \Rightarrow \neg H\neg\alpha}(\to R)$$

$$\frac{\dfrac{\dfrac{\dfrac{\dfrac{\dfrac{\dfrac{\dfrac{\alpha \Rightarrow \alpha}{\bullet\alpha \Rightarrow P\alpha}(\dagger R) \quad \bot \Rightarrow \bot}{\bullet\alpha \star \neg P\alpha \Rightarrow \bot}(\to R)}{\alpha \star \circ\neg P\alpha \Rightarrow \bot}(\text{Dual}_{\bullet\circ})}{\circ\neg P\alpha \Rightarrow \neg\alpha}(\to R)}{\neg P\alpha \Rightarrow H\neg\alpha}(HR)}{\dfrac{\neg P\alpha \star \neg H\neg\alpha \Rightarrow \bot}{\neg H\neg\alpha \Rightarrow \neg\neg P\alpha}(\to R)} \quad \bot \Rightarrow \bot (\to L) \quad \neg\neg P\alpha \Rightarrow P\alpha}{\neg H\neg\alpha \Rightarrow P\alpha}$$

For (15), it suffices to show $\vdash \top \Rightarrow \vdash ?\alpha \vee \neg ?\alpha$, first, one has $\vdash ?\alpha \wedge \neg ?\alpha \Rightarrow \bot$:

$$
\cfrac{
\cfrac{
\cfrac{
\cfrac{
\cfrac{
\cfrac{
\cfrac{
\cfrac{
\cfrac{\cfrac{\cfrac{\alpha \Rightarrow \alpha}{\S?\alpha \Rightarrow \alpha}\ (?\mathrm{L})}{\S^2?\alpha \Rightarrow \alpha}\ (4)}{\S?\alpha \Rightarrow ?\alpha}\ (?\mathrm{R}) \qquad \bot \Rightarrow \bot
}{\S?\alpha \star \neg?\alpha \Rightarrow \bot}\ (\rightarrow \mathrm{R})
}{\S^2?\alpha \star \neg?\alpha \Rightarrow \bot}\ (4)
}{\S?\alpha \star \S\neg?\alpha \Rightarrow \bot}\ (\mathrm{Dual}_\S)
}{\S(?\alpha \wedge \neg?\alpha) \star \S(?\alpha \wedge \neg?\alpha) \Rightarrow \bot}\ (\wedge\mathrm{L} \times 2)
}{\S(?\alpha \wedge \neg?\alpha) \Rightarrow \bot}\ (\mathrm{IA1})
}{?\alpha \wedge \neg?\alpha \Rightarrow \bot}\ (\mathrm{T})
}{}
$$

Then, one has $\vdash \neg(?\alpha \wedge \neg?\alpha) \Rightarrow ?\alpha \vee \neg?\alpha$:

$$
\cfrac{
\cfrac{\cfrac{\neg?\alpha \Rightarrow \neg?\alpha}{\neg?\alpha \Rightarrow \neg?\alpha \vee ?\alpha}\ (\vee\mathrm{R}) \quad \cfrac{\neg\neg?\alpha \Rightarrow ?\alpha}{\neg\neg?\alpha \Rightarrow \neg?\alpha \vee ?\alpha}\ (\vee\mathrm{R})}{\neg?\alpha \vee \neg\neg?\alpha \Rightarrow \neg?\alpha \vee ?\alpha}\ (\vee\mathrm{L}) \qquad \neg(?\alpha \wedge \neg?\alpha) \Rightarrow \neg?\alpha \vee \neg\neg?\alpha
}{\neg(?\alpha \wedge \neg?\alpha) \Rightarrow \neg?\alpha \vee ?\alpha}\ (\mathrm{Cut})
$$

Finally, one has

$$
\cfrac{
\cfrac{\cfrac{?\alpha \wedge \neg?\alpha \Rightarrow \bot}{\neg\bot \Rightarrow \neg(?\alpha \wedge \neg?\alpha)}\ (\mathrm{Prop29\ (1)}) \qquad \top \Rightarrow \neg\bot}{\top \Rightarrow \neg(?\alpha \wedge \neg?\alpha)}\ (\mathrm{Cut}) \qquad \neg(?\alpha \wedge \neg?\alpha) \Rightarrow \neg?\alpha \vee ?\alpha
}{\top \Rightarrow \neg?\alpha \vee ?\alpha}\ (\mathrm{Cut})
$$
$\qquad\qquad\qquad\qquad\qquad\qquad\qquad\qquad\qquad\qquad\qquad\qquad\qquad\qquad \Box$

Theorem 30 (Soundness and Completeness) rG *is sound and complete with respect to* **rAL**.

Proof. The proof method is similar to Theorem 14. Note that the completeness result can be obtained by the FMP result in the later part as well. $\qquad\Box$

Lemma 31 (Conservative Extension) *For any sequent* $\alpha \Rightarrow \beta$ *in* G, $\vdash_G \alpha \Rightarrow \beta$ *iff* $\vdash_{rG} \alpha \Rightarrow \beta$.

Proof. By Theorem 14, $\vdash_G \alpha \Rightarrow \beta$ iff $\models_{\mathbf{AL}} \alpha \Rightarrow \beta$. While by Theorem 30, $\vdash_{rG} \alpha \Rightarrow \beta$ iff $\models_{\mathbf{rAL}} \alpha \Rightarrow \beta$. Hence one suffices to show $\models_{\mathbf{AL}} \alpha \Rightarrow \beta$ iff $\models_{\mathbf{rAL}} \alpha \Rightarrow \beta$. The left to the right direction is easy since every rIA1 × tqBa.t is a IA1 × tqBa.t. Conversely assume that $\not\models_{\mathbf{AL}} \alpha \Rightarrow \beta$, then by Lemma 23, one obtains a rIA1 × tqBa.t M such that $\not\models_M \alpha \Rightarrow \beta$. Therefore $\not\models_{\mathbf{rAL}} \alpha \Rightarrow \beta$.$\Box$

In what follows, we are going to prove the FMP of rG.

Definition 32 Let $\mathcal{T}$ be a set of formulas, a formula structure Γ is a $\mathcal{T}$-formula structure if all formulas appearing in it belong to $\mathcal{T}$. Let $\mathcal{FS}(\mathcal{T})$ be the set of all $\mathcal{T}$-formula structure. Let the notation $c(\mathcal{T})$ denotes the closure of $\mathcal{T}$ under $(\top, \bot, \wedge, \vee, \neg)$ and subformulas. A sequent $\Gamma \Rightarrow \beta$ is a $\mathcal{T}$-sequent if all formulas appearing in it belong to $\mathcal{T}$. One has $\vdash \Gamma \Rightarrow_{\mathcal{T}} \beta$ if there is a derivation of $\Gamma \Rightarrow_{\mathcal{T}} \beta$ s.t. all sequents appearing in it are $\mathcal{T}$-sequent.

Lemma 33 (Interpolation) *If* $\vdash \Gamma[\Delta] \Rightarrow_\mathcal{T} \beta$, *then there is a* $\gamma \in \mathcal{T}$ *such that* $\vdash \Delta \Rightarrow_\mathcal{T} \gamma$ *and* $\vdash \Gamma[\gamma] \Rightarrow_\mathcal{T} \beta$ *additionally for* $\Delta = \odot\Delta'$ *and* $\odot = \{\S, \bullet, \circ\}$ $\vdash \odot\gamma \Rightarrow_\mathcal{T} \gamma$.

Proof. We proceed by induction on the length of proof of $\Gamma[\Delta] \Rightarrow_\mathcal{T} \beta$. The proof for axioms is obvious. Take $(\bot)$ as an example, that is $\Gamma[\bot] \Rightarrow \alpha$. Since $\bot \in \mathcal{T}$, then $\bot$ is the required interpolant. Assuming that the end sequent is obtained by an arbitrary rule (R), let us consider the following cases, others can be treated similarly:

(R)=($\wedge$L). Assume the premise is $\vdash \Gamma[\Delta'[\alpha_i]] \Rightarrow_\mathcal{T} \beta$ and the conclusion is $\vdash \Gamma[\Delta'[\alpha_1 \wedge \alpha_2]] \Rightarrow_\mathcal{T} \beta$. Then by the induction hypothesis, there is a $\gamma \in \mathcal{T}$ such that (1) $\vdash \Delta'[\alpha_i] \Rightarrow_\mathcal{T} \gamma$, $\vdash \Gamma[\gamma] \Rightarrow_\mathcal{T} \beta$ and $\vdash \odot\gamma \Rightarrow_\mathcal{T} \gamma$. Then from (1) by ($\wedge$L), one obtains $\vdash \Delta'[\alpha_1 \wedge \alpha_2] \Rightarrow_\mathcal{T} \gamma$. Therefore γ is the required interpolant. The ($\vee$R), ($\rightarrow$R), ($\dagger$L), (?R), (HR), (GR), (Com) cases can be treated similarly.

(R)=($\wedge$R). Assume the premises are $\vdash \Gamma[\Delta] \Rightarrow_\mathcal{T} \alpha$ and $\vdash \Gamma[\Delta] \Rightarrow_\mathcal{T} \beta$. By induction hypothesis, there are $\gamma_1, \gamma_2 \in \mathcal{T}$ such that (1) $\vdash \Delta \Rightarrow_\mathcal{T} \gamma_1$, (2) $\vdash \Gamma[\gamma_1] \Rightarrow_\mathcal{T} \alpha$, (3) $\vdash \Delta \Rightarrow_\mathcal{T} \gamma_2$, (4) $\vdash \Gamma[\gamma_2] \Rightarrow_\mathcal{T} \beta$, (5) $\vdash \odot\gamma_1 \Rightarrow_\mathcal{T} \gamma_1$ and (6) $\vdash \odot\gamma_2 \Rightarrow_\mathcal{T} \gamma_2$. By applying ($\wedge$L) to (2) and (4), one has (7) $\vdash \Gamma[\gamma_1 \wedge \gamma_2] \Rightarrow_\mathcal{T} \alpha$ and (8) $\vdash \Gamma[\gamma_1 \wedge \gamma_2] \Rightarrow_\mathcal{T} \beta$. Next, by applying ($\wedge$R) to (7) and (8), one has $\vdash \Gamma[\gamma_1 \wedge \gamma_2] \Rightarrow_\mathcal{T} \alpha \wedge \beta$. Again by applying ($\wedge$R) to (1) and (3), one has $\vdash \Delta \Rightarrow_\mathcal{T} \gamma_1 \wedge \gamma_2$. Finally, we apply ($\wedge$L) and ($\wedge$R) to (5) and (6), one has $\vdash \odot(\gamma_1 \wedge \gamma_2) \Rightarrow_\mathcal{T} \gamma_1 \wedge \gamma_2$. Therefore, $\gamma_1 \wedge \gamma_2$ is the required interpolant. The ($\vee$L) case can be treated similarly.

(R)=($\rightarrow$L). Assume the premises are $\vdash \Delta' \Rightarrow_\mathcal{T} \alpha$ and $\vdash \Gamma[\sigma] \Rightarrow_\mathcal{T} \beta$, and the conclusion is $\vdash \Gamma[\Delta' \star \alpha \rightarrow \sigma] \Rightarrow_\mathcal{T} \beta$. Subcase (i): if $\Delta' \star \alpha \rightarrow \sigma$ is contained in Δ i.e. $\Delta = \Delta''[\Delta' \star \alpha \rightarrow \sigma]$, then premises are $\vdash \Delta' \Rightarrow_\mathcal{T} \alpha$ and $\vdash \Gamma'[\Delta''[\sigma]] \Rightarrow_\mathcal{T} \beta$. By induction hypothesis and ($\rightarrow$L), one has the interpolant for $\Delta''[\sigma]$ is the required interpolant. Subcase (ii): if Δ is contained in $\Delta' \star \alpha \rightarrow \sigma$, then $\Delta' = \Delta''[\Delta]$ or $\Delta = \alpha \rightarrow \sigma$. If $\Delta' = \Delta''[\Delta]$, by induction hypothesis and ($\rightarrow$L), one has the interpolant for Δ in $\Delta''[\Delta] \Rightarrow_\mathcal{T} \alpha$ is the required interpolant. If $\Delta = \alpha \rightarrow \sigma$, then $\alpha \rightarrow \sigma$ is the required interpolant. Subcase (iii): if Δ and $\Delta' \star \alpha \rightarrow \sigma$ are independent, then the premises are $\vdash \Delta' \Rightarrow_\mathcal{T} \alpha$ and $\vdash \Gamma'[\sigma][\Delta] \Rightarrow_\mathcal{T} \beta$, by induction hypothesis and ($\rightarrow$L), one has the interpolant for Δ in $\Gamma'[\sigma][\Delta] \Rightarrow_\mathcal{T} \beta$ is the required interpolant. The ($\cdot$L), ($\cdot$R), (?L), (HL) and (GL) cases can be treated similarly.

(R)=($\dagger$R). Assume the premise is $\vdash \Gamma \Rightarrow_\mathcal{T} \alpha$, then the conclusion is $\vdash \odot\Gamma \Rightarrow_\mathcal{T} \dagger\alpha$. Subcase (i): $\Gamma = \Gamma'[\Delta]$, then by induction hypothesis, there is a $\gamma \in \mathcal{T}$ such that (1) $\vdash \Delta \Rightarrow_\mathcal{T} \gamma$, (2) $\vdash \Gamma'[\gamma] \Rightarrow_\mathcal{T} \alpha$ and (3) $\vdash \odot\gamma \Rightarrow_\mathcal{T} \gamma$. By applying ($\dagger$R) to (2), one has $\vdash \odot\Gamma'[\gamma] \Rightarrow_\mathcal{T} \dagger\alpha$. Therefore, γ is the required interpolant. Subcase (ii): $\Delta = \odot\gamma$, then $\dagger\alpha$ is the required interpolant.

(R)=(Dual$_\S$). Assume the premise is $\vdash \S\Gamma_1 \star \Gamma_2 \Rightarrow_\mathcal{T} \bot$, then the conclusion is $\vdash \Gamma_1 \star \S\Gamma_2 \Rightarrow_\mathcal{T} \bot$. Subcase (i): Δ is contained in Γ_1 or Γ_2, then by induction hypothesis and (Dual), one has the interpolant for Δ is the required interpolant. Subcase (ii): $\Delta = \Gamma_1 \star \S\Gamma_2$, then $\bot$ is the required interpolant. Subcase (iii): $\Delta = \S\Gamma_2$, then by induction hypothesis, there is a $\gamma \in \mathcal{T}$ such that (1) $\vdash \S\Gamma_2 \Rightarrow_\mathcal{T} \gamma$, (2) $\vdash \Gamma_2 \star \gamma \Rightarrow_\mathcal{T} \bot$ and (3) $\vdash \S\gamma \Rightarrow_\mathcal{T} \gamma$. By (T) and ($\rightarrow$L) with

$\vdash \bot \Rightarrow_{\mathcal{T}} \bot$ on (3), one has (4) $\vdash \S\gamma \star \neg\gamma \Rightarrow_{\mathcal{T}} \bot$. Further by (Dual) and ($\to$R), one has (5) $\vdash \S\neg\gamma \Rightarrow_{\mathcal{T}} \neg\gamma$. By ($\to$R) on (2), one has (6) $\vdash \Gamma_2 \Rightarrow_{\mathcal{T}} \neg\gamma$. By (Cut) on (5) and (6), one has $\vdash \S\Gamma_2 \Rightarrow_{\mathcal{T}} \neg\gamma$. One has (7) $\vdash \gamma \star \neg\gamma \Rightarrow_{\mathcal{T}} \bot$. By (Cut) on (7) and $\vdash \Gamma_1 \Rightarrow_{\mathcal{T}} \gamma$, one has $\vdash \Gamma_1 \star \neg\gamma \Rightarrow_{\mathcal{T}} \bot$, then $\neg\gamma$ is the required interpolant. The (Dual$_{\bullet\circ}$), (Dual$_{\circ\bullet}$) and (IA1) cases can be treated similarly.

(R)=(T). Assume the premise is $\vdash \Gamma[\odot\Delta'] \Rightarrow_{\mathcal{T}} \beta$, then the conclusion is $\vdash \Gamma[\Delta'] \Rightarrow_{\mathcal{T}} \beta$. Subcase (i): Δ is contained in Δ', then by induction hypothesis, there is a $\gamma \in \mathcal{T}$ such that (1) $\vdash \Delta \Rightarrow_{\mathcal{T}} \gamma$, (2) $\vdash \Gamma[\odot\Delta'[\gamma]] \Rightarrow_{\mathcal{T}} \beta$ and (3) $\vdash \odot\gamma \Rightarrow_{\mathcal{T}} \gamma$. Next, by applying (T) to (2), one has $\vdash \Gamma[\Delta'[\gamma]] \Rightarrow_{\mathcal{T}} \beta$. Therefore, γ is the required interpolant. Subcase (ii): Δ' is contained in Δ, then the interpolant for Δ is the required interpolant. Subcase (iii): Δ' and Δ are independent, then the interpolant for Δ is the required interpolant.

(R)=(4). Assume the premise is $\vdash \Gamma[\odot\Delta'] \Rightarrow_{\mathcal{T}} \beta$, then the conclusion is $\vdash \Gamma[\odot^2\Delta'] \Rightarrow_{\mathcal{T}} \beta$. Subcase (i): Δ is contained in Δ' but $\Delta \neq \odot\Delta'$, then the interpolant for Δ is the required interpolant. Subcase (ii): $\Delta = \odot\Delta'$, then by induction hypothesis, there is a $\gamma \in \mathcal{T}$ such that (1) $\vdash \Delta \Rightarrow_{\mathcal{T}} \gamma$, (2) $\vdash \Gamma[\gamma] \Rightarrow_{\mathcal{T}} \beta$ and (3) $\vdash \odot\gamma \Rightarrow_{\mathcal{T}} \gamma$. By applying (Cut) to (2) and (3), one has $\vdash \Gamma[\odot\gamma] \Rightarrow_{\mathcal{T}} \beta$, then γ is the required interpolant. Subcase (iii): Δ' is contained in Δ or independent, then the interpolant for Δ is the required interpolant.

(R)=(Cut). Assume the premises are (1) $\vdash \Delta' \Rightarrow_{\mathcal{T}} \alpha$ and (2) $\vdash \Gamma[\alpha] \Rightarrow_{\mathcal{T}} \beta$, then the conclusion is $\vdash \Gamma[\Delta'] \Rightarrow_{\mathcal{T}} \beta$. Subcase (i): Δ is contained in Δ' i.e. $\Delta' = \Delta''[\Delta]$, then by induction hypothesis and (Cut), one has the interpolant for the $\vdash \Delta''[\Delta] \Rightarrow_{\mathcal{T}} \alpha$ is the required interpolant. Subcase (ii): Δ' is contained in Δ i.e. $\Delta = \Delta''[\Delta']$, then by induction hypothesis and (Cut), one has the interpolant for the $\vdash \Gamma[\Delta''[\alpha]] \Rightarrow_{\mathcal{T}} \beta$ is the required interpolant. Subcase (iii): Δ and Δ' are independent, then $\Gamma[\alpha] = \Gamma'[\alpha][\Delta]$, by induction hypothesis and (Cut), the interpolant for the Δ in $\Gamma'[\alpha][\Delta]$ is the required interpolant. $\qquad \square$

Remark 34 The above interpolation lemma's additional content: $\vdash \odot\gamma \Rightarrow_{\mathcal{T}} \gamma$ where $\odot = \{\S, \bullet, \circ\}$ is specially designed for solving the case of (R)=(4) when $\Delta = \odot\Delta'$. Such a lemma is rooted in Buszkowski's work [8] towards finite embeddability property and FMP of nonassociative Lambek calculus and its various lattice extensions. Lin [19,20] further studied this lemma to prove some non-classical modal logics' FMP or SFMP.

Definition 35 A formula α is a letter if $\alpha \in \mathbf{Var} \cup \{\top, \bot\}$ or $\alpha = \dagger\beta$ for some $\dagger \in \mathcal{F}$ where $\dagger \in \{!, P, F, ?, G, H\}$. Let $\mathcal{L}_e$ be the set of all letters. A formula α is called a literal if $\alpha \in \mathcal{L}_e$ or $\alpha = \neg\beta$ for some $\beta \in \mathcal{L}_e$. The set of all literals under language $\mathcal{L}$ is denoted by $\mathcal{L}_i$. A formula α is in *disjunction normal form* (DNF) if α is of the form $\bigvee_{i<l}\bigwedge_{j<k}\beta_{i,j}$ where $\beta_{i,j} \in \mathcal{L}_i$ and $l, k > 0$.

Let $\mathcal{T}$ be a set of formulas such that $\mathcal{T} = c(\mathcal{T})$. Suppose that $\mathcal{T}_{li} \subseteq \mathcal{T}$ be the set of all literals in $\mathcal{T}$. We say $\mathcal{T}$ is finitely based if $\mathcal{T}_{li}$ is finite. For any formula $\alpha \in \mathcal{T}$, there is a DNF formula $\beta \in \mathcal{T}$ such that α is equivalent to β under distributive, De morgan and double negation laws. If one omits the repetition of literals, then one has a unique formula in DNF which is equivalent to α. We denote the unique DNF formula corresponding to α by $df_{\mathcal{T}}(\alpha)$. Let

$df(\mathcal{T}) = \{df_{\mathcal{T}}(\alpha) \mid \alpha \in \mathcal{T}\}$. If $\mathcal{T}_{li}$ is finite, then $df(\mathcal{T})$ is finite.

Corollary 36 *If $\vdash \Gamma[\Delta] \Rightarrow_{\mathcal{T}} \beta$, then there is a $\gamma \in df(\mathcal{T})$ such that $\vdash \Delta \Rightarrow_{\mathcal{T}} \gamma$ and $\vdash \Gamma[\gamma] \Rightarrow_{\mathcal{T}} \beta$.*

Proof. Follows from the Lemma 33 and the definitions of $df(\mathcal{T})$. □

Definition 37 We define $\leq_{\mathcal{T}}$ on $\mathcal{FS}(\mathcal{T})$ as follows: for $\Delta_1, \Delta_2 \in \mathcal{FS}(\mathcal{T})$, $\Delta_1 \leq_{\mathcal{T}} \Delta_2$ iff for any context $\Gamma[-]$ and formula $\varphi \in \mathcal{T}$, if $\Gamma[\Delta_2] \Rightarrow_{\mathcal{T}} \varphi$, then $\Gamma[\Delta_1] \Rightarrow_{\mathcal{T}} \varphi$.

Let $\Delta_1 \approx_{\mathcal{T}} \Delta_2$ be $\Delta_1 \leq_{\mathcal{T}} \Delta_2$ and $\Delta_2 \leq_{\mathcal{T}} \Delta_1$, then $\approx_{\mathcal{T}}$ is an equivalence relation. Let $[\alpha]_{\mathcal{T}} = \{\Delta \mid \Delta \approx_{\mathcal{T}} \alpha \,\&\, \Delta \in \mathcal{FS}(\mathcal{T})\}$ for any $\alpha \in \mathcal{T}$. Let $[\mathcal{T}] = \{[\alpha]_{\mathcal{T}} \mid \alpha \in \mathcal{T})\}$. Since $[\alpha]_{\mathcal{T}} = [df_{\mathcal{T}}(\alpha)]_{\mathcal{T}}$ and the number of $[df_{\mathcal{T}}(\alpha)]_{\mathcal{T}}$ is finite, $[\mathcal{T}]$ is finite.

Lemma 38 *For any $\Delta \in \mathcal{FS}(\mathcal{T})$, there is a $\alpha \in df(\mathcal{T})$ such that $\Delta \approx_{\mathcal{T}} \alpha$.*

Proof. For any $\Gamma[-]$ and $\gamma \in \mathcal{T}$, assume that $\vdash \Gamma[\Delta] \Rightarrow_{\mathcal{T}} \gamma$. By corollary 36, there is a $\beta_j \in df(\mathcal{T})$ such that $\vdash \Delta \Rightarrow_{\mathcal{T}} \beta_j$ and $\vdash \Gamma[\beta_j] \Rightarrow_{\mathcal{T}} \gamma$. Obviously, the number of β_j is finite. Let δ be all the conjunctions of β_j. Clearly $\delta \in \mathcal{T}$. By (∧R) and (∧L), one has (1) $\vdash \Delta \Rightarrow_{\mathcal{T}} \delta$ and (2) $\vdash \Gamma[\delta] \Rightarrow_{\mathcal{T}} \gamma$. Then $df_{\mathcal{T}}(\delta) \in df(\mathcal{T})$. Let $\theta = df_{\mathcal{T}}(\delta)$, then one has (3) $\vdash \Delta \Rightarrow_{\mathcal{T}} \theta$ and (4) $\vdash \Gamma[\theta] \Rightarrow_{\mathcal{T}} \gamma$. Therefore, the assumption $\vdash \Gamma[\Delta] \Rightarrow_{\mathcal{T}} \gamma$ implies (4) $\vdash \Gamma[\theta] \Rightarrow_{\mathcal{T}} \gamma$, then one has $\theta \leq_{\mathcal{T}} \Delta$. Further, assume that (5) $\vdash \Gamma'[\theta] \Rightarrow_{\mathcal{T}} \sigma$ for some context $\Gamma'[-]$ and formula $\sigma \in \mathcal{T}$. By applying (Cut) to (3) and (5), one has $\vdash \Gamma'[\Delta] \Rightarrow_{\mathcal{T}} \sigma$. Therefore, one has $\Delta \leq_{\mathcal{T}} \theta$. Consequently, one has $\Delta \approx_{\mathcal{T}} \theta$. □

Definition 39 Let $\mathbb{Q} = ([\mathcal{T}], \wedge^*, \vee^*, \neg^*, !^*, P^*, F^*, H^*, G^*, \cdot^*, \rightarrow^*, \perp^*, \top^*)$ be the quotient algebra of $[\mathcal{T}]$ where all operations are defined as follows: for any $[\alpha]_{\mathcal{T}}, [\beta]_{\mathcal{T}} \in [\mathcal{T}]$, $\odot = \{\S, \bullet, \circ\}$ and their corresponding formulas $\dagger = \{!, P, F\}$,

(1) $\top^* = [\top]_{\mathcal{T}}$;

(2) $\perp^* = [\perp]_{\mathcal{T}}$;

(3) $[\alpha]_{\mathcal{T}} \wedge^* [\beta]_{\mathcal{T}} = [\alpha \wedge \beta]_{\mathcal{T}}$;

(4) $[\alpha]_{\mathcal{T}} \vee^* [\beta]_{\mathcal{T}} = [\alpha \vee \beta]_{\mathcal{T}}$;

(5) $\dagger^*[\alpha]_{\mathcal{T}} = [\gamma]_{\mathcal{T}}$ s.t. $\gamma \approx_{\mathcal{T}} \odot\alpha$;

(6) $H^*[\alpha]_{\mathcal{T}} = [\gamma_1 \vee \ldots \vee \gamma_n]_{\mathcal{T}}$ s.t. $F^*[\gamma_i]_{\mathcal{T}} \leq^* [\alpha]_{\mathcal{T}}$ for any $i \in \{1, \ldots, n\}$;

(7) $G^*[\alpha]_{\mathcal{T}} = [\gamma_1 \vee \ldots \vee \gamma_n]_{\mathcal{T}}$ s.t. $P^*[\gamma_i]_{\mathcal{T}} \leq^* [\alpha]_{\mathcal{T}}$ for any $i \in \{1, \ldots, n\}$;

(8) $?^*[\alpha]_{\mathcal{T}} = [\gamma_1 \vee \ldots \vee \gamma_n]_{\mathcal{T}}$ s.t. $!^*[\gamma_i]_{\mathcal{T}} \leq^* [\alpha]_{\mathcal{T}}$ for any $i \in \{1, \ldots, n\}$;

(9) $[\alpha]_{\mathcal{T}} \cdot^* [\beta]_{\mathcal{T}} = [\gamma]_{\mathcal{T}}$ s.t. $\gamma \approx_{\mathcal{T}} \alpha \star \beta$;

(10) $[\alpha]_{\mathcal{T}} \rightarrow^* [\beta]_{\mathcal{T}} = [\gamma_1 \vee \ldots \vee \gamma_n]_{\mathcal{T}}$ s.t. $[\alpha]_{\mathcal{T}} \wedge^* [\gamma_i]_{\mathcal{T}} \leq^* [\beta]_{\mathcal{T}}$ for any $i \in \{1, \ldots, n\}$.

Note that the symbols "$\dagger$" and "$\odot$" in (6) are in one-to-one correspondence similar to Definition 27. We define $[\alpha]_{\mathcal{T}} \leq^* [\beta]_{\mathcal{T}}$ as $[\alpha]_{\mathcal{T}} \wedge^* [\beta]_{\mathcal{T}} = [\alpha]_{\mathcal{T}}$.

Lemma 40 *All the operations defined in Definition 39 are well-defined.*

Proof. We only provide the cases for operations of $\dagger^* \in \{!^*, P^*, F^*\}$, others can be treated similarly. Clearly, by the definition of $\dagger^*[\alpha]_\mathcal{T}$ and Lemma 38, $[\gamma]_\mathcal{T}$ exists and is unique. Further, let $[\alpha_1]_\mathcal{T} = [\alpha_2]_\mathcal{T}$, one can show that $\dagger^*[\alpha_1]_\mathcal{T} = \dagger^*[\alpha_2]_\mathcal{T}$. Since $\alpha_1 \in [\alpha_1]_\mathcal{T}$, then $\alpha_1 \in [\alpha_2]_\mathcal{T}$. By the definition of the equivalence class, one has $\alpha_1 \approx_\mathcal{T} \alpha_2$. Given any context $\Gamma[-]$ and formula $\beta \in \mathcal{T}$. Assume that $\vdash \Gamma[\odot\alpha_1] \Rightarrow_\mathcal{T} \beta$, then one has $\vdash \Gamma[\odot\alpha_2] \Rightarrow_\mathcal{T} \beta$. Therefore, one has $\odot\alpha_2 \leq_\mathcal{T} \odot\alpha_1$. By similar argument, one has $\odot\alpha_1 \leq_\mathcal{T} \odot\alpha_2$. Consequently, one has $\odot\alpha_1 \approx_\mathcal{T} \odot\alpha_2$. Assume $\dagger^*[\alpha_1]_\mathcal{T} = [\gamma]_\mathcal{T}$ such that $\gamma \approx_\mathcal{T} \odot\alpha_1$, then $\gamma \approx_\mathcal{T} \odot\alpha_2$ and $[\gamma]_\mathcal{T} = \dagger^*[\alpha_2]_\mathcal{T}$. Therefore $\dagger^*[\alpha_1]_\mathcal{T} = \dagger^*[\alpha_2]_\mathcal{T}$. $\square$

Lemma 41 *The following conditions are equivalent for all $\alpha, \beta \in \mathcal{T}$:*

$$(1)\ \alpha \leq_\mathcal{T} \beta;\ (2)\ \vdash \alpha \Rightarrow_\mathcal{T} \beta;\ (3)\ [\alpha]_\mathcal{T} \leq^* [\beta]_\mathcal{T}.$$

Proof. For (1) and (2), assume $\vdash \alpha \Rightarrow_\mathcal{T} \beta$. Given any context $\Gamma[-]$ and formula $\varphi \in \mathcal{T}$, assume that $\vdash \Gamma[\beta] \Rightarrow_\mathcal{T} \varphi$. By (Cut) one has $\vdash \Gamma[\alpha] \Rightarrow_\mathcal{T} \varphi$. Therefore, one has $\alpha \leq_\mathcal{T} \beta$. Conversely, assume $\alpha \leq_\mathcal{T} \beta$. Since $\vdash \beta \Rightarrow_\mathcal{T} \beta$, then one has $\vdash \alpha \Rightarrow_\mathcal{T} \beta$. For (2) and (3), Assume $[\alpha]_\mathcal{T} \leq^* [\beta]_\mathcal{T}$, then one has $[\alpha]_\mathcal{T} \wedge^* [\beta]_\mathcal{T} = [\alpha]_\mathcal{T}$. Since $[\alpha]_\mathcal{T} \wedge^* [\beta]_\mathcal{T} = [\alpha \wedge \beta]_\mathcal{T}$, then one has $[\alpha \wedge \beta]_\mathcal{T} = [\alpha]_\mathcal{T}$. By the definition of the equivalence class, one has $\alpha \wedge \beta \approx_\mathcal{T} \alpha$. Further, one has $\alpha \leq_\mathcal{T} \alpha \wedge \beta$ and $\alpha \wedge \beta \leq_\mathcal{T} \beta$. Therefore one has $\alpha \leq_\mathcal{T} \beta$. Conversely, assume $\alpha \leq_\mathcal{T} \beta$, then one has $\vdash \alpha \Rightarrow_\mathcal{T} \beta$. By $(\wedge R)$, one has $\vdash \alpha \wedge \beta \Leftrightarrow_\mathcal{T} \alpha$. Therefore, one has $[\alpha \wedge \beta]_\mathcal{T} = [\alpha]_\mathcal{T} = [\alpha]_\mathcal{T} \wedge^* [\beta]_\mathcal{T}$. Therefore, $[\alpha]_\mathcal{T} \leq^* [\beta]_\mathcal{T}$. Consequently, $\alpha \leq_\mathcal{T} \beta$ iff $[\alpha]_\mathcal{T} \leq^* [\beta]_\mathcal{T}$. $\square$

Lemma 42 *For any $[\alpha]_\mathcal{T}, [\beta]_\mathcal{T} \in [\mathcal{T}]$, $\dagger^* \in \{!^*, P^*, F^*\}$, the following conditions hold for $\mathbb{Q}$:*

(Adj$_1$) $P^*[\alpha]_\mathcal{T} \leq^* [\beta]_\mathcal{T}$ iff $[\alpha]_\mathcal{T} \leq^* G^*[\beta]_\mathcal{T}$;

(Adj$_2$) $F^*[\alpha]_\mathcal{T} \leq^* [\beta]_\mathcal{T}$ iff $[\alpha]_\mathcal{T} \leq^* H^*[\beta]_\mathcal{T}$;

(Adj$_3$) $!^*[\alpha]_\mathcal{T} \leq^* [\beta]_\mathcal{T}$ iff $[\alpha]_\mathcal{T} \leq^* ?^*[\beta]_\mathcal{T}$;

(Res) $[\alpha]_\mathcal{T} \cdot^* [\beta]_\mathcal{T} \leq^* [\gamma]_\mathcal{T}$ iff $[\alpha]_\mathcal{T} \leq^* [\beta]_\mathcal{T} \to^* [\gamma]_\mathcal{T}$;

(Com) $[\alpha]_\mathcal{T} \cdot^* [\beta]_\mathcal{T} = [\beta]_\mathcal{T} \cdot^* [\alpha]_\mathcal{T}$;

(Mon) If $[\alpha]_\mathcal{T} \leq^* [\beta]_\mathcal{T}$, then $\dagger^*[\alpha]_\mathcal{T} \leq^* \dagger^*[\beta]_\mathcal{T}$;

(N) $\dagger^*\bot^* = \bot^*$;

(K) $\dagger^*([\alpha]_\mathcal{T} \vee^* [\beta]_\mathcal{T}) = \dagger^*[\alpha]_\mathcal{T} \vee^* \dagger^*[\beta]_\mathcal{T}$;

(T) $[\alpha]_\mathcal{T} \leq^* \dagger^*[\alpha]_\mathcal{T}$;

(4) $\dagger^* \dagger^* [\alpha]_\mathcal{T} \leq^* \dagger^*[\alpha]_\mathcal{T}$;

(IA1) If $!^*[\alpha]_\mathcal{T} \cdot^* !^*[\alpha]_\mathcal{T} = \bot^*$, then $!^*[\alpha]_\mathcal{T} = \bot^*$;

(Dual$_\S$) If $!^*[\alpha]_\mathcal{T} \cdot^* [\beta]_\mathcal{T} = \bot^*$, then $[\alpha]_\mathcal{T} \cdot^* !^*[\beta]_\mathcal{T} = \bot^*$;

(Dual$_{\bullet\circ}$) If $P^*[\alpha]_\mathcal{T} \cdot^* [\beta]_\mathcal{T} = \bot^*$, then $[\alpha]_\mathcal{T} \cdot^* F^*[\beta]_\mathcal{T} = \bot^*$;

(Dual$_{\circ\bullet}$) If $F^*[\alpha]_\mathcal{T} \cdot^* [\beta]_\mathcal{T} = \bot^*$, then $[\alpha]_\mathcal{T} \cdot^* P^*[\beta]_\mathcal{T} = \bot^*$.

Proof. We only provide the proofs for the following cases. (Com) and (N) are easy to check, (Res), (Adj$_i$) $i \in \{2, 3\}$ are similar to (Adj), and (IA1), (Dual$_{\bullet\circ}$)

and (Dual$_{o\bullet}$) are similar to (Dual$_\S$).

(Adj) Assume that $P^*[\alpha]_\mathcal{T} \leq^* [\beta]_\mathcal{T}$. Let $P^*[\alpha]_\mathcal{T} = [\gamma]_\mathcal{T}$ such that $\vdash \bullet\alpha \approx_\mathcal{T} \gamma$, then by Lemma 41 one has $\bullet\alpha \Rightarrow_\mathcal{T} \beta$. Let $G^*[\beta]_\mathcal{T} = [\delta \vee \ldots \vee \delta_n]_\mathcal{T}$ s.t. $P^*[\delta_i]_\mathcal{T} \leq^* [\beta]_\mathcal{T}$ for any $i \in \{1, \ldots, n\}$. By ($\vee$R), one has $\vdash \alpha \Rightarrow_\mathcal{T} \delta \vee \ldots \vee \delta_n \vee \alpha$. Clearly, α is one of the δ_i. Therefore, $[\alpha]_\mathcal{T} \leq^* G^*[\beta]_\mathcal{T}$. Assume that $[\alpha]_\mathcal{T} \leq^* G^*[\beta]_\mathcal{T}$. Then one has $\vdash \alpha \Rightarrow_\mathcal{T} \delta \vee \ldots \vee \delta_n$. By ($\vee$R), one has $\vdash \bullet(\delta \vee \ldots \vee \delta_n) \Rightarrow_\mathcal{T} \beta$. By applying (Cut), one has $\vdash \bullet\alpha \Rightarrow_\mathcal{T} \beta$. Therefore, one has $P^*[\alpha]_\mathcal{T} \leq^* [\beta]_\mathcal{T}$.

(Mon) Assume $[\alpha]_\mathcal{T} \leq^* [\beta]_\mathcal{T}$. Let $\dagger^*[\alpha]_\mathcal{T} = [\gamma_1]_\mathcal{T}$ and $\dagger^*[\beta]_\mathcal{T} = [\gamma_2]_\mathcal{T}$ such that $\odot\alpha \approx_\mathcal{T} \gamma_1$ and $\odot\beta \approx_\mathcal{T} \gamma_2$. It suffices to prove that $\vdash \gamma_1 \Rightarrow_\mathcal{T} \gamma_2$. By Lemma 41, one obtains $\vdash \alpha \Rightarrow_\mathcal{T} \beta$. By Lemma 41 and Definition 37, $\vdash \odot\beta \Rightarrow_\mathcal{T} \gamma_2$. Hence by (Cut), one obtains $\vdash \odot\alpha \Rightarrow_\mathcal{T} \gamma_2$. Therefore $\vdash \gamma_1 \Rightarrow_\mathcal{T} \gamma_2$.

(K) Clearly, $[\alpha]_\mathcal{T} \leq^* [\alpha]_\mathcal{T} \vee^* [\beta]_\mathcal{T}$ and $[\beta]_\mathcal{T} \leq^* [\alpha]_\mathcal{T} \vee^* [\beta]_\mathcal{T}$. By (Mon), one obtains $\dagger^*[\alpha]_\mathcal{T} \leq^* \dagger^*([\alpha]_\mathcal{T} \vee^* [\beta]_\mathcal{T})$ and $\dagger^*[\beta]_\mathcal{T} \leq^* \dagger^*([\alpha]_\mathcal{T} \vee^* [\beta]_\mathcal{T})$. Then $\dagger^*[\alpha]_\mathcal{T} \vee^* \dagger^*[\beta]_\mathcal{T} \leq^* \dagger^*([\alpha]_\mathcal{T} \vee^* [\beta]_\mathcal{T})$. Let $[\gamma_1]_\mathcal{T} = \dagger^*[\alpha]_\mathcal{T}$, $[\gamma_2]_\mathcal{T} = \dagger^*[\beta]_\mathcal{T}$ and $[\gamma_3]_\mathcal{T} = \dagger^*[\alpha \vee \beta]_\mathcal{T}$ s.t. $\gamma_1 \approx_\mathcal{T} \odot\alpha$, $\gamma_2 \approx_\mathcal{T} \odot\beta$ and $\gamma_3 \approx_\mathcal{T} \odot(\alpha \vee \beta)$. By Lemma 41, one has $\vdash \odot\alpha \Rightarrow_\mathcal{T} \gamma_1$ and $\vdash \odot\beta \Rightarrow_\mathcal{T} \gamma_2$. By ($\vee$R) and ($\vee$L), one has $\vdash \odot(\alpha \vee \beta) \Rightarrow_\mathcal{T} \gamma_1 \vee \gamma_2$. By $\gamma_3 \approx_\mathcal{T} \odot(\alpha \vee \beta)$ and Lemma 41, one has $\dagger^*([\alpha]_\mathcal{T} \vee^* [\beta]_\mathcal{T}) \leq^* \dagger^*[\alpha]_\mathcal{T} \vee^* \dagger^*[\beta]_\mathcal{T}$.

(T) Let $\dagger^*[\alpha]_\mathcal{T} = [\gamma]_\mathcal{T}$ such that $\gamma \approx_\mathcal{T} \odot\alpha$. It suffices to prove that $\vdash \alpha \Rightarrow_\mathcal{T} \gamma$. Clearly, $\vdash \odot\alpha \Rightarrow_\mathcal{T} \gamma$. Then by (T), one obtains $\vdash \alpha \Rightarrow_\mathcal{T} \gamma$.

(4) Let $\dagger^*[\alpha]_\mathcal{T} = [\theta_1]_\mathcal{T}$ and $\dagger^*[\theta_1]_\mathcal{T} = [\theta_2]_\mathcal{T}$ such that $\theta_1 \approx_\mathcal{T} \odot\alpha$ and $\theta_2 \approx_\mathcal{T} \odot\theta_1$. It suffices to show $\theta_2 \leq_\mathcal{T} \theta_1$ by Lemma 41. Assume $\vdash \Gamma[\theta_1] \Rightarrow_\mathcal{T} \varphi$ for some context $\Gamma[-]$ and $\varphi \in \mathcal{T}$. Hence $\vdash \Gamma[\odot\alpha] \Rightarrow_\mathcal{T} \varphi$. Then by (4), one obtains $\vdash \Gamma[\circ(\odot\alpha)] \Rightarrow_\mathcal{T} \varphi$. Whence $\vdash \Gamma[\odot\theta_1] \Rightarrow_\mathcal{T} \varphi$. Therefore $\vdash \Gamma[\theta_2] \Rightarrow_\mathcal{T} \varphi$.

(Dual$_\S$) Assume $!^*[\alpha]_\mathcal{T} \cdot^* [\beta]_\mathcal{T} = \perp^*$. Let $!^*[\alpha]_\mathcal{T} = [\gamma_1]_\mathcal{T}$ and $!^*[\beta]_\mathcal{T} = [\gamma_2]_\mathcal{T}$ such that $\gamma_1 \approx_\mathcal{T} \S\alpha$ and $\gamma_2 \approx_\mathcal{T} \S\beta$. By Definition 39 and Lemma 41, one has $\vdash \S\alpha \star \beta \Rightarrow_\mathcal{T} \perp$. By (Dual$_\S$) rule, one has $\vdash \alpha \star \S\beta \Rightarrow_\mathcal{T} \perp$. Therefore, one has $[\alpha]_\mathcal{T} \cdot^* P^*[\beta]_\mathcal{T} = \perp^*$.

$\square$

Lemma 43 *The following conditions hold for* $\mathbb{Q}$: *for any* $\dagger \in \{!, P, F\}$,

(1) If $\dagger\alpha \in \mathcal{T}$, *then* $\dagger^*[\alpha]_\mathcal{T} = [\dagger\alpha]_\mathcal{T}$;   *(4) If* $?\alpha \in \mathcal{T}$, *then* $?^*[\alpha]_\mathcal{T} = [?\alpha]_\mathcal{T}$;

(2) If $H\alpha \in \mathcal{T}$, *then* $H^*[\alpha]_\mathcal{T} = [H\alpha]_\mathcal{T}$;   *(5) If* $\alpha \cdot \beta \in \mathcal{T}$, *then* $[\alpha]_\mathcal{T} \cdot^* [\beta]_\mathcal{T} = [\alpha \cdot \beta]_\mathcal{T}$;

(3) If $G\alpha \in \mathcal{T}$, *then* $G^*[\alpha]_\mathcal{T} = [G\alpha]_\mathcal{T}$;   *(6) If* $\alpha \rightarrow \beta \in \mathcal{T}$, *then* $[\alpha]_\mathcal{T} \rightarrow^* [\beta]_\mathcal{T} = [\alpha \rightarrow \beta]_\mathcal{T}$.

Proof. We only provide the proofs for (1), others can be treated similarly. Assume $\dagger^*[\alpha]_\mathcal{T} = [\gamma]_\mathcal{T}$ such that $\gamma \in \mathcal{T}$ and $\odot\alpha \approx_\mathcal{T} \gamma$. It suffices to show $\gamma \approx_\mathcal{T} \dagger\alpha$. Assume that $\vdash \Gamma[\gamma] \Rightarrow_\mathcal{T} \varphi$ for some context $\Gamma[-]$ and $\varphi \in \mathcal{T}$. Then $\vdash \Gamma[\odot\alpha] \Rightarrow_\mathcal{T} \varphi$. By ($\dagger$L), $\vdash \Gamma[\dagger\alpha] \Rightarrow_\mathcal{T} \varphi$. Assume $\vdash \Gamma[\dagger\alpha] \Rightarrow_\mathcal{T} \varphi$. Clearly $\vdash \odot\alpha \Rightarrow_\mathcal{T} \dagger\alpha$. Then by (Cut), $\vdash \Gamma[\odot\alpha] \Rightarrow_\mathcal{T} \varphi$. Thus $\vdash \Gamma[\gamma] \Rightarrow_\mathcal{T} \varphi$. $\square$

Remark 44 By Lemma 43, one has $\neg^*[\alpha]_\mathcal{T} = [\neg\alpha]_\mathcal{T}$ and $?^*[\alpha]_\mathcal{T} = [?\alpha]_\mathcal{T}$. Since $\mathcal{T}$ is closed under $\neg$, then the De Morgan rule and double negation rule hold for $\mathbb{Q}$. (5) $!^*[\alpha]_\mathcal{T} \leq^* ?^*!^*[\alpha]_\mathcal{T}$ can be obtained by (Dual$_\S$) in Lemma 42.

Corollary 45 *Let $\mathcal{T}$ be finitely based, then $\mathbb{Q}$ is a finite rIA1 × tqBa.t.*

Lemma 46 *If $\nvdash_{rG} \alpha \Rightarrow_{\mathcal{T}} \beta$, then $\not\models_{\mathbb{Q}} [\alpha]_{\mathcal{T}} \leq^* [\beta]_{\mathcal{T}}$.*

Proof. Let $\mathcal{T}$ be the smallest set containing α, β s.t. $\mathcal{T} = c(\mathcal{T})$, assume $\nvdash_{rG} \alpha \Rightarrow \beta$, then $\nvdash_{rG} \alpha \Rightarrow_{\mathcal{T}} \beta$. Construct $\mathbb{Q} = ([\mathcal{T}], \wedge^*, \vee^*, \neg^*, !^*, P^*, F^*, \cdot^*, \rightarrow^* , \perp^*, \top^*)$ as above and an assignment $\sigma : \mathbf{Var} \longrightarrow [\mathcal{T}]$ such that $\sigma(p) = [p]_{\mathcal{T}}$. By induction on the complexity of the formula, one can easily prove that $\hat{\sigma}(\delta) = [\delta]_{\mathcal{T}}$ by Definition 39 and Lemma 43. Assume that $\models_{\mathbb{Q},\sigma} [\alpha]_{\mathcal{T}} \leq^* [\beta]_{\mathcal{T}}$, then by Lemma 41, one has $\vdash \alpha \Rightarrow_{\mathcal{T}} \beta$, which contradicts to our initial assumption. Therefore, if $\nvdash_{rG} \alpha \Rightarrow_{\mathcal{T}} \beta$, then $\not\models_{\mathbb{Q}} [\alpha]_{\mathcal{T}} \leq^* [\beta]_{\mathcal{T}}$.   □

Corollary 47 *If $\nvdash_{rG} \alpha \Rightarrow \beta$, then $\not\models_{\mathbf{rAL}} \mu(\alpha) \leq \mu(\beta)$ where μ is a mapping from $\mathbf{Var}$ to a finite rIA1 × tqBa.t.*

Theorem 48 (FMP) rG *has FMP.*

Theorem 49 (Decidability) rG *is decidable.*

Theorem 50 (FMP) G *has FMP.*

Proof. Since rG is a conservative extension of G by Lemma 31 and rG has FMP by Theorem 48, then G has FMP.   □

Theorem 51 (Decidability) G *is decidable.*

5   Conclusion and Future Work

In this paper, we study a logic L of algebra IA1×tqBa.t by algebraic proof theory for temporal and open information. We present the axiomatic system H and sequent calculus G with soundness and completeness proved. By showing a different kind of interpolation lemma, we prove a conservative extension rG FMP result of G and thus decidability of rG and G. One of our future works is to see whether there is a feasible way to rewrite the calculus into a context-free language. If so, then a decidable algorithm can be possibly established for G.

References

[1] An, A., J. Stefanowski, S. Ramanna, C. J. Butz, W. Pedrycz, G. Wang et al., "Rough sets, fuzzy sets, data mining and granular computing," Springer Berlin Heidelberg, 2007.

[2] Banerjee, M., *Rough sets and 3-valued lukasiewicz logic*, Fundamenta Informaticae **31** (1997), pp. 213–220.

[3] Banerjee, M. and M. Chakraborty, *Rough algebra*, Bulletin of the Polish Academy of Sciences-Mathematics **41** (1993), pp. 293–298.

[4] Banerjee, M. and M. K. Chakraborty, *Rough sets through algebraic logic*, Fundamenta Informaticae **28** (1996), pp. 211–221.

[5] Banerjee, M., S. Ju, M. A. Khan and L. Tang, *Open world models: a view from rough set theory*, in: *Facets of Uncertainties and Applications: ICFUA, Kolkata, India, December 2013*, Springer, 2015, pp. 77–86.

[6] Bello, R. and R. Falcon, *Rough sets in machine learning: a review*, Thriving Rough Sets: 10th Anniversary-Honoring Professor Zdzisław Pawlak's Life and Legacy & 35 Years of Rough Sets (2017), pp. 87–118.

[7] Bialynicki-Birula, A. and H. Rasiowa, *On the representation of quasi-boolean algebras*, Journal of Symbolic Logic **22** (1957).

[8] Buszkowski, W., *Interpolation and fep for logics of residuated algebras*, Logic Journal of IGPL **19** (2011), pp. 437–454.

[9] Cattaneo, G., D. Ciucci and D. Dubois, *Algebraic models of deviant modal operators based on de morgan and kleene lattices*, Information Sciences **181** (2011), pp. 4075–4100.

[10] Chan, C.-C., J. W. Grzymala-Busse and W. P. Ziarko, "Rough sets and current trends in computing," Springer Berlin Heidelberg, 2008.

[11] Ewald, W. B., *Intuitionistic tense and modal logic*, The Journal of Symbolic Logic **51** (1986), pp. 166–179.

[12] Galatos, N., P. Jipsen, T. Kowalski and H. Ono, "Residuated lattices: an algebraic glimpse at substructural logics," Elsevier, 2007.

[13] Girard, J.-Y., *Linear logic*, Theoretical computer science **50** (1987), pp. 1–101.

[14] Hardegree, G. M., *Material implication in orthomodular (and boolean) lattices.*, Notre Dame Journal of Formal Logic **22** (1981), pp. 163–182.

[15] Ju, S. and H. Liu, *The logical structure of open sets*, in: *Logic and Cognition: The Progress in Logical Study in China* (2003), pp. 596–603.

[16] Khan, M. A., M. Banerjee and S. Panda, *Logics for temporal information systems in rough set theory*, ACM Transactions on Computational Logic **24** (2023), pp. 1–29.

[17] Liang, C. and D. Miller, *Kripke semantics and proof systems for combining intuitionistic logic and classical logic*, Annals of Pure and Applied Logic **164** (2013), pp. 86–111.

[18] Lin, T., N. Cercone and A. Wasilewska, *Topological rough algebras*, Rough Sets and Data Mining: Analysis of Imprecise Data (1997), pp. 411–425.

[19] Lin, Z., *Non-associative lambek calculus with modalities: interpolation, complexity and fep*, Logic Journal of the IGPL **22** (2014), pp. 494–512.

[20] Lin, Z., M. K. Chakraborty and M. Ma, *Residuated algebraic structures in the vicinity of pre-rough algebra and decidability*, Fundamenta Informaticae **179** (2021), pp. 239–274.

[21] Moisil, G. C., *Recherches sur l'algèbre de la logique*, Ann. Sci. Univ. Jassy **22** (1935), pp. 1–117.

[22] Monteiro, L. F., *Axiomes indépendants pour les algebres de lukasiewicz trivalentes*, Bulletin mathématique de la Société des Sciences Mathématiques et Physiques de la République Populaire Roumaine **7** (1963), pp. 199–202.

[23] Pawlak, Z., *Rough sets*, International journal of computer & information sciences **11** (1982), pp. 341–356.

[24] Quine, W. V., *Natural kinds*, in: *Essays in honor of Carl G. Hempel: A tribute on the occasion of his sixty-fifth birthday*, Springer, 1969 pp. 5–23.

[25] Saha, A., J. Sen and M. K. Chakraborty, *Algebraic structures in the vicinity of pre-rough algebra and their logics*, Information Sciences **282** (2014), pp. 296–320.

[26] Sen, J., "Some embeddings in linear logic and related issues," Ph.D. thesis, University of Calcutta, India (2001).

[27] Sen, J. and M. K. Chakraborty, *A study of interconnections between rough and 3-valued łukasiewicz logics*, Fundamenta Informaticae **51** (2002), pp. 311–324.

[28] Wittgenstein, L., "Philosophical investigations," John Wiley & Sons, 2010.

[29] Zhang, Q., Q. Xie and G. Wang, *A survey on rough set theory and its applications*, CAAI Transactions on Intelligence Technology **1** (2016), pp. 323–333.

Assumable Logic Programming

Zhizheng Zhang [1]

School of Computer Science and Engineering
Southeast University
Nanjing, China

Abstract

In the case of incomplete information, a common reasoning framework is an assumption-based two-step deductive reasoning process where intelligent agents make assumptions first and then use them to build their belief sets. This paper proposes a logic programming paradigm ALP (Assumable Logic Programming) that is developed as a knowledge representation and reasoning tool for designing intelligent agents capable of performing the framework. ALP extends ASP (Answer Set Programming) with a new modal operator used to precede a literal in rules bodies, and thus allows for the representation of assumption knowledge. We define the language of ALP, discuss the relation of ALP to ASP, and show that some problems that are beyond the power of ASP can be solved by ALP through cases study and some theoretical results.

Keywords: Logic programming, answer set, assumption-making, belief-building.

1 Introduction

In the case of incomplete information, a popular reasoning framework is two-stage where intelligent agents make assumptions first and then use them to build their belief sets through deductive reasoning [15]. Many existing formalisms can be considered as approaches for this framework. Examples include Assumption-Based Truth Maintenance System introduced in [6], [26], and [7], Probabilistic Assumption-Based Model and language proposed in [17] and [1], Poole's Default Theory in [22] and [23], Supposition-based logic in [3] etc. Some efforts are made to explore the way of identifying assumptions in reasoning, in which assumptions are not given explicitly. For example, [5] explores the derivation of assumptions to explain observed events, [24] presents an approach to hypothetical planning that involves generating assumptions about actions that can not be derived from the knowledge-base etc. Besides, many studies show that assumption-based reasoning is closely related to the topics like argumentation [4], action reasoning [18], planning [21], contextual reasoning [15], defeasible reasoning [11], default reasoning [16] [12] etc.

---

[1] Email: seu_zzz@seu.edu.cn. The work was supported by the Pre-research Key Laboratory Fund for Equipment (Grant No.6142101210205).

This paper presents a new logic programming formalism for the two-stage framework. Specifically, we propose Assumable Logic Programming (ALP) language, where an assumption operator : is introduced to precede a literal in rules bodies, and thus allow to represent an assumption e as : e to express "e does not yield contradiction" or "e is acceptable" or "e is assumable" or "Assuming e" or "e is possible" etc., that can be used for designing intelligent agents whose behaviors of assumption making and belief building are defined on the answer-set based deductive reasoning.

The rest of the paper will introduce ALP formally and is organized as follows. In the next section, ASP and its extensions are briefly introduced as background knowledge for the self-contained requirement and as objects compared with ALP. In section 3, we introduce the syntax and semantics of the ALP program. In section 4, some properties and the relation of ALP to ASP are given. We conclude in section 5 with some further discussion.

We will restrict our discussion in this paper to propositional programs. However, as usual in answer set programming, we admit rule schemata containing variables bearing in mind that these schemata are just convenient representations for the set of their ground instances.

2   ASP and Its Extensions

2.1   Answer Set Program

Follow the description of ASP from [10]. A regular ASP program is a collection of rules of the form

$$l_1 \ or \ ... \ or \ l_k \leftarrow l_{k+1}, ..., l_m, not \ l_{m+1}, ..., not \ l_n \tag{1}$$

where the ls are literals, not denotes default negation, or is epistemic disjunction. The left-hand side of a rule is called the $head$ and the right-hand side is called the $body$. A rule is called a fact if its body is empty and its head contains only one literal, and a rule is called a constraint if its head is empty.

A collection of literals is consistent if it does not contain both a literal l and its contrary $\bar{l}$. Let M be a consistent collection of literals, r be an ASP rule of the form (1), the notion of satisfiability denoted by $\models_{ASP}$ is defined below.

- (M satisfies r's head). $M \models_{ASP} l_1 \ or \ ... \ or \ l_k$ if for some $1 \leq i \leq k, l_i \in M$

- (M satisfies r's body). $M \models_{ASP} l_{k+1}, ..., l_m, not \ l_{m+1}, ..., not \ l_n$ if for all $k+1 \leq i \leq m, l_i \in M$ and for all $m+1 \leq i \leq n, l_i \notin M$

- (M satisfies rule r). $M \models_{ASP} r$ if whenever $M \models_{ASP} l_{k+1}, ..., l_m, not \ l_{m+1}, ..., not \ l_n$, it holds that $M \models_{ASP} l_1 \ or \ ... \ or \ l_k$

M satisfies an ASP program Π if M satisfies every rule in Π, then M is called a *model* of Π. M is an answer set of Π iff it is the least model (in the sense of set inclusion) that satisfies Π^M, where Π^M is G-L reduct of Π with respect to M achieved by two rules:

- delete all rules whose bodies are not satisfied by M.

- delete $not \ l$ in the bodies of the remaining rules.

$AS(\Pi)$ is used to denote the set of all answer sets of an ASP program Π.

2.2 CR-Prolog

CR-Prolog extends the regular ASP with a purpose of representing indirect exceptions to defaults ([10]). Follow the description of CR-Prolog from [2] and [10], a CR-Prolog program is a collection of regular ASP rules or consistency-restoring rules (CR-rule) of the form

$$l_1 \text{ or } ... \text{ or } l_k \xleftarrow{+} l_{k+1}, ..., l_m, not\ l_{m+1}, ..., not\ l_n$$

where the ls are literals, not denotes negation as failure, or is epistemic disjunction. And, $\leq$ is a partial order defined on sets of CR-rules in the program. This partial order is often referred to as a preference relation based on the set-theoretic inclusion ($R_1 \leq R_2$ iff $R_1 \subset R_2$) or defined by the cardinality of the corresponding sets ($R_1 \leq R_2$ iff $|R_1| \subset |R_2|$).

The set of regular ASP rules of a CR-Prolog program Ω is denoted by Ω^r; By $\alpha(r)$ we denote a regular rule obtained from a consistency-restoring rule r by replacing $\xleftarrow{+}$ by $\leftarrow$, and α can be expanded in a standard way to a set R of CR-rules, i.e., $\alpha(R) = \{\alpha(r)|r \in R\}$.

A minimal (with respect to the preference relation of the program) collection R of CR-rules of Ω such that $\Omega^r \cup R$ is consistent (i.e., has an answer set) is called an abductive support of Ω. Then, a set M is called an answer set of Ω if it is an answer set of a regular program $\Omega \cup \alpha(R)$ for some abductive support R of Ω. We use $AS_\subset(\Omega)$ and $AS_\sharp(\Omega)$ to denote the collection of answer sets of Ω w.r.t. the preference relation on set-theoretic inclusion and the cardinality respectively.

2.3 Abductive ASP

We consider two versions of the abductive answer set programs. In [13], an abductive logic program (ABLP93) Γ is defined as a pair $< P, \mathcal{A} >$ where P is a regular ASP program and $\mathcal{A}$ is a set of literals from the language of P called abducibles. G a ground literal represents a positive observation. A set S is a belief set of Γ with respect to E if S is an answer set of $P \cup E$ where $E \subseteq \mathcal{A}$. S is called $\mathcal{A}$ minimal if there is no belief set T of Γ such that $T \cap \mathcal{A} \subset S \cap \mathcal{A}$. A set E is an explanation of G with respect to Γ if G is true in a belief set S of Γ such that $E = S \cap \mathcal{A}$. An explanation E of G is minimal if no $E' \subset E$ is an explanation of G. E is a minimal explanation of G iff S is an $\mathcal{A}$ minimal belief set of $< P \cup \{\leftarrow not\ G\}, \mathcal{A} >$.

In [14], an abductive logic program (ABLP95) Γ is defined as a pair $< P, \mathcal{A} >$ where both P and $\mathcal{A}$ are regular ASP programs. G a ground literal represents a positive observation. A pair (E, F) is a explanation of G with respect to Γ if

(i) $G \subseteq M$ for $\forall M \in AS((P - F) \cup E)$

(ii) $(P - F) \cup E)$ is consistent

(iii) $E \subseteq (\mathcal{A} - P)$ and $F \subseteq \mathcal{A} \cap P$

On the other hand, a pair (E, F) is an anti-explanation of G with respect to Γ if

  (i) $G \not\subseteq M$ for $\forall M \in AS((P - F) \cup E)$

  (ii) $(P - F) \cup E$ is consistent

  (iii) $E \subseteq (\mathcal{A} - P)$ and $F \subseteq \mathcal{A} \cap P$

An (anti-)explanation (E, F) of G is called minimal if for any (anti-)explanation (E', F') of G, $E' \subseteq E$ and $F' \subseteq F$ imply $E' = E$ and $F' = F$.

3 Assumable Logic Program

3.1 Syntax

An ALP rule r is written as

$$l_1 \text{ or } ... \text{ or } l_k \leftarrow e_1, ..., e_m : e_{m+1}, ..., e_n \tag{2}$$

where the ls are literals in propositional logic language, es are extended literals that are propositional literals possibly preceded by default negation not, : is called assumption operator. $head(r)$ is used to denote the left-hand side of r where or is an epistemic disjunction. $body(r)$ is used to denote the right-hand side of r. $e_1, ..., e_m$ is called the precondition of r and denoted by $pbody(r)$. $e_{m+1}, ..., e_n$ is called the assumption of r and denoted by $assump(r)$. As in usual logic programming, a rule is called a fact if its body is empty (equivalent to containing only a literal $\top$) and its head contains only one literal, and a rule is called a constraint if its head is empty (equivalent to containing only a literal $\bot$). An ALP rule is called assumption-free if its assumption is empty, otherwise it is called an assumption rule. Sometimes, we use $head(r) \leftarrow body(r)$ or $head(r) \leftarrow pbody(r) : assump(r)$ to denote r. $lit(r)$ is used to denote the set of propositional logic literals appearing in r. r can be read as *When assump(r) is possible, head(r) is believed if pbody(r) is believed or head(r) is believed if pbody(r) is believed, when assump(r) do not yield a contradiction* etc.

An ALP program is a collection of ALP rules. $lit(\Pi)$ is used to denote the set of propositional logic literals appearing in Π. $alit(\Pi)$ is used to denote the set of extended literals appearing in the assumptions of rules in Π. For convenient description, sometimes an ALP program Π is written as a pair (Π^D, Π^W) in which Π^W is the set of assumption-free ALP rules in Π and Π^D is the set of assumption rules in Π. An ALP program is default negation free if there is no default negation not appearing in the program. We say an ALP program is an ALP^{-not} program if it is default negation free.

It is clear that an assumption-free ALP rule is a regular ASP rule, and an assumption-free ALP program is an ASP program that can be dealt with by ASP solvers like DLV ([8]), CLASP ([9]).

For the well-known example "Normally, birds can fly" of default knowledge, its ALP encoding is the rule:

$$canfly \leftarrow bird : canfly$$

This rule defines quite a natural reading *If it is acceptable that a given bird can fly then we believe that the bird can fly* or *If assuming that a bird can fly does not cause inconsistency, then believe that it can fly* of the above statement.

A statement "Assuming a bird is not sick, normally it can fly" can then be encoded in an ALP rule:

$$canfly \leftarrow bird : canfly, not\ sick$$

3.2 Semantics

3.2.1 Satisfiability

Let M be a consistent set of literals, r be an ALP rule of the form (2), the notion of satisfiability denoted by $\models_{\text{ALP}}$ is defined below.

- $M \models_{\text{ALP}} l$ if $l \in M$
- $M \models_{\text{ALP}} not\ l$ if $l \notin M$
- $M \models_{\text{ALP}} head(r)$ if $\exists 1 \leq i \leq k, M \models_{\text{ALP}} l_i$
- $M \models_{\text{ALP}} pbody(r)$ if $\forall 1 \leq i \leq m, M \models_{\text{ALP}} e_i$
- $M \models_{\text{ALP}} assump(r)$ if $\forall m + 1 \leq i \leq n, M \models_{\text{ALP}} e_i$
- $M \models_{\text{ALP}} body(r)$ if $M \models_{\text{ALP}} pbody(r)$ and $M \models_{\text{ALP}} assump(r)$.
- $M \models_{\text{ALP}} r$ if whenever $M \models_{\text{ALP}} body(r)$, $M \models_{\text{ALP}} head(r)$.

We say M is a model of an ALP program Π, denoted by $M \models_{\text{ALP}} \Pi$, if we have $M \models_{\text{ALP}} r$ for $\forall r \in \Pi$. A set M of literals is inconsistent if it contains a literal l and its contrary $\bar{l}$.

It is easily verified that $\{bird, canfly\}$ is a model of the rules $canfly \leftarrow bird : canfly$ and $canfly \leftarrow bird : canfly, not\ sick$ mentaned above.

3.2.2 Foundmental Principles

Here, we present three fundamental principles for reasoning within the framework.

(i) **Consistency of Assumption Making.** This principle tells that the assumptions that can be established must make the theory (program) consistent, that is, in assuming a formula, the agent reasons and behaves as if it is a fact and will not cause conflicts.

(ii) **Rationality of Belief Building on Assumptions.** This principle tells that the agent's beliefs are obtained by reasoning within the scope of assumptions, that is, the belief set cannot exceed the results of reasoning based on assumptions and the given theory (program).

(iii) **Consistency between Assumptions and Beliefs.** This principle says that an agent's assumptions and beliefs must be consistent.

The three principles are natural but abstract. Below, we will provide specific technical methods to implement these principles. Next, we first provide a method for assumption-making, and then define a method for belief-building.

3.2.3   Assumptiom Making

The notion of *Assumption Set* of an ALP program is viewed as the result of assumption-making within the framework by the principle of **Consistency of Assumption Making**.

Definition 3.1 Given an ALP program Π, an arbitrary set $A \subseteq lit(\Pi)$, A is an assumption set of Π if and only if

$$A \in AS\big(\Pi^{(A)} \cup \overleftarrow{A}\big)$$

where $AS()$ denotes the set of all answer sets as mentioned in subsection 2.1, and $\Pi^{(A)}$ is a program obtained by

$$\Pi^{(A)} = \{head(r) \leftarrow pbody(r) | r \in \Pi \ and \ A \models_{ALP} assump(r)\}$$

and $\overleftarrow{A}$ is used to denote the fact rules set $\{l \leftarrow |l \in A\}$.

$ASS(\Pi)$ is used to denote the collection of all assumption sets of an ALP program Π.

Theorem 3.2 *For an ALP program Π, an assumption set of Π is a model of Π.* [2]

Example 3.3 Consider Π_1 that consists of two rules:

$$canfly \leftarrow bird : canfly$$

$$bird \leftarrow$$

There are four consistent sets of literals: $A_{11} = \emptyset$, $A_{12} = \{bird\}$, $A_{13} = \{canfly\}$, and $A_{14} = \{bird, canfly\}$. We have

$$\Pi_1^{(A_{11})} = \Pi_1^{(A_{12})} = \{bird \leftarrow\}$$

$$\Pi_1^{(A_{13})} = \Pi_1^{(A_{14})} = \{canfly \leftarrow bird. \quad bird \leftarrow .\}$$

Then,

$$\Pi_1^{(A_{11})} \cup \overleftarrow{A}_{11} = \{bird \leftarrow .\}$$

$$\Pi_1^{(A_{12})} \cup \overleftarrow{A}_{12} = \{bird \leftarrow .\}$$

$$\Pi_1^{(A_{13})} \cup \overleftarrow{A}_{13} = \{canfly \leftarrow bird. \quad canfly \leftarrow . \quad bird \leftarrow .\}$$

$$\Pi_1^{(A_{14})} \cup \overleftarrow{A}_{14} = \{canfly \leftarrow bird. \quad canfly \leftarrow . \quad bird \leftarrow .\}$$

Thus,

$$AS\big(\Pi_1^{(A_{11})} \cup \overleftarrow{A}_{11}\big) = AS\big(\Pi_1^{(A_{12})} \cup \overleftarrow{A}_{12}\big) = \{bird\}$$

---

[2] Due to space limitations, the proofs of the theorems are given in the full version of this paper.

$$AS(\Pi_1^{(A_{13})} \cup \overleftarrow{A}_{13}) = AS(\Pi_1^{(A_{14})} \cup \overleftarrow{A}_{14}) = \{bird, canfly\}$$

Then,

$$A_{11} \notin AS(\Pi_1^{(A_{11})} \cup \overleftarrow{A}_{11}) \quad A_{12} \in AS(\Pi_1^{(A_{12})} \cup \overleftarrow{A}_{12})$$

$$A_{13} \notin AS(\Pi_1^{(A_{13})} \cup \overleftarrow{A}_{13}) \quad A_{14} \in AS(\Pi_1^{(A_{14})} \cup \overleftarrow{A}_{14})$$

By Definition 3.1, both $A_{12} = \{bird\}$ and $A_{14} = \{bird, canfly\}$ are assumption sets of Π_1.

Example 3.4 Consider Π_2 that consists of two rules:

$$canfly \leftarrow bird : canfly, not\ sick.$$

$$bird \leftarrow .$$

There are eight consistent sets of literals: $A_{21} = \emptyset$, $A_{22} = \{bird\}$, $A_{23} = \{canfly\}$, $A_{24} = \{sick\}$, $A_{25} = \{bird, canfly\}$, $A_{26} = \{bird, sick\}$, $A_{27} = \{canfly, sick\}$, and $A_{28} = \{bird, canfly, sick\}$. We have

$$\Pi_2^{(A_{21})} = \Pi_3^{(A_{22})} = \Pi_2^{(A_{24})} = \Pi_2^{(A_{26})} = \Pi_2^{(A_{27})} = \Pi_2^{(A_{28})} = \{bird \leftarrow .\}$$

$$\Pi_2^{(A_{23})} = \Pi_2^{(A_{25})} = \{canfly \leftarrow bird. \quad bird \leftarrow .\}$$

Then,

$$\Pi_2^{(A_{21})} \cup \overleftarrow{A}_{21} = \{bird \leftarrow .\}$$

$$\Pi_2^{(A_{22})} \cup \overleftarrow{A}_{22} = \{bird \leftarrow .\}$$

$$\Pi_2^{(A_{23})} \cup \overleftarrow{A}_{23} = \{canfly \leftarrow bird. \quad bird \leftarrow . \quad canfly \leftarrow .\}$$

$$\Pi_2^{(A_{24})} \cup \overleftarrow{A}_{24} = \{bird \leftarrow . \quad sick \leftarrow .\}$$

$$\Pi_2^{(A_{25})} \cup \overleftarrow{A}_{25} = \{canfly \leftarrow bird. \quad bird \leftarrow . \quad canfly \leftarrow .\}$$

$$\Pi_2^{(A_{26})} \cup \overleftarrow{A}_{26} = \{bird \leftarrow . \quad sick \leftarrow .\}$$

$$\Pi_2^{(A_{27})} \cup \overleftarrow{A}_{27} = \{bird \leftarrow . \quad sick \leftarrow . \quad canfly \leftarrow .\}$$

$$\Pi_2^{(A_{28})} \cup \overleftarrow{A}_{28} = \{bird \leftarrow . \quad sick \leftarrow . \quad canfly \leftarrow .\}$$

Thus,

$$AS(\Pi_2^{(A_{21})} \cup \overleftarrow{A}_{21}) = AS(\Pi_2^{(A_{22})} \cup \overleftarrow{A}_{22}) = \{\{bird\}\}$$

$$AS(\Pi_2^{(A_{23})} \cup \overleftarrow{A}_{23}) = AS(\Pi_2^{(A_{25})} \cup \overleftarrow{A}_{25}) = \{\{bird, canfly\}\}$$

$$AS(\Pi_2^{(A_{24})} \cup \overleftarrow{A}_{24}) = AS(\Pi_2^{(A_{26})} \cup \overleftarrow{A}_{26}) = \{\{bird, sick\}\}$$

$$AS(\Pi_2^{(A_{27})} \cup \overleftarrow{A}_{27}) = AS(\Pi_2^{(A_{28})} \cup \overleftarrow{A}_{28}) = \{\{bird, canfly, sick\}\}$$

Then,

$$A_{21} \notin AS(\Pi_2^{(A_{21})} \cup \overleftarrow{A}_{21}), \quad A_{23} \notin AS(\Pi_2^{(A_{23})} \cup \overleftarrow{A}_{23})$$

$$A_{24} \notin AS(\Pi_2^{(A_{24})} \cup \overleftarrow{A}_{24}), \quad A_{27} \notin AS(\Pi_2^{(A_{27})} \cup \overleftarrow{A}_{27})$$

$$A_{22} \in AS(\Pi_2^{(A_{22})} \cup \overleftarrow{A}_{22}), \quad A_{25} \in AS(\Pi_2^{(A_{25})} \cup \overleftarrow{A}_{25})$$

$$A_{26} \in AS(\Pi_2^{(A_{26})} \cup \overleftarrow{A}_{26}), \quad A_{28} \in AS(\Pi_2^{(A_{28})} \cup \overleftarrow{A}_{28})$$

Hence, A_{22}, A_{25}, A_{26}, and A_{28} are assumption sets of Π_2.

Just as the examples above show, in the scenario of incomplete informa-tion, different assumption sets may be generated. Some of them satisfy the assumption body of more rules or satisfy more assumption literals, while some of them satisfy less (in the sense of set inclusion or cardinality). Based on this observation, we define several strategies of assumption making while keeping the principles unchanged.

Definition 3.5 Given an ALP program Π, A is an assumption set of Π,

(i) $\max_{\subseteq}^{R}$ **Strategy**: A is a $\max_{\subseteq}^{R}$ assumption set of Π if there is no assump-tion set A' of Π such that $\Pi^{(A)} \subset \Pi^{(A')}$.

(ii) $\min_{\subseteq}^{R}$ **Strategy**: A is a $\min_{\subseteq}^{R}$ assumption set of Π if there is no assumption set A' of Π such that $\Pi^{(A)} \supset \Pi^{(A')}$.

(iii) $\max_{\#}^{R}$ **Strategy**: A is a $\max_{\#}^{R}$ assumption set of Π if there is no assump-tion set A' of Π such that $|\Pi^{(A)}| < |\Pi^{(A')}|$.

(iv) $\min_{\#}^{R}$ **Strategy**: A is a $\min_{\#}^{R}$ assumption set of Π if there is no assumption set A' of Π such that $|\Pi^{(A)}| > |\Pi^{(A')}|$.

(v) $\max_{\subseteq}^{L}$ **Strategy**: A is a $\max_{\subseteq}^{L}$ assumption set of Π if there is no assump-tion set A' of Π such that $\{e \in alit(\Pi)|A \models_{\mathrm{ALP}} e\} \subset \{e \in alit(\Pi)|A' \models_{\mathrm{ALP}} e\}$.

(vi) $\min_{\subseteq}^{L}$ **Strategy**: A is a $\min_{\subseteq}^{L}$ assumption set of Π if there is no assumption set A' of Π such that $\{e \in alit(\Pi)|A \models_{\mathrm{ALP}} e\} \supset \{e \in alit(\Pi)|A' \models_{\mathrm{ALP}} e\}$.

(vii) $\max_{\#}^{L}$ **Strategy**: A is a $\max_{\#}^{L}$ assumption set of Π if there is no assumption set A' of Π such that $|\{e \in alit(\Pi)|A \models_{\mathrm{ALP}} e\}| < |\{e \in alit(\Pi)|A' \models_{\mathrm{ALP}} e\}|$.

(viii) $\min_{\#}^{L}$ **Strategy**: A is a $\min_{\#}^{L}$ assumption set of Π if there is no assumption set A' of Π such that $|\{e \in alit(\Pi)|A \models_{\mathrm{ALP}} e\}| > |\{e \in alit(\Pi)|A' \models_{\mathrm{ALP}} e\}|$.

Reconsider the program Π_1, $A_{12} = \{bird\}$ is both a $\min_{\subseteq}^{R}$ assumption set and a $\min_{\subseteq}^{L}$ assumption set, $A_{14} = \{bird, canfly\}$ is both a $\max_{\subseteq}^{R}$ assumption set and a $\max_{\subseteq}^{L}$ assumption set and so on.

For Π_2, $A_{26} = \{bird, sick\}$ is both a $\min_{\subseteq}^{R}$ assumption set and a $\min_{\subseteq}^{L}$ assumption set, and $A_{25} = \{bird, canfly\}$ is a $\max_{\subseteq}^{R}$ assumption set, a $\max_{\subseteq}^{L}$, and a $\max_{\#}^{L}$ assumption set and so on.

3.2.4 Belief Building

The notion of *assumable answer set* of an ALP program is viewed as the result of belief-building over assumption-making within the framework by the principles of **Rationality of Belief Building on Assumptions** and **Consistency between Assumptions and Beliefs.**

Definition 3.6 Given an ALP program Π, an arbitrary set $M \subseteq lit(\Pi)$, M is an assumable answer set of Π if and only if there exists an assumption set A of Π such that

(i) $M \in AS(\Pi^{(A)})$, and

(ii) $M \subseteq A$

We say that M is an assumable answer set of Π on the assumption set A, and that (M, A) is a view of Π.

$AAS(\Pi)$ is used to denote the collection of all assumable answer sets of an ALP program Π. $VIEW(\Pi)$ is used to denote the collection of all views of an ALP program Π.

Theorem 3.7 *For an ALP program* $\Pi = (\Pi^D, \Pi^W)$, *let* M *be an assumable answer set of* Π

$$M \models_{\text{ALP}} \Pi^W$$

Example 3.8 Continue Π_1 mentioned above, let us consider $M_1 = \{bird\}$ and $M_2 = \{bird, canfly\}$. We have

$$AS(\Pi_1^{(A_{12})}) = \{bird\} \qquad AS(\Pi_1^{(A_{14})}) = \{bird, canfly\}$$

Thus,

$$M_1 \in AS(\Pi_1^{(A_{12})}) \text{ and } M_1 \subseteq A_{12}$$

$$M_2 \in AS(\Pi_1^{(A_{14})}) \text{ and } M_2 \subseteq A_{14}$$

Hence, $\{bird\}$ is an assumable answer set of Π_1 on the assumption set A_{12}, $\{bird, canfly\}$ is also an assumable answer set of Π_1 on the assumption set A_{14}.

Example 3.9 Continue Π_2 mentioned above, we have

$$AS(\Pi_2^{(A_{22})}) = AS(\Pi_2^{(A_{26})}) = AS(\Pi_2^{(A_{28})}) = \{bird\}$$

$$AS(\Pi_2^{(A_{25})}) = \{bird, canfly\}$$

Thus,

$$\{bird\} \in AS(\Pi_2^{(A_{22})}) \text{ and } \{bird\} \subseteq A_{22}$$

$$\{bird\} \in AS(\Pi_2^{(A_{26})}) \text{ and } \{bird\} \subseteq A_{26}$$

$$\{bird\} \in AS(\Pi_2^{(A_{28})}) \text{ and } \{bird\} \subseteq A_{28}$$

$$\{bird, canfly\} \in AS(\Pi_2^{(A_{25})}) \text{ and } M_2 \subseteq A_{25}$$

	$\max_C^R$ **VIEW**	$\min_C^R$ **VIEW**	$\max_{\sharp}^R$ **VIEW**	$\min_{\sharp}^R$ **VIEW**
Π_1	({bird, canfly},{bird, canfly})	({bird},{bird})	({bird, canfly},{bird, canfly})	({bird},{bird})
Π_2	({bird, canfly},{bird, canfly})	({bird},{bird}) ({bird},{bird, sick}) ({bird},{bird, canfly, sick})	({bird, canfly},{bird, canfly})	({bird},{bird}) ({bird},{bird, sick}) ({bird},{bird, canfly, sick})

	$\max_C^L$ **VIEW**	$\min_C^L$ **VIEW**	$\max_{\sharp}^L$ **VIEW**	$\min_{\sharp}^L$ **VIEW**
Π_1	({bird, canfly},{bird, canfly})	({bird},{bird})	({bird, canfly},{bird, canfly})	({bird},{bird})
Π_2	({bird, canfly},{bird, canfly})	({bird},{bird, sick})	({bird, canfly},{bird, canfly})	({bird},{bird, sick})

Fig. 1. Views of Π_1 and Π_2

Hence, $\{bird\}$ is an assumable answer set of Π_2 on the assumption set A_{22} or A_{26} or A_{28}, $\{bird, canfly\}$ is also an assumable answer set of Π_2 on the assumption set A_{25}.

Definition 3.10 Given an ALP program Π, M is called $\max_X^Y(\min_X^Y)$ assumable answer set of Π if (M, A) is a view of Π and A is a $\max_X^Y(\min_X^Y)$ assumption set of Π. Correspondingly, the pair (M, A) is called a $\max_X^Y(\min_X^Y)$ view of Π. where $Y \in \{R, L\}$ and $X \in \{\subset, \sharp\}$.

$\max_X^Y(\min_X^Y)$-$AAS(\Pi)$ is used to denote the collection of all $\max_X^Y(\min_X^Y)$ assumable answer sets of an ALP program Π. $\max_X^Y(\min_X^Y)$-$VIEW(\Pi)$ is used to denote the collection of all $\max_X^Y(\min_X^Y)$ views of an ALP program Π, where $Y \in \{R, L\}$ and $X \in \{\subset, \sharp\}$. Figure 1 shows the views of Π_1 and Π_2 under different assumption-making strategies.

The intuitions of max and min strategies of assumption making are direct: max means that the reasoner is positive/optimistic/ credulous/brave in making assumptions. min is just the opposite. Let us consider Π_1 and Π_2 again. A positive reasoner's view is $(\{bird, canfly\}, \{bird, canfly\})$ such that its belief set is $\{bird, canfly\}$, and a cautious reasoner just has a belief set $\{bird\}$.

4   Some Properties and Relations

4.1   Nonmonotonicity

Obviously, ALP is an extension of ASP and ALP, it's reasoning is nonmonotonic. However, the following theorem and example show that nonmonotonicity does not solely stem from default negation and epistemic disjunctive operators. Assumption operators can also lead to nonmonotonicity. For convenient description, an ALP not containing default negations is marked by ALP^{-not}.

Theorem 4.1 *ALP^{-not}-based reasoning is nonmonotonic.*

Theorem 4.1 can be demonstrated by the following examples.

Example 4.2 Π_3 is an ALP^{-not} program containing one rule:

$$p \leftarrow : p$$

Π_3 has two views: $(\emptyset, \emptyset)$ and $(\{p\}, \{p\})$. Consider Π_3' that is obtained from Π_3

by adding a fact:

$$p \leftarrow : p$$

$$\neg p \leftarrow$$

Clearly, Π'_3 has only one view $(\{\neg p\}, \{\neg p\})$.

4.2 Relation to ASP

Now, let us consider the relationship between ASP and ALP.

Theorem 4.3 *For an assumption-free ALP program* Π:

$$AAS(\Pi) = Z_X^Y\text{-}AAS(\Pi) = AS(\Pi)$$

where $Z \in \{max, min\}$, $X \in \{\subset, \sharp\}$, *and* $Y \in \{R, L\}$.

Theorem 4.3 tells that an ALP program containing no assumption rules degenerates into an ASP program. In other words, ALP is an smooth extension of ASP.

Define a mapping η frome an ASP program to an ALP program, identifies an ASP rule r:

$$l_1 \ or \ ... \ or \ l_k \leftarrow l_{k+1}, ..., l_m, not \ l_{m+1}, ..., not \ l_n$$

with the ALP rule $\eta(r)$:

$$l_1 \ or \ ... \ or \ l_k \leftarrow l_{k+1}, ..., l_m \ : not \ l_{m+1}, ..., not \ l_n$$

Then we have

Theorem 4.4 *For any ASP program* Π, *if* S *is an answer set of* Π, *then* (S, S) *is a view of* $\eta(\Pi)$.

Now, let us focus on the relationship among *not* and :. The most obvious difference is that the assumption operator : works during the assumption-making stage, while *not* operates during the belief-building stage.

It seems that the assumption rule *On the assumption of* α, β *is believed if* γ *is believed* can also be coded into an ASP rule $\beta \leftarrow \gamma, not \ \neg\alpha$ where *not* $\neg\alpha$ is used to express α *is assumable* or *it is consistent to assume* α. However, the following cases demonstrate the difference between ASP encodings and ALP encodings of the assumptions.

First of all, let us see the difference by observing an example that contains:

- An assumption: *p if it is consistent to assume p.*
- A constraint: *p is impossible.*

Intuitively, the constraint is a denial of p such that the assumption is blocked, thus the result is $\emptyset$. If the case is modeled by an ASP program

$$p \leftarrow not \ \neg p$$

$$\leftarrow p$$

where the assumption p is made through $not\ \neg p$. There is no solution because the ASP program is unsatisfiable. If the case is represented by an ALP program

$$p \leftarrow: p$$

$$\leftarrow p$$

where the assumption p is made through $: p$. There is an assumable answer set $\emptyset$ as expected.

Now, let us consider another case with two assumptions:

- p if it is consistent to assume r, and
- q if it is consistent to assume $\neg r$.

If they are represented as an ASP program

$$p \leftarrow not\ \neg r$$

$$q \leftarrow not\ r$$

The result is its answer set $\{p, q\}$. Meanwhile, if they are represented as an ALP program

$$p \leftarrow: r$$

$$q \leftarrow: \neg r$$

There are three assumable answer sets $\{p\}$, $\{q\}$, and $\emptyset$. Among them, both $\{p\}$ and $\{q\}$ are max_C^R (and $max_\sharp^R$) assumable answer sets, and $\emptyset$ is a min_C^R (and $min_\sharp$) assumable answer set. Consider that r and its contrary $\neg r$ cannot appear in one world, the results given by the ALP program should be more praised than that of the ASP program.

Another seeming ASP-based encoding of *it is consistent to assume α* is *not not α*. But, it is easy to verify that the encoding of the second case in this way

$$p \leftarrow not\ not\ r$$

$$q \leftarrow not\ not\ \neg r$$

has only one answer set $\emptyset$.

Now, we focus on the the difference of not and $:$ in representing defaults. Consider an example of indirect exceptions to defaults.

Example 4.5 Consider a theory contains two laws:

- Birds generally have the ability to fly.
- Having the ability to fly means having wings.

and an observations: A bird without wings. [3]

---

[3] The wings of New Zealand's kiwis have degenerated beyond sight.

Consider a regular ASP representation Π_4 of the example:

$$fly \leftarrow bird, not \neg fly. \quad wing \leftarrow fly. \quad bird \leftarrow . \quad \neg wing \leftarrow .$$

Consider an ALP representation Π_5 of the example:

$$fly \leftarrow bird : fly. \quad wing \leftarrow fly. \quad bird \leftarrow . \quad \neg wing \leftarrow .$$

Π_4 is unsatisfiable, while Π_5 provides a reasonable answer $\{bird, \neg wing\}$. The representation of indirect exceptions seems to be beyond the power of ASP, which led to the development of a extension of ASP called CR-Prolog [10]. ALP seems to be adept at representing indirect exceptions to defaults. Next subsection will shows that a CR-Prolog program can be converted into an ALP program.

As shown in the examples above, assumption operators bring richer knowledge representation by realize the two-stage framework that some beliefs are built on making assumptions.

4.3 Relation to CR-Prolog

Define a mapping β from a CR-prolog to ALP, identifies a CR-rule r:

$$l_1 \ or \ ... \ or \ l_k \overset{+}{\leftarrow} l_{k+1}, ..., l_m, not \ l_{m+1}, ..., not \ l_n \tag{3}$$

with an ALP rule $\beta(r)$:

$$l_1 \ or \ ... \ or \ l_k \leftarrow l_{k+1}, ..., l_m, not \ l_{m+1}, ..., not \ l_n : apply_r \tag{4}$$

where $apply_r$ is used to denote the fresh atom obtained from a CR-rule r. Besides, β identifies a regular ASP rule in the CR-prolog program with itself.

Theorem 4.6 *For any CR-Prolog program Ω*

$$AS_\star(\Omega) = min_\star^R\text{-}AAS(\beta(\Omega))$$

where $\star \in \{\subset, \sharp\}$.

Example 4.7 Consider a simple CR-Prolog program Ω that contains only one CR-rule r:

$$a \overset{+}{\leftarrow}$$

It is easy to see Ω has only one answer set $\emptyset$. Then, an ALP program $\beta(\Omega)$ is

$$a \leftarrow : apply_r$$

whose $min_\sharp^R$ and $min_\subset^R$ assumable answer set is also $\emptyset$.

4.4 Relation to Abductive ASP

Define a mapping θ, for an ABLP93 program $\Gamma =< P, \mathcal{A} >$, $\theta(\Gamma)$ is an ALP program:

$$P \cup \{l \leftarrow : l | l \in \mathcal{A}\}$$

We have

Theorem 4.8 *For the ABLP93 program* $\Gamma =< P, \mathcal{A} >$, S *is a* $\min_C^R$ *assumable answer set of* $\theta(\Gamma)$ *if and only if* S *is a* $\mathcal{A}$-*minimal belief set of* Γ.

Example 4.9 Consider an abductive logic program $\Gamma_1 =< P, \mathcal{A} >$ in [27]:

- P: $p \leftarrow not\ a$
- $\mathcal{A}$: a

The program has one $\mathcal{A}$-minimal belief set $\{p\}$. $\theta(\Gamma_1)$ is an ALP

$$p \leftarrow not\ a$$

$$a \leftarrow: a$$

Both $\{a\}$ and $\{p\}$ are its assumable answer sets, only $\{p\}$ is its $\min_C^R$ assumable answer set as expected.

Theorem 4.10 *For the ABLP93 program* $\Gamma =< P, \mathcal{A} >$ *and a positive observation* G.

(i) *If* S *is a* $\min_C^R$ *assumable answer set of the ALP program* $\theta(\Gamma) \cup \{\leftarrow not\ G\}$, *then* $S \cap \mathcal{A}$ *is a credulous explanation of* G *with respect to* Γ.

(ii) *If* E *is a credulous explanation of* G *with respect to* Γ, *then there exists a* $\min_C^R$ *assumable answer set* S *of the ALP program* $\theta(\Gamma) \cup \{\leftarrow not\ G\}$ *such that* $E = S \cap \mathcal{A}$.

Example 4.11 Continue to consider Γ_1 mentioned in Example 4.9, given a positive observation $G = p$. Clearly, $\emptyset$ is a credulous explanation of p with respect to Γ_1. Now, we have $\theta(\Gamma_1) \cup \{\leftarrow not\ G\}$:

$$p \leftarrow not\ a. \quad a \leftarrow: a. \quad \leftarrow not\ p.$$

that has a $\min_C^R$ assumable answer set $\emptyset$ such that $E = \emptyset$.

Define a mapping θ', for the ABLP95 program $\Gamma =< P, \mathcal{A} >$, $\theta'(\Gamma)$ is an ALP program:

$$(P - \mathcal{A}) \cup \{head(r) \leftarrow body(r), apply_r | r \in (\mathcal{A} - P)\} \cup$$

$$\{apply_r \leftarrow: apply_r | r \in (\mathcal{A} - P)\} \cup$$

$$\{head(r) \leftarrow body(r), not\ block_r | r \in (\mathcal{A} \cap P)\} \cup$$

$$\{block_r \leftarrow: block_r | r \in (\mathcal{A} \cap P)\}$$

where both $apply_r$ and $block_r$ are used to denote the fresh atoms obtained from $r \in \mathcal{A}$. We have

Theorem 4.12 *For the ABLP95 program* $\Gamma =< P, \mathcal{A} >$ *and a positive observation* G.

(i) *If S is a min_C^R assumable answer set of the ALP program $\theta'(\Gamma) \cup \{\leftarrow \ not\ G\}$, then there exists a minimal explanation of G with respect to Γ is*

$$(\{r|apply_r \in S\}, \{r|block_r \in S\})$$

(ii) *If (E, F) is a minimal explanation of G with respect to Γ, then there exists a min_C^R assumable answer set S of the ALP program $\theta'(\Gamma) \cup \{\leftarrow \ not\ G\}$ such that*

$$E = \{r|apply_r \in S\}$$
$$F = \{r|block_r \in S\}$$

Example 4.13 Consider an abductive logic program $\Gamma_2 =< P, \mathcal{A} >$ and a positive observation G:

- P:

$$fly \leftarrow bird$$
$$bird \leftarrow penguin$$
$$penuin \leftarrow$$

- $\mathcal{A}$:

$$(1).\ fly \leftarrow bird$$
$$(2).\ \neg fly \leftarrow penguin$$

- G: $\neg fly$

Thus, $\theta'(\Gamma_2)$ is:

$$bird \leftarrow penguin$$
$$penuin \leftarrow$$
$$fly \leftarrow bird, not\ block_1$$
$$\neg fly \leftarrow penguin, apply_2$$
$$block_1 \leftarrow: block_1$$
$$apply_2 \leftarrow: apply_2$$

Then, $\theta'(\Gamma_2) \cup \{\leftarrow \ not\ \neg fly\}$ has a min_C^R assumable answer set

$$\{apply_2,\ block_1,\ \neg fly,\ penguin,\ bird\}$$

by which

$$E = \{\neg fly \leftarrow penguin\}$$
$$F = \{fly \leftarrow bird\}$$

such that (E, F) is a minimal explanation of $\neg fly$ with respect to Γ_2.

Theorem 4.14 *For the ABLP95 program $\Gamma =< P, \mathcal{A} >$ and a positive observation G.*

(i) *If S is a min_C^R assumable answer set of $\theta'(\Gamma) \cup \{\leftarrow G\}$, then there exists a minimal anti-explanation of G with respect to Γ is*

$$(\{r|apply_r \in S\}, \{r|block_r \in S\})$$

(ii) *If (E, F) is a minimal anti-explanation of G with respect to Γ, then there exists a min_C^R assumable answer set S of $\theta'(\Gamma) \cup \{\leftarrow G\}$ such that*

$$E = \{r|apply_r \in S\}, F = \{r|block_r \in S\}$$

Example 4.15 Continue to consider the abductive logic program Γ_2 used in the Example 4.13. By the Theorem 4.14, the anti-explanation of fly with respect to Γ_2 is the min_C assumable answer set of the ALP program

$$\theta'(\Gamma_2) \cup \{\leftarrow fly\}$$

Obviously, a min_C assumable answer set of the program is $\{penguin, bird\}$ that tells neither (1) nor (2) is used, and the corresponding minimal anti-explanation (E, F) of fly is

$$E = \emptyset \qquad F = \{fly \leftarrow bird\}$$

.

5   Conclusion

This paper introduces ALP that extends logic programming with an operator : to express the notion of assumption in logic programming. ALP can be viewed as a tool to design the intelligent agent capable of assumption-based reasoning that is a framework of many intelligent behaviors in the case of incomplete information. Three fundamental principles of reasoning within the framework are proposed to form a pattern of reasoning of assumptions and beliefs. By these principles, ALP provides an approach to reasoning by using the answer set-based reasoning in both assumption-making and belief-building, which makes the existing ASP solvers able to facilitate the implementation of the ALP solver. Several strategies of assumption-making are given in the definition of the semantics of ALP. Those strategies depict the attitude of agents to assumptions and therefore provide options for designing a variety of intelligent agents. The preliminary exploration results on the relationship between ALP and ASP, ALP and CR-Prolog, and ALP and abductive ASP show that ALP provides a more general way to model the problems with defaults, and exceptions, and to solve the explanation problems. Due to space limitations, the algorithm and its complexity in solving ALP programs are given in the full version of this paper.

Future work includes more properties of the ALP languages and applications. The first next step is to explore the power of ALP by studying the relation of ALP to other nonmonotonic logics such as circumscription[19], autoepistemic logic[20], default logic[25] and its variants, and so on.

References

[1] Anrig, B., R. Haenni, J. Kohlas and N. Lehmann, *Assumption-based modeling using abel*, in: *Proceedings of ECSQARU-FAPR*, 1997, pp. 171–182.

[2] Balduccini, M. and M. Gelfond, *Logic programs with consistency-restoring rules*, in: *AAAI Technical Report SS-03-05*, 2003.

[3] Besnard, P. and P. Siegel, *Supposition-based logic for automated nonmontonic reasoning*, in: *CADE*, 1988.

[4] Bondarenko, A., F. Toni and R. A. Kowalski, *An assumption-based framework for non-monotonic reasoning*, in: *LPNMR-93*, 1993, pp. 171–189.

[5] Cox, P. T. and T. Pietrzykowski, *Causes for events: their computation and applications*, in: *International Conference on Automated Deduction*, Springer, 1986, pp. 608–621.

[6] de Kleer, J., *An assumption-based tms*, Artificial Intelligence **28** (1986), pp. 127–162.

[7] de Kleer, J., *A general labeling algorithm for assumption-based truth maintenance*, in: *Proceedings of AAAI-88*, 1988, pp. 188–192.

[8] Faber, W., G. Pfeifer, N. Leone, T. Dell'armi and G. Ielpa, *Design and implementation of aggregate functions in the dlv system*, Theory Pract. Log. Program. **8** (2008), pp. 545–580.

[9] Gebser, M., B. Kaufmann and T. Schaub, *Conflict-driven answer set solving: From theory to practice*, Artif. Intell. **187-188** (2012), pp. 52–89.

[10] Gelfond, M. and Y. Kahl, *Knowledge representation, reasoning, and the design of intelligent agents: The answer-set programming approach* (2014).

[11] Giannikis, G. K. and A. Daskalopulu, *Defeasible reasoning with e-contracts*, in: *2006 IEEE/WIC/ACM International Conference on Intelligent Agent Technology*, 2006, pp. 690–694.

[12] Giannikis, G. K. and A. Daskalopulu, *The representation of e-contracts as default theories*, in: *International Conference on Industrial, Engineering and Other Applications of Applied Intelligent Systems*, Springer, 2007, pp. 963–973.

[13] Inoue, K. and C. Sakama, *Transforming abductive logic programs to disjunctive programs*, in: *ICLP-93*, 1993, p. 335.

[14] Inoue, K. and C. Sakama, *Abductive framework for nonmonotonic theory change.*, , **95**, Citeseer, 1995, pp. 204–210.

[15] Jago, M., *Modelling assumption-based reasoning using contexts*, in: *Proceedings of workshop on Context Representation and Reasoning (CRR'05)*, 2005.

[16] Kaminski, M., *A comparative study of open default theories*, Artif. Intell. **77** (1995), pp. 285–319.

[17] Kohlas, J. and P.-A. Monney, *Probabilistic assumption-based reasoning*, in: *Uncertainty in Artificial Intelligence*, 1993, pp. 485–491.

[18] Kowalski, R. A. and F. Sadri, *Reconciling the event calculus with the situation calculus*, J. Log. Program. **31** (1997), pp. 39–58.

[19] McCarthy, J., *Circumscription - a form of non-monotonic reasoning*, Artif. Intell. **13** (1980), pp. 27–39.

[20] Moore, R. C., *Semantical considerations on nonmonotonic logic*, in: *International Joint Conference on Artificial Intelligence*, 1985.

[21] Pellier, D. and H. Fiorino, *Multi-agent assumption-based planning*, in: *IJCAI-05*, 2005, pp. 1717–1718.

[22] Poole, D. L., *A logical framework for default reasoning*, Artif. Intell. **36** (1988), pp. 27–47.

[23] Poole, D. L., *Who chooses the assumptions*, in: *Abductive Reasoning* (1997).

[24] Reichgelt, H. and N. Shadbolt, *A specification tool for planning systems*, in: *ECAI-1990*, 1990, pp. 541–546.

[25] Reiter, R., *A logic for default reasoning*, Artif. Intell. **13** (1980), pp. 81–132.

[26] Reiter, R. and J. de Kleer, *Foundations of assumption-based truth maintenance systems: Preliminary report*, in: *Proceedings of AAAI-87*, 1987.

[27] Sakama, C. and K. Inoue, *Abductive logic programming and disjunctive logic programming: their relationship and transferability*, J. Log. Program. **44** (2000), pp. 75–100.

BTPK-based interpretable method for NER tasks based on Talmudic Public Announcement Logic

Yulin Chen[a] Beishui Liao [a] [1]   Bruno Bentzen [a] Bo Yuan[a]
Zelai Yao [a]   Haixiao Chi [a] Dov Gabbay[b]

[a] *Zhejiang University, Zheda Road 38, Hangzhou, 310028, China*

[b] *King's College London, Strand London WC2R 2LS, United Kingdom London, 695014,*

Abstract

As one of the basic tasks in natural language processing (NLP), named entity recognition (NER) is an important basic tool for downstream tasks of NLP, such as information extraction, syntactic analysis, machine translation and so on. The internal operation logic of the current name entity recognition model is black-box to the user, so the user has no basis to determine which name entity makes more sense. Therefore, a user-friendly explainable recognition process would be very useful for many people. In this paper, we propose a novel interpretable method, BTPK (Binary Talmudic Public Announcement Logic model), to help users understand the internal recognition logic of the name entity recognition tasks based on Talmudic Public Announcement Logic. BTPK model can also capture the semantic information in the input sentences, that is, the context dependency of the sentence. We observed the public announcement of BTPK presents the inner decision logic of Bidirectional Recurrent Neural Networks (BRNNs), and the explanations obtained from a BTPK model show us how BRNNs essentially handle NER tasks.

Keywords: named entity recognition, interpretable, Talmudic Public Announcement Logic

1 Introduction

Named Entity Recognition (NER) is an information extraction task aimed at classifying words in unstructured text [3,6]. Due to their ability to establish dependencies in neighboring words, Bidirectional Recurrent Neural Networks (BRNNs) have demonstrated excellent performance in many NER tasks [9]. Despite the advantages of such deep learning based methods, their inherent black box nature makes them unable to explain decision results [5,10]. In application areas where NER technology provides extensive underlying support such as health-care or autonomous driving, a transparent internal decision system is critical for the system reliability and user trust. Many interpretable works have been carried out for RNN [7,11,8], However, there are few research efforts on the explainability of BRNNs in NER tasks, although models with explainability are crucial [2].

---

[1] corresponding author.

Talmudic public announcement logic (TPK) is one of the modal logics that serves as a good formalism for representing decisions depending on the future [1]. Regarding past and future information as context information, TPK is able to model the context relationship and represent the implicit logic in context through modal logic. Some work has been done in exporting logical insights derived from the Talmud to AI, including the description of active historical databases, the study of retroactive update, etc.In this paper, we use the Talmudic public announcement logic (TPK) model [1] as a tool to explain the process of NER and bring transparency to the RNN-based models, since the reversible and modifiable recognition process in NER is very much in line with the problem that TPK is trying to deal with. We propose a new binary TPK model (called BTPK) based on the original TPK model, which can deal with actions depending on future determinations by public announcements [1]. By modifying the accessibility relation in a temporal tree structure, the public announcement at a future state will tell which path should be chosen. Thus, a logical explanation can be obtained for any trained BRNN based on BTPK.

We summarize our main contributions as follows: (1) We propose a BTPK-based learning method based on the original TPK model and apply it to a BRNN model to obtain logic explanationa for a BRNN.(2) We carry out a case study on real dataset to show how BRNN handle NER tasks, as well as to explore the potential for further reasoning on a BTPK model.

2   Preliminary

The original TPK model is a deterministic Talmudic $\mathbf{K}$ frame based on a time-action tree structure. The time-action model is a tree structure with a set of states $S = \{s_0, s_1, s_2, s'_2, \dots\}$ (s_0 is the root), and a set of actions $A = \{a_1, a_2, a_3, \dots\}$. The elements of A are actions moving the agent from any state to a new one. This corresponds to a successor function R_1 (denoted by $\rightarrow$), and can be written in the form $s_0 R_1 s_1$. A time-action sequence has the form of $s_0 a_1 a_2 \dots a_n$. In scenarios where the present course of action is indeterminate and the subsequent state is uncertain, the concept of a public announcement is proposed as a means of resolving the ambiguity. This entails providing clarification regarding the previous undetermined path and identifying the accurate successor to the decision point.

A deterministic TPK model [1] can be defined as a 6-tuple $(S, R_1, R, \rho, s_0, \pi)$ where (S, R_1, s_0) is a tree with root s_0 and successor relation R_1, R is the transitive closure of R_1, ρ is the public announcement function and π is an assignment for each atom q, such that $s \models q$ iff $s \in \pi(q)$. D is the distance from the root, and if $s'_3 \rho s_3$ then $D(s_3) = D(s'_3) + 1$. Let $M = \langle S, A, R, \rho, \pi \rangle$ be a TPK model and $s \in S$, then the semantics of a Deterministic TPK model can be defined as follows: As for the relation R_1: $M, t \models \Box A$ iff $\forall s : t R_1 s \rightarrow M, s \models A$; $M, t \models YA$ iff $\forall s : s R_1 t \rightarrow M, s \models A$. As for the relation ρ: $M, t \models \boxminus A$ iff $\forall s : t \rho s \rightarrow M, s \models A$; $M, t \models \mathbb{Y}A$ iff $\forall s : s \rho t \rightarrow M, s \models A$. D_n is a time constant: $t \models D_n$ iff the distance of t from s_o is n.

3   Approach

The overall framework is illustrated in Figure 1. Initially, a Bidirectional Recurrent Neural Network (BRNN) learner is trained on a set of training data to obtain a well-trained BRNN model. The public announcements of the sequences are computed through an analysis of the backward and forward hidden states. Section 3.1 provides the definition of public announcement within the binary TPK model (BTPK). In Section 3.3, the BTPK models of BRNNs are generated.

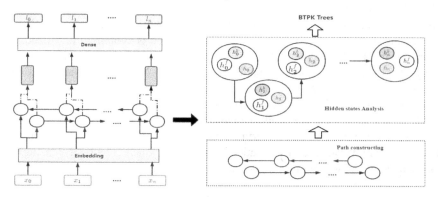

Fig. 1. The illustration of the framework of BTPK-based learning. h_i in the blue circles represents the hidden state of a BRNN, h_i^f in the white circles represents the hidden states in a forward RNN and h_i^b in the green circles represents the hidden states in a backward RNN, where i is the element order. l_i in yellow box denotes the output of the BRNN.

3.1   Definitions

Definition 3.1 (**Task definition**) *We regard NER as a sequence labeling problem, whose input includes a set of sequences and labels. For any sequence* $W = (w_1, w_2, ..., w_n)$, *the corresponding labels are* $Y = (y_1, y_2, ..., y_n)$, *where* w_i *denotes an entity in the sequence, and* y_i *comes from BIO tagging schema for labeling elements from the sentence.*

We view the final output of one RNN as the final option, so there are at most two options for each entity in the BRNN. According to the original TPK model, we define a binary TPK (BTPK) model as follows.

Definition 3.2 (**BTPK**) A binary TPK (BTPK) model is defined as a $T =< V, E >$ with public announcements P and height $|H|$, where $|V|$ is the order, $V = V_1 \cup V_2$, $|E|$ is the size and E is represented by the successor relation R_1. Height $|H|$ denotes the depth of a tree.

Based on the above definitions, let each node of the same height be annotated by the possible label of a named entity, we can construct a tree that represents all options in BRNN to recognize a sequence.

3.2   Path construction

For any sequence $W = (w_1, w_2, ..., w_n)$ and the corresponding labels $Y = (y_1, y_2, ..., y_n)$, we can map the bi-directional hidden states to the path in a BTPK model $T' = < V, E >$, where the hidden states of BRNN constitute the vertices of BTPK. We present the mapping from a BRNN network to a BTPK model of height $|H| = n + 1$ as follows:

$$V_1 = \left\{ h_1^f, h_2^f, ..., h_n^f \right\} \qquad V_2 = \left\{ h_1^b, h_2^b, ..., h_n^b \right\}$$

$$H = \{w_0, w_1, w_2, ..., w_n\} \qquad n \leq |E| < |V| * (|V| - 1) \qquad (1)$$

where h_i^f means the hidden state (feature vector) of the ith element w_i in forward RNN, and h_i^b means the feature vector of the ith element w_i in backward RNN, w_0 denotes the start. V is the vertices of the graph, which is composed of V_1 and V_2, where V_1 and V_2 denote the vertices of forward RNN and backward RNN, respectively. H is denoted by the elements in the sequence. As mentioned above, for all $x, y \in V$, $xy \in E$ iff xR_1y or yR_1x, written as $x < y$ or $y < x$. Unlike standard binary trees, the size $|E|$ is greater than or equal to n because there may be loops in a BTPK model, since there are public announcements in the tree.

It is important to note that the primary methodology for constructing the path of BTPK models involves the identification of branches, which entails recognizing the points in the model where different decisions may be made and subsequently indicating the potential outcomes of these decisions at the corresponding nodes. This study aims to identify the branches of trees in the BTPK model through the distinct masking of forward and backward hidden states.

3.3   Public announcement extraction

The present study involves the construction of the branches and paths of BTPK models through the utilization of trained BRNNs. Nevertheless, the process of path construction solely achieves the representation of knowledge pertaining to concealed states of trained Bidirectional Recurrent Neural Networks (BRNNs), while the decision-making logic that is implicitly involved remains undisclosed. Consequently, we conduct a further examination of the correlation between these concealed states and utilize the technique of public announcement to represent said correlation.

The challenge of identifying public announcements includes identifying the key factor that determine the predicted label of a given entity. The public announcements in BTPK models are derived from the forward hidden states, backward hidden states, and BRNN hidden states that combine the forward and backward ones, as illustrated in Figure 1. Firstly, we split the input sequences into a series of grams such that

$$G_i = \{w_j, ... w_k\} \qquad (0 \leq j \leq k \leq n) \qquad (2)$$

where G_i denotes the i-th gram, and n denotes the length of an input sequence. w_j and w_k denote the j-th words and k-th words in an input sequence, respectively. Then, we select grams one by one for intervention, where the selected gram will be mapped to the corresponding prototypes in the feature space, as shown in Figure 2. To be precise, the hidden states of the chosen n-grams are set to zero through physical intervention. The detection of public announcements is achieved through an analytical process that involves the examination of both the original hidden states and the invented hidden states, drawing inspiration from the techniques employed by the Millers' methods [4].

4 Experiments and Discussion

In this section, we will first introduce our experimental setup. Secondly, we will show how to generate a BTPK model from a trained BRNN in a real-world instance. Thirdly, we try to reason about ambiguous entities through TPK semantics.

4.1 Experiment setup

This paper trains BRNNs on a Chinese public NER dataset **CBVM**, which is available on GITHUB and includes 7 label categories. We extracted 8791 available sequences from it, including 7814 train samples and 977 test samples. As for training parameters, we set $batch\_size = 32$, $learning\_rate = 0.0001$. Hidden states and embedding dimensions are fixed at 128.

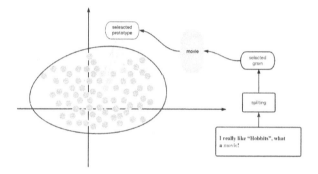

Fig. 2. The illustration of the identification of the public announcement in $task_{103}$.

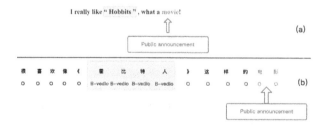

Fig. 3. The illustration of the public announcement in $task_{103}$. The original sequences are in Chinese, we also show them in English to help readers understand them.

4.2 Generating a BTPK model

For an arbitary input sequence, we can generate a global explanation for the decision process of a trained BRNN using the BTPK model. The hidden state analysis for **Example 4.1** is demonstrated in Figure 2. Firstly, we split the input sequences into a series of grams and each gram is composed of one or more words. Then, the selected gram will be mapped to the corresponding prototypes in feature space, where the selected gram (words) are "movie" in Figure 2. According to our experimental results, we find that the predicted label of entities "Hobbits" will not be "video" when we fix the hidden states of the selected n-grams to 0 by physical intervention. Therefore, it's possible to derive the conclusion that "movie" is the public announcement for the predicted label of entities as "Hobbits", as shown in Figure 3. Based on the above knowledge, a BTPK model is established in Figure 4.

Example 4.1   *Consider the sentence* $task_{103}$ =*{I really like "Hobbits", what a movie". }*

4.3   Semantic ambiguity explanation

Intuitively, we consider the human-readable explanations which consist of public announcements and a natural language template. To generate user-friendly explanations for those without background in name entity tagging, we consider "B_book" and "I_book" to be "book", "B_video" and "I_video" to be "video" and "B_music" and "I_music" to be "music". Generally, we have the following explanations for **Example 4.1, Question**: Why is "Hobbits" recognized as a video name rather than other labels (*e.g.*, book name or music name)?

Explanation: Because "movie" (public announcement) appears in the following words, it is more reasonable to recognize it as "video".

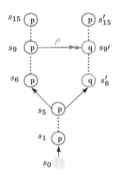

Fig. 4: BTPK model.

In this example, the explanation is obtained by a logic reasoning process of a BTPK model M in Figure 4. Let axiom q denote that the entity is recognized as "vedio", axiom p denote that the entity is recognized as other labels. From semantics of TPK, "Hobbits" is correctly recognized iff $(M, s'_6 \vDash q)$ *and* $(M, s'_7 \vDash q)$ *and* $(M, s'_8 \vDash q)$ *and* $(M, s'_9 \vDash q)$, where s'_i denotes the states in height $|H| = i$. When the system gets words from s_1 and goes forward to s_{15}, the path can be represented as $M, s_1 \vDash \Box p$. But there is a public announcement $s_{159}\rho s'_9$, so the system will go back to s'_6 and then go forward to the end state s'_{15}, generating a new path on the right branch of the tree. The new path can be denoted by $(M, s'_6 \vDash q)$ *and* $(M, s'_7 \vDash q)$ *and* $(M, s'_8 \vDash q)$ *and* $(M, s'_9 \vDash q)$ $(M, s_5 \vDash \Upsilon p)$ *and* $(M, s'_{10} \vDash \Box p)$, which is the ground truth of this sequence. Thus, the recognition process of the entity can be presented in a logical way by the BTPK model, and the public announcement illustrates how to go back to a more reasonable state.

5   Conclusions and future work

We proposed a new BTPK-based interpretable method for NER tasks, which can effectively and logically capture the semantics in the context and give explanations in form of trees to show the internal mechanisms of BRNN models. We apply the BTPK-based interpretable method to a trained BRNN model to obtain logic explanations. In addition, we also demonstrate how to reason on a BTPK model to understand the inner decision making path of a trained BRNN. For future work, we plan to combine the BTPK-based interpretable method (as in this work) with transfer learning for cross-lingual NER tasks.

References

[1] Abraham, M., I. Belfer, D. M. Gabbay and U. Schild, *Future determination of entities in talmudic public announcement logic*, Journal of Applied logic **11** (2013), pp. 63–90.

[2] Agarwal, O., Y. Yang, B. C. Wallace and A. Nenkova, *Interpretability analysis for named entity recognition to understand system predictions and how they can improve*, Computational Linguistics **47** (2021), pp. 117–140.

[3] Alvi, M. Z. H., S. Zaman, J. R. Saurav, S. Haque, M. S. Islam and M. R. Amin, *B-NER: A novel bangla named entity recognition dataset with largest entities and its baseline evaluation*, IEEE Access **11** (2023), pp. 45194–45205.
URL https://doi.org/10.1109/ACCESS.2023.3267746

[4] Cornish, T. A. O., *Mill's methods for complete intelligent data analysis*, in: X. Liu, P. R. Cohen and M. R. Berthold, editors, *Advances in Intelligent Data Analysis, Reasoning about Data, Second International Symposium, IDA-97, London, UK, August 4-6, 1997, Proceedings*, Lecture Notes in Computer Science **1280** (1997), pp. 65–76.
URL https://doi.org/10.1007/BFb0052830

[5] Fu, J., P. Liu and G. Neubig, *Interpretable multi-dataset evaluation for named entity recognition*, arXiv preprint arXiv:2011.06854 (2020).

[6] Fu, Y., N. Lin, Z. Yang and S. Jiang, *Towards malay named entity recognition: an open-source dataset and a multi-task framework*, Connect. Sci. **35** (2023).
URL https://doi.org/10.1080/09540091.2022.2159014

[7] Hou, B.-J. and Z.-H. Zhou, *Learning with interpretable structure from gated rnn*, IEEE transactions on neural networks and learning systems **31** (2020), pp. 2267–2279.

[8] Krakovna, V. and F. Doshi-Velez, *Increasing the interpretability of recurrent neural networks using hidden markov models*, arXiv preprint arXiv:1606.05320 (2016).

[9] Li, J., A. Sun, J. Han and C. Li, *A survey on deep learning for named entity recognition*, IEEE Transactions on Knowledge and Data Engineering **34** (2022), pp. 50–70.

[10] Lin, B. Y., D.-H. Lee, M. Shen, R. Moreno, X. Huang, P. Shiralkar and X. Ren, *Triggerner: Learning with entity triggers as explanations for named entity recognition*, arXiv preprint arXiv:2004.07493 (2020).

[11] Wisdom, S., T. Powers, J. Pitton and L. Atlas, *Interpretable recurrent neural networks using sequential sparse recovery*, arXiv preprint arXiv:1611.07252 (2016).

An Interpretable Method for Biosignal-based Gesture Recognition (Extended Abstract)

Sheng Wei

ZLAIRE
Zhejiang University

Beishui Liao

ZLAIRE
Zhejiang University

Abstract

Biosignal-based human-computer interaction has received widespread attention due to its non-invasive nature and high accuracy, but its interpretability needs to be enhanced if it is to be applied to human health-related fields such as rehabilitation. To address this problem, we fuse two different biosignals to obtain a more comprehensive information representation and propose an interpretable feature selection and classification method to enhance its interpretability while achieving higher recognition accuracy.

Keywords: Biosignal, Hand Gesture Recognition, Feature Select, Classification.

1 Introduction

Human-computer interaction (HCI) is an important way for people to use artificial intelligence and is widely used in many fields. As the most flexible part of human body, hand gestures can express a large number of different meanings[6], and they have the advantages of various types of gestures, rich meanings, little influence by external environment, and a wide range of application scenarios, etc. Gesture recognition-based interaction has become an important means of HCI and is one of the current research hot spots in this field, so HCI based on gesture recognition is of great significance[11].

However, due to the poor interpretability of artificial intelligence algorithms, it is still not accepted by users in many fields related to human life and property safety. For example, in rehabilitation treatment and assessment, as a field closely related to human health, good interpretability is an important factor for patient acceptance of the treatment and assessment method[10]. Therefore, we need to improve the interpretability of the method. There are some model-agnostic interpretation methods: Variable Importance(VI)[8], Local interpretable model-agnostic explanations (LIME)[9], SHapley Additive exPlanations(SHAP)[17], etc. These methods are separated from the machine learning model, and their greatest advantage is their flexibility, which can be applied to different types of models. However, it is not enough to achieve high-precision interpretable HCI only by using those interpretable methods. Therefore, we can improve the accuracy and interpretability of HCI by fusing multimodal biosignal data representing different information to improve the comprehensive representation and accuracy of human body information[18], and enhance the trust of user data sources, combined with interpretable algorithms to achieve user-acceptable HCI based on biosignals. In this way, we can pave the way for the application of biosignals in the field of rehabilitation treatment and assessment.

To solve this problem, we use surface Electromyography (sEMG) signals[7] and A-mode Ultrasound (AUS) signals[16] to characterize different human body information as signal sources for gesture recognition. By fusing them, we can obtain richer human body information. Also, we propose a weighted soft voting algorithm based on feature selection to achieve feature weight acquisition while completing feature selection. And we perform back-end fusion by the soft voting algorithm to improve the classification accuracy and explain the process and results of prediction.

The structure of this paper is organized as follows. In the next section, we briefly describe the research process of biosignal-based interpretable gesture recognition. Then, in Section 3, we present the interpretable feature selection and classification method proposed in this paper. Finally, we conclude this paper with a final remark.

2   Sequence of processes

2.1   Choice of sensing modality

In the selection of multi-modal bio-signals, sEMG signal-based HCI has been studied extensively and is a very mature HCI interface[5], but it still has problems such as insensitivity to fine movements and susceptibility to muscle fatigue, so we hope to find other modal signals that can complement its shortcomings, and try to fuse them to obtain a more comprehensive and robust information representation. At the same time, to make biosignal-based HCI more user-friendly, we need to choose HCI signal sources that are interpretable and non-intrusive.

In terms of the signal nature, the sEMG signal is a non-smooth electrical signal generated by the electrical signal of human muscle contraction flooding to the skin surface, reflecting the electrophysiological information generated by the nerve signal of the subject's motor intention after muscle amplification, which is a temporal dimension but lacks the spatial information of individual channels[3]. While the AUS signal reflects the morphological information of the muscle-muscle interface and the muscle-skeleton interface of the subject[15]. It is spatial dimensional information but lacks temporal information due to its low sampling frequency. The AUS signal and the sEMG signal reflect the characteristics of the muscle in different states from different dimensions, so they have strong complementary properties for gesture recognition[14]. Therefore, we consider adding the AUS signal, which can obtain deep muscle information, to complement the information of the sEMG signal, to achieve more accurate and robust gesture recognition.

2.2 Feature extraction and fusion

Feature extraction is an important method to get useful information hidden in the signal and remove unnecessary parts and interferences. With the in-depth research related to biosignal feature extraction, a large number of time domain, frequency domain and time-frequency domain features have been widely used in sEMG and AUS signal feature extraction[1,13]. Most of the manual features have clear meanings and can be easily understood and interpreted by users, and different features can be extracted and selected for different purposes to improve the interpretability and accuracy of the algorithm.

For manual discrete features, some methods can be used to rank them and choose some of the most important features for feature fusion, which can improve their interpretability as well as efficiency and accuracy[4]. In addition to using manual discrete features to combine and downscale them into the classifier for training, discrete features can also be two-dimensionalized and then trained[1]. It is also possible to combine traditional manual features with deep features based on deep learning[2], as they represent concrete and abstract features of signals respectively, and their fusion can improve the efficiency and interpretability of feature information extraction.

2.3 Design of classification algorithms

For the interpretability of traditional machine learning algorithms. Some methods such as Linear Discriminant Analysis (LDA) and decision trees have good interpretability, but their accuracy is low in some cases. To solve this problem, ensemble learning methods such as soft voting algorithms can be used to make full use of the characteristics of those interpretable methods to improve the accuracy and ensure their interpretability. For the machine learning algorithms with weak interpretability, some post-explanation models can be used to explain them.

For the interpretability of deep learning algorithms, we can try to build an interpretable multi-modal biosignal neural network model. We can simulate and explain the relationship between biosignals and gesture movements based on the relevant knowledge of physiology, morphology and anatomy of human forearm muscle tissues. And we should consider the signal features as well as the relevant muscle features to design the network structure when building the network model, instead of using the same feature extraction and classification network model for different signals[12]. At the same time, since the influence of related muscles is also different for different gestures, an attention mechanism can be added to emphasize the effective region information, which can improve both the accuracy and its interpretability, and try to combine the attention mechanism and game theory to achieve the interpretable attention mechanism with better effect.

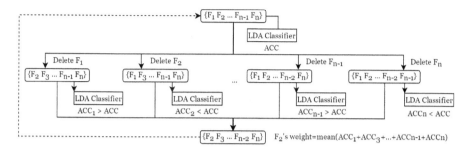

Fig. 1. An example of automatic feature selection strategy.

3   Methods

Since manual features and traditional machine learning algorithms have better interpretability than deep learning methods, this paper focuses on the research related to feature selection and ensemble learning.

Since multiple features are often extracted at the same time in feature extraction of biosignals, but not all features can be combined to achieve the best results, we want to design an interpretable feature selection and classification method to improve accuracy and interpretability, which consists of two main parts, feature automatic selection strategy and weighted soft voting classification.

3.1   Automatic feature selection strategy

The purpose of feature selection is to find the best combination of features for recognition among all discrete features, reduce the negative impact of negative traits, and improve the model efficiency. Therefore, an automatic feature selection method needs to be designed. Since the LDA classifier is an interpretable linear classifier and does not require a hyper-parameter setting, we use it as the classifier used in this method. Also, to prevent data leakage, our feature selection method is used only on the training set, which is divided into training-training subsets and training-test subsets according to a ratio, and finally, the resulting feature combinations and their weights are applied to the whole training and test sets.

In this paper, a generic feature automatic selection method is proposed, whose flow chart is shown in Fig. 1. The main steps are as follows:

- Step 1. Combine all n features to get the feature set F, which is fed into the LDA classifier to get the classification accuracy, which we called ACC;

- Step 2. The individual features in the feature set are sequentially deleted to obtain n feature subsets F_{in} each of size $n - 1$, which are sequentially fed into the LDA classifier to obtain the classification accuracy of each subset;

- Step 3. If the accuracy of a subset of features after deletion of a feature is higher than the accuracy before deletion, it means that the feature harms the task. So, these features with an accuracy higher than the ACC are removed from the set of features in descending order, until the target number

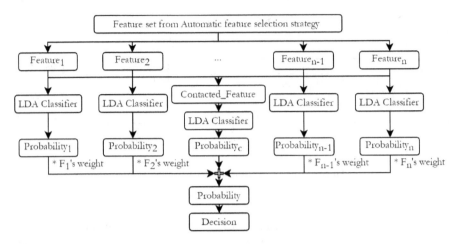

Fig. 2. The process of the weighted soft voting algorithm.

of dimensions is reached or all the features with accuracy higher than the previous level have been removed to get the feature set F_i;

- Step 4. Record the accuracy of each sub-feature set and accumulate it to each feature weight contained in it;

- Step 5. Repeat the above four steps until the feature set of the target dimension is obtained.

After the above steps, the combination of features with the best recognition effect and the weights of each feature can be obtained. In a multimodal fusion scenario, we can achieve more efficient modal fusion and gesture recognition by performing feature selection for each modality separately. In addition, the interpretability of the algorithm can be improved by showing the accuracy changes of each feature subset during feature selection.

3.2 Weighted soft voting algorithm

In previous studies, researchers have often combined multiple features and downscaled them before feeding them into the classifier for training[13], because the fused features contain information on multiple features, and in this way, we can often obtain better recognition results than single features. Therefore, we designed a weighted soft voting algorithm to fuse multiple features from the feature set.

First, we input each discrete feature from the automatic feature selection strategy into the classifier for training separately and get their respective probability matrices. In the next step, we splice each discrete feature after normalization to obtain its concatenated features and feed them into the classifier as well to obtain their probability matrices. Finally, we use the weights of each feature obtained in the automatic feature selection strategy to weigh and sum their associated probability matrices ($Probability_c$), then obtain the final probability matrix to make a judgment of the gesture category. The process of the weighted soft voting algorithm is shown in Fig. 2. Since both the extracted features and the LDA classifier are interpretable, the results obtained from them are still highly interpretable after weighted summing.

Since different features represent various meanings, they perform differently in the recognition of different gestures, and back-end fusion using a soft voting algorithm can more fully exploit the recognition advantages of distinct features on their dominant gestures. At the same time, since different features perform differently in gesture recognition, by weighting the feature weights obtained from the automatic feature selection strategy to the likelihood matrix of each feature, the influence of effective features can be further strengthened, and that of ineffective features would be weaken compared with the direct use of all the discrete features. Thus our method can achieve a more robust gesture recognition with higher accuracy and interpretability.

3.3   Evaluation methods

To evaluate the accuracy improvement of the method. We evaluate our model using four evaluation metrics that are common in previous work: accuracy, precision, recall, and F1 score, calculated as follows:

$$Accuracy = \frac{TP + FN}{TP + FP + TN + FN} \tag{1}$$

$$Precision = \frac{TP}{TP + FP} \tag{2}$$

$$Recall = \frac{TP}{TP + FN} \tag{3}$$

$$F_1 Score = 2 * \frac{Precision * Recall}{Precision + Recall} \tag{4}$$

where TP is true positive, FP is false positive, FN is false negative, and TN is true negative.

Because biological data collection is more difficult, it tends to be smaller. To verify the robustness of the method, with a certain amount of data, we can use the division of different training and testing sets ratios to verify its robustness. For example, the data measured in the first small period in the dataset is used as the training set, and at the same time, more parts of the data in the later part of the dataset are used as the test set, if the results are still satisfactory, it means that the method has high robustness.

4   Final Remark

In this paper, we propose an interpretable feature selection and classification method based on biosignals represented by sEMG signals with AUS signals, which obtains the weights of each feature while using a backward deletion strategy for feature selection, and then feeds the automatically selected feature set and fused characteristics into the classifier separately for training, and sums up all the possibility matrices after weighting them to obtain the final result.

This method can make full use of the information on each feature to improve its accuracy and robustness. Because the whole process is interpretable, it can be better understood by users and improve its acceptability in practical applications. At the same time, the method is general and may provide ideas and insights for pattern recognition of multiple features.

References

[1] Chen, G. Q., W. L. Wang, Z. Wang, H. H. Liu, Z. L. Zang and W. K. Li, *Two-dimensional discrete feature based spatial attention capsnet for semg signal recognition*, Applied Intelligence **50** (2020), pp. 3503–3520.

[2] Fajardo, J. M., O. Gomez and F. Prieto, *Emg hand gesture classification using handcrafted and deep features*, Biomedical Signal Processing and Control **63** (2021).

[3] Fang, Y. F., D. L. Zhou, K. R. Li, Z. J. Ju and H. H. Liu, *Attribute-driven granular model for emg-based pinch and fingertip force grand recognition*, IEEE Transactions on Cybernetics **51** (2021), pp. 789–800.

[4] Guo, N., C. Liu, C. Li, T. Lu, L. Wen and Q. Zeng, *Explainable feature-based hierarchical approach to predict remaining process time*, Ruan Jian Xue Bao/Journal of Software (in Chinese) (2023).

[5] Huang, Y. J., K. B. Chen, X. M. Zhang, K. Wang and J. Ota, *Joint torque estimation for the human arm from semg using backpropagation neural networks and autoencoders*, Biomedical Signal Processing and Control **62** (2020).

[6] Lee, K. S. and M. C. Jung, *Ergonomic evaluation of biomechanical hand function*, Saf Health Work **6** (2015), pp. 9–17.

[7] Li, Y., G. Chai, C. Lu and Z. Tang, *On-line semg hand gesture recognition based on incremental adaptive learning*, Computer Science **46** (2019), pp. 274–279.

[8] Murray, K. and M. M. Conner, *Methods to quantify variable importance: implications for the analysis of noisy ecological data*, Ecology (Durham) **90** (2009), pp. 348–355.

[9] Ribeiro, M. T., S. Singh and C. Guestrin, *"why should i trust you?": Explaining the predictions of any classifier*, in: *Proceedings of the 22nd ACM SIGKDD International Conference on Knowledge Discovery and Data Mining*, pp. 1135–1144.

[10] Shafivulla, M., *Semg based human computer interface for physically challenged patients*, 2016 International Conference on Advances in Human Machine Interaction (Hmi) (2016), pp. 171–174.

[11] Sun, Y., C. Xu, G. Li, W. Xu, J. Kong, D. Jiang, B. Tao and D. Chen, *Intelligent human computer interaction based on non redundant emg signal*, Alexandria Engineering Journal **59** (2020), pp. 1149–1157.

[12] Wei, S., Y. Zhang and H. Liu, *A multimodal multilevel converged attention network for hand gesture recognition with hybrid semg and a-mode ultrasound sensing*, IEEE transactions on cybernetics **PP** (2022).

[13] Wei, S., Y. Zhang, J. Pan and H. Liu, *A novel preprocessing approach withsoft voting forhand gesture recognition witha-mode ultrasound sensing*, in: *15th International Conference on Intelligent Robotics and Applications, ICIRA 2022, August 1, 2022 - August 3, 2022*, Lecture Notes in Computer Science (including subseries Lecture Notes in Artificial Intelligence and Lecture Notes in Bioinformatics) **13458 LNAI**, pp. 363–374.

[14] Xia, W., Y. Zhou, X. Yang, K. He and H. Liu, *Toward portable hybrid surface electromyography/a-mode ultrasound sensing for human–machine interface*, IEEE Sensors Journal **19** (2019), pp. 5219–5228.

[15] Yang, X., Z. Chen, N. Hettiarachchi, J. Yan and H. Liu, *A wearable ultrasound system for sensing muscular morphological deformations*, IEEE Transactions on Systems, Man, and Cybernetics: Systems **51** (2021), pp. 3370–3379.

[16] Zhang, Y. H., D. Jiang and A. Demosthenous, *Design of a cmos analog front-end for wearable a-mode ultrasound hand gesture recognition*, Prime 2022: 17th International Conference on Phd Research in Microelectronics and Electronics (2022), pp. 97–100.

[17] Zheng, S., Y. Cao and M. Yoshikawa, *Secure shapley value for cross-silo federated learning*, Proceedings of the VLDB Endowment **16** (2023), pp. 1657–1670.

[18] Zou, Y., L. Cheng and Z. Li, *A multimodal fusion model for estimating human hand force: Comparing surface electromyography and ultrasound signals*, IEEE Robotics & Automation Magazine (2022), pp. 2–16.

Design and Implementation of Legal Question Answering Robot

Xiaotong Fang

China University of Mining and Technology
Jiangsu
P.R. China

Abstract

This paper presents the development of a legal question-answering robot and its human-computer interaction system. The objective is to provide legal consultation services through an intelligent robot, aiming to reduce costs and workload for legal professionals. The research includes the construction of a legal question-answering dataset through web scraping, fine-tuning experiments with Chinese pre-trained language models, and the development of a human-computer interaction system using the Pepper robot. The system successfully addresses the unequal distribution of legal resources and demonstrates the potential of AI technology in improving the accessibility of legal resources and meeting the demand for legal consultation services.

Keywords: question-answering system, pre-trained language model, human-computer interaction.

1   Introduction

In recent times, ChatGPT [6], developed by OpenAI, has demonstrated remarkable natural language generation abilities, showcasing its ability to comprehend user intentions and provide excellent responses. The underlying technology of large-scale pre-trained language models used by ChatGPT for knowledge storage provides robust support for generating high-quality answers in question-answering systems.

AI-powered legal consultation can directly assist clients in organizing and analyzing legal issues and provide them with alternative solutions to choose from. Moreover, the cost of the entire consultation process is almost negligible, and even if a certain fee is charged, it is much lower than the standard charges in the legal industry, well within the clients' affordability range.

Our work makes several contributions. First, we complete the professional dataset in the field of legal question-answering. We utilize web crawling techniques to extract a large amount of manually annotated legal question-answering data from the Chinese Legal Service website and perform effective data cleaning to obtain the dataset. Second, we conduct fine-tuning experiments on multiple open-source Chinese pre-trained language models using the legal question-answering dataset. Compared to conventional retrieval-based legal question-answering models, our research provides references and solutions for improving the effectiveness of generative pre-trained Chinese legal question-answering models. Finally, we develop a human-machine interaction system for legal question-answering based on the Pepper robot [9], integrating the best-performing legal question-answering model.

2   Data collection

To construct a legal question-answering dataset, this paper utilized the Selenium library in Python to automate data crawling from the Chinese Legal Service website (12348 China Legal Net) [1]. The Chinese Legal Service website, established by the Ministry of Justice of the People's Republic of China, integrates over 1.39 million legal service personnel and offers legal consultation services. Its consultation service section contains a substantial amount of high-quality, manually annotated legal question-answering data. In total, approximately 350,000 raw data entries were collected from the Chinese Legal Service website, spanning the time range from 2018 to 2022.

The dataset obtained through Selenium crawling contains four fields: "title" for the legal question-answering title, "question" for the legal question, "answer" for the legal answer, and "time" for the question timestamp.

This paper implements data cleaning techniques to enhance the quality of the legal question-answering dataset. After applying the necessary data cleaning steps to the original dataset, a final legal question-answering dataset consisting of 336,760 data entries is obtained.

Statistical analysis was conducted on the length of legal question-answer data. It was observed that the legal questions generally have shorter lengths, with the majority of them being less than 200 characters. On the other hand, the legal answers tend to be longer, with a majority falling within the range of 51 to 500 characters.

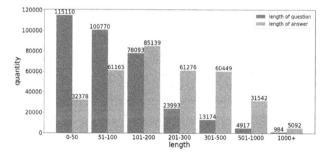

Fig. 1. Length of questions and answers in the Legal Question Answering Dataset.

---

[1] Source: Chinese Legal Service website http://www.12348.gov.cn/

3 Experiment

Currently, deep learning-based encoder-decoder architectures have become the mainstream approach for natural language generation. Most existing legal question-answering systems are built upon methods such as knowledge graphs and BERT [10,12], which employ query-based approaches to retrieve legal answers. In this paper, we compare and test the performance of three generative language models, namely Chinese GPT-3, ChatGLM [11] based on GLM [3], and PromptCLUE based on T5 [8]. We fine-tune these models using a Chinese legal question-answering dataset, evaluating their effectiveness in the context of legal question-answering tasks.

LLM models are mostly derived from transformer. PromptCLUE based on T5 adopts an encoder-decoder architecture, while Chinese GPT-3 uses a decoder-only architecture. ChatGLM, based on GLM, utilizes the Prefix-LM architecture. Regarding performance on few-shot and zero-shot tasks, autoregressive language models like GPT have demonstrated promising results. These models are trained by generating the next word in a sequence given the preceding words. They have been widely employed in text generation and question answering. Prefix-LM is a variant of the encoder-decoder architecture. By controlling the mask matrix, the visibility of input tokens can be managed, allowing the model to handle both natural language understanding and generation tasks effectively.

The Chinese GPT-3 model is based on GPT-3 [2] and is pretrained using a large amount of Chinese unsupervised data and downstream task data. Currently, the Alibaba DAMO Academy has only released the specific parameters for GPT-3 base, GPT-3 large, GPT-3 1.3B, and GPT-3 2.7B.

In this paper, fine-tuning is performed on GPT-3 large and GPT-3 1.3B. The hyperparameters used for fine-tuning are presented in Table 1. The fine-tuning of GPT-3 large is conducted on a server RTX 3090, while the fine-tuning of GPT-3 1.3B is performed on a server with Tesla A100.

Hyperparameter	GPT-3 large	GPT-3 1.3B
Bbatch size	16	8
Epochs	20	10
Warmup steps	200	200
Learning rate	0.0003	0.0001
Optimizer	AdamW	AdamW

Table 1
The hyperparameters of the fine-tuning experiment of GPT-3

The ChatGLM-6B is implemented using P-Tuning v2 [5] for tuning, which reduces the amount of parameters that need to be optimized to 0.1%. Additionally, model quantization and Gradient Checkpoint methods are employed, enabling the model to run with a minimum requirement of just 7GB of GPU memory. It is p-tuned on a server with an RTX 3090. During the p-tuning process, the quantization bit is set to 4. Under this configuration, the model parameters of INT4 are frozen. The loss curve is depicted in Fig. 2.

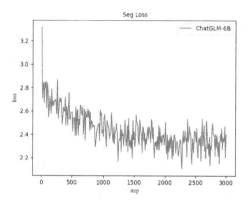

Fig. 2. Loss of ChatGLM-6B.

In this paper, fine-tuning is also performed on PromptCLUE. To fine-tune the PromptCLUE, the legal question-answering dataset is used. The experimental setup includes a batch size of 8 and a maximum sequence length of 512.

The inference experiments are conducted. It is found that GPT-3 large, due to its relatively smaller model size, performs poorly in legal question answering. The parameters and scale of the GPT-3 1.3B model are larger but the logical coherence in question answering still remains weak, primarily due to computational resource limitations.

"question":"Sign a contract with a company, but hold a subsidiary position at the same time, only pay one salary. The social security provident fund is paid in accordance with the minimum base, which is inconsistent with the salary. Now the company wants to transfer more positions, want to leave. Can I get compensation? Or do I have to quit to get compensation?" ,
"answer":"First, the amount of compensation shall be different according to different positions. If there is no competition restriction, the amount of compensation can be negotiated by both parties with 28% to 30% of the monthly wage standard as the compensation standard. Article 24 of the Labor Contract Law stipulates : An employer and a worker may agree in the labor contract to keep confidential the business secrets of the employer and matters related to intellectual property rights."

Fig. 3. Inference result of GPT-3 1.3B (1).

During inference testing of GPT-3 1.3B, it is observed that the model occasionally exhibits a tendency to evade questions, leading to a decline in the question-answering efficiency of GPT-3 1.3B.

"question":"Li Mou (female) and Wang Mou (male) are remarried, Wang Mou died in March 2020, Wang Mou has personal real estate before marriage, may I ask Li Mou if there is the right to inherit Wang Mou real estate, found that Wang mou has given the property to his biological children, in this case Li Mou can still inherit?",
"answer":"I suggest you explain the situation more specifically so that we can better analyze the answer for you

Fig. 4. Inference result of GPT-3 1.3B (2).

The ChatGLM-6B demonstrates better performance in legal question answering. However, upon comparison with annotated data provided by China Legal Service Network, significant discrepancies were found between ChatGLM-6B and the annotated data.

```
"question": "The other party defaults on wages, and now the business license breach has been executed, and the
legal person has no (mediation agreement legal person guarantees) but only limits high consumption. The
equipment was moved during the execution of my application, and now it's gone missing, and I still don't pay you
back.",
"answer":"Hello, according to your statement, the other party has failed to perform the debt, you can apply for
enforcement according to Article 123 of the Civil Procedure Law.According to the provisions of Article 122 of
the Provisions of the Supreme People's Court on Several Issues concerning the Enforcement Work of the People's
Courts, enforcement cases shall be accepted by the enforcement court. You can apply to the enforcement court for
enforcement."
```

Fig. 5. Inference result of ChatGLM-6B.

Finally, in this paper, PromptCLUE is tested through inference experiments, and it is observed that PromptCLUE is capable of providing answers based on specific legal clauses. However, the majority of the responses lack direct answers to the questions.

```
"question": "On April 30, 2021, my father's right forearm was amputated due to an accident at a factory. There
are nearly centenarians and two college students in the family. He is still hospitalized in the hospital, and
has been hospitalized for more than two months without greeting or paying medical bills. I need your help.",
"answer": "According to Article 14 of the Regulations on Industrial Injury Insurance, an employee shall be
identified as an industrial injury under any of the following circumstances :(1) The employee is injured by an
accident due to work during working hours and in the workplace;(2) being injured in an accident while engaged
in preparatory or concluding work related to work in the workplace before and after working hours;(3) being
```

Fig. 6. Inference result of PromptCLUE.

This study evaluates the generative legal question-answering model using a comprehensive assessment of Rouge-1, Rouge-2, Rouge-L [4], and Bleu-4 [7] metrics. These metrics measure the similarity between the generated answer and the reference answer. Specifically, Rouge-1 and Rouge-2 evaluate similarity based on n-gram overlap, Rouge-L utilizes longest common subsequence (LCS), and Bleu-4 quantifies similarity through 4-gram precision with a brevity penalty. The final results are presented in Table 2.

Model	Rouge-1	Rouge-2	Rouge-L	Bleu-4
ChatGLM-6B p-tuning	21.6	5.1	15.7	4.1
GPT-3 large	33.3	5.7	16.9	17.8
GPT-3 1.3B	37.5	17.5	26.4	24.7
PromptCLUE	50.1	25.5	36.4	30.3

Table 2
Experimental results of the Legal Question-answering model: evaluation using Rouge-1, Rouge-2, Rouge-L and Bleu-4 metrics

It is found that PromptCLUE achieves the best performance in the evaluation system used in this study, closely followed by GPT-3 1.3B. Due to its smaller model size and fewer parameters, GPT-3 large performs less effectively than GPT-3 1.3B. Surprisingly, the results of ChatGLM-6B p-turning are worse than those of GPT-3 large. Inference results reveal that ChatGLM-6B outperforms GPT-3 large and even GPT-3 1.3B. This unexpected result may be attributed to the insufficient parameter updates caused by freezing INT4 during P-Tuning. Ultimately, PromptCLUE is chosen as the legal question-answering model to provide legal advisory services in this study.

4 Interactive System for the Pepper Robot

This paper develops a human-machine interactive legal question-answering system. Building upon the Pepper robot, a legal question-answering interaction framework was designed as illustrated in Fig. 7. PromptCLUE was converted to ONNX [1] format and deployed on a server using the Flask framework to provide legal question-answering services for the robot.

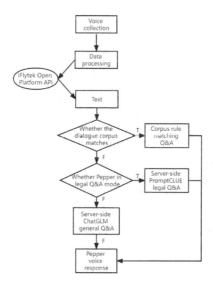

Fig. 7. Flow chart of Pepper question-answering.

5 Conclusion

To develop a legal question-answering robot, this paper conducted research in legal question-answering dataset construction, legal question-answering model training, and the development of the Pepper human-machine interaction system. Fine-tuning experiments were conducted using legal question-answering data on Chinese pre-trained language models to create a generative legal question-answering model. By comparing the experimental results, it was observed that PromptCLUE achieved excellent performance in our legal question-answering task.

References

[1] Bai, J., F. Lu, K. Zhang et al., *Onnx: Open neural network exchange*, GitHub repository (2019), p. 54.

[2] Brown, T., B. Mann, N. Ryder, M. Subbiah, J. D. Kaplan, P. Dhariwal, A. Neelakantan, P. Shyam, G. Sastry, A. Askell et al., *Language models are few-shot learners*, Advances in neural information processing systems **33** (2020), pp. 1877–1901.

[3] Du, Z., Y. Qian, X. Liu, M. Ding, J. Qiu, Z. Yang and J. Tang, *Glm: General language model pretraining with autoregressive blank infilling*, in: *Proceedings of the 60th Annual Meeting of the Association for Computational Linguistics (Volume 1: Long Papers)*, 2022, pp. 320–335.

[4] Lin, C.-Y., *Rouge: A package for automatic evaluation of summaries*, in: *Text summarization branches out*, 2004, pp. 74–81.

[5] Liu, X., K. Ji, Y. Fu, W. L. Tam, Z. Du, Z. Yang and J. Tang, *P-tuning v2: Prompt tuning can be comparable to fine-tuning universally across scales and tasks*, arXiv preprint arXiv:2110.07602 (2021).

[6] Ouyang, L., J. Wu, X. Jiang, D. Almeida, C. Wainwright, P. Mishkin, C. Zhang, S. Agarwal, K. Slama, A. Ray et al., *Training language models to follow instructions with human feedback*, Advances in Neural Information Processing Systems **35** (2022), pp. 27730–27744.

[7] Papineni, K., S. Roukos, T. Ward and W.-J. Zhu, *Bleu: a method for automatic evaluation of machine translation*, in: *Proceedings of the 40th annual meeting of the Association for Computational Linguistics*, 2002, pp. 311–318.

[8] Raffel, C., N. Shazeer, A. Roberts, K. Lee, S. Narang, M. Matena, Y. Zhou, W. Li and P. J. Liu, *Exploring the limits of transfer learning with a unified text-to-text transformer*, The Journal of Machine Learning Research **21** (2020), pp. 5485–5551.

[9] Robotics, S., *Pepper - the world's first social humanoid robot*, 2023, June 2. URL https://www.softbankrobotics.com/emea/en/pepper

[10] Wu, J., J. Liu and X. Luo, *Few-shot legal knowledge question answering system for covid-19 epidemic*, in: *2020 3rd International Conference on Algorithms, Computing and Artificial Intelligence*, 2020, pp. 1–6.

[11] Zeng, A., X. Liu, Z. Du, Z. Wang, H. Lai, M. Ding, Z. Yang, Y. Xu, W. Zheng, X. Xia et al., *Glm-130b: An open bilingual pre-trained model*, arXiv preprint arXiv:2210.02414 (2022).

[12] Zhang, N. N. and Y. Xing, *Questions and answers on legal texts based on bert-bigru*, Journal of Physics: Conference Series **1828** (2021), p. 012035 (9pp).

Modal Lambek Calculus with Primary Assumptions: Decidability and Context-freeness

Zhe Lin [1]   Xinshu Wang

Department of Philosophy, Xiamen University
Xiamen, China

Abstract

In this paper we study modal Lambek calculus with primary assumptions. We prove that modal Lambek calculus with primary assumptions is decidabe. Further we show that the categorial grammars based on modal Lambek calculus with transitive primary assumptions are context-free, which partially provides the answer to long term open problem concerning the generatic capacity of Lambek calculus with primary assumptions.

Keywords: modal Lambek calculus, primary assumption, decidability, context-freeness.

1   Introduction

Lambek calculus (L) was introduced by Lambek in 1958 [8] as a syntactic calculus for categorial grammars. He provided a sequent system for this calculus and proved the cut-elimination, which yields the decidability. Pentus [9] shows that the categorial grammars based on it are context-free. As a type logic for categorial grammars with application to linguistic analysis, not only the pure logic are studied but addtional non-logical assumptions are considered. Buszkowski [2,4] shows that L with assumptions is undecidable and generates all r.e. language while its weaker version(Nonassociative L) is decidable in P-time and generates context-free languages [3]. Bulińska[1] studies L enriched with simple assumptions: $p \Rightarrow q$ where p, q are propositional variables and shows decidability and the fact that categorial grammars based on L with such assumptions are context-free. Foret [6] studies similar problem by another method. Dudakov, Karlov, Kuznetsov and Fofanova show a certain subexponential extension of the Lambek calculus (related to primary assumptions) are decidable and their R-total derivability are context-free [5]

The main contribution of the present paper is that we study Lambek calculus with modalities (introduced by Jäger [7]) and enriched with primary

---

[1]   Email address: pennyshaq@163.com

assumptions: $p_1, ..., p_n \Rightarrow q$. We prove that these logics enriched with finitely many primary assumptions are decidable. We impose some restrictions on the set of Φ. Φ is transitive i.e. it is closed under (Cut), while the formula in the right side of a sequent would not be any of the antecedent $p_1, ..., p_n$. For example, if $p_1, p_2, p_3 \Rightarrow q \in \Phi$ and $p_4, q \Rightarrow r \in \Phi$, then $p_4, p_1, p_2, p_3 \Rightarrow r \in \Phi$, while if $p_1, p_2, p_3 \Rightarrow q \in \Phi$ and $p_4, q \Rightarrow p_3 \in \Phi$, then $p_4, p_1, p_2, p_3 \Rightarrow p_3 \notin \Phi$. We show that categorial grammars based on modal Lambek calculus with a set of transitive primary assumptions are context-free. One movitivation of considering the transitive primary assumptions is that in real applications, the assumptions under consideration are simple and untransitive for instance $VP \Leftrightarrow PN \backslash S$ (the type of verb phrase conjunction).

This paper is organized as follows. In next section, we introduce L with modalities enriched primary assumptions($\mathrm{L}_\diamond(\Phi)$). We show their corresponding sequent systems are all decidable. In section 3, we consider the type grammar based on these logics, and show they are context-free.

2   Modal Lambek Calculus with Primary Assumptions : decidability

In this section, we consider Lambek calculus with modal operators enriched with primary assumptions (denoted by $\mathrm{L}_\diamond(\Phi)$). First, we recall some basic notions for Lambek calculus with modal operators (denoted by $\mathrm{L}_\diamond$).

Definition 2.1 The set of formulas (terms) $\mathcal{F}$ is defined inductively as follows:

$$\mathcal{F} \ni \alpha ::= p \mid \alpha \cdot \beta \mid \alpha \backslash \beta \mid \alpha / \beta \mid \Diamond \alpha \mid \blacksquare \alpha$$

Definition 2.2 Let , and $\circ$ be structural counterparts for $\cdot$ and $\Diamond$ respectively. The set of all formula structures $\mathcal{FS}$ is defined inductively as follows:

$$\mathcal{FS} \ni \Gamma ::= \alpha \mid (\Gamma, \Gamma) \mid \circ\Gamma$$

A *sequent* is an expression of the form $\Gamma \Rightarrow \alpha$ where Γ is a formula structure and α is a formula. A *context* is a formula structure $\Gamma[-]$ with a designated position $[-]$ which can be filled with a formula structure. In particular, a single position $[-]$ is a context. For instance $\Gamma[\Delta]$ is the formula structure obtained from $\Gamma[-]$ by substituting Δ for $-$. By $f(\Gamma)$ we mean the formula obtained from Γ by replacing all structure operations by their relevant formula connectives.

Definition 2.3 The sequent system for $\mathrm{DFL}_\diamond(\Phi)$ consists of a finite set of primary assumptions Φ and the following axiom and rules:

(1) Axiom:

$$\alpha \Rightarrow \alpha \quad (\mathrm{Id})$$

(2) Connective rules:

$$\frac{\Gamma[(\alpha, \beta)] \Rightarrow \gamma}{\Gamma[\alpha \cdot \beta] \Rightarrow \gamma} (\cdot\mathrm{L}) \qquad \frac{\Gamma \Rightarrow \alpha \quad \Delta \Rightarrow \beta}{(\Gamma, \Delta) \Rightarrow \alpha \cdot \beta} (\cdot\mathrm{R})$$

$$\frac{\Gamma[\beta] \Rightarrow \gamma \quad \Delta \Rightarrow \alpha}{\Gamma[(\Delta, \alpha \backslash \beta)] \Rightarrow \gamma} (\backslash\mathrm{L}) \quad \frac{(\alpha, \Gamma) \Rightarrow \beta}{\Gamma \Rightarrow \alpha \backslash \beta} (\backslash\mathrm{R}) \quad \frac{\Gamma[\beta] \Rightarrow \gamma \quad \Delta \Rightarrow \alpha}{\Gamma[(\alpha / \beta, \Delta)] \Rightarrow \gamma} (/\mathrm{L}) \quad \frac{(\Gamma, \beta) \Rightarrow \alpha}{\Gamma \Rightarrow \alpha / \beta} (/\mathrm{R})$$

(3) Modal rules:

$$\frac{\Gamma[\circ\alpha] \Rightarrow \beta}{\Gamma[\Diamond\alpha] \Rightarrow \beta} (\Diamond\mathrm{L}) \qquad \frac{\Gamma \Rightarrow \alpha}{\circ\Gamma \Rightarrow \Diamond\alpha} (\Diamond\mathrm{R}) \qquad \frac{\Gamma[\alpha] \Rightarrow \beta}{\Gamma[\circ\blacksquare\alpha] \Rightarrow \beta} (\blacksquare\mathrm{L}) \qquad \frac{\circ\Gamma \Rightarrow \alpha}{\Gamma \Rightarrow \blacksquare\alpha} (\blacksquare\mathrm{R})$$

(4) Cut rule:

$$\frac{\Delta \Rightarrow \alpha \quad \Gamma[\alpha] \Rightarrow \beta}{\Gamma[\Delta] \Rightarrow \beta} (\mathrm{Cut})$$

(5) Structural rules:

$$\frac{\Gamma[\Delta_1, (\Delta_2, \Delta_3)] \Rightarrow \beta}{\Gamma[(\Delta_1, \Delta_2), \Delta_3] \Rightarrow \beta}(As_1) \qquad \frac{\Gamma[(\Delta_1, \Delta_2), \Delta_3] \Rightarrow \beta}{\Gamma[\Delta_1, (\Delta_2, \Delta_3)] \Rightarrow \beta}(As_2)$$

A sequent $\Gamma \Rightarrow \alpha$ is provable in $L_\diamond$, notation $\vdash_{L_\diamond} \Gamma \Rightarrow \alpha$, if there is a derivation of $\Gamma \Rightarrow \alpha$ in $L_\diamond$. We write $\vdash_{L_\diamond} \alpha \Leftrightarrow \beta$ if $\vdash_{L_\diamond} \alpha \Rightarrow \beta$ and $\vdash_{L_\diamond} \beta \Rightarrow \alpha$.

Theorem 2.4 *(Cut elimination)* $\vdash_{L_\diamond(\Phi)} \Gamma \Rightarrow \beta$ *iff* $\vdash_{L_\diamond(\Phi)} \Gamma \Rightarrow \beta$ *without any application of* (Cut).

Proof. Assume that there is a subderivation of $\Gamma \Rightarrow \beta$ ended with an application of (Cut) as follows:

$$\frac{\vdash \Delta \Rightarrow \alpha \quad \vdash \Sigma[\alpha] \Rightarrow \beta}{\vdash \Sigma[\Delta] \Rightarrow \beta}(Cut)$$

We suffice to show that if $\Delta \Rightarrow \alpha$ and $\Sigma[\alpha] \Rightarrow \beta$ are both provable in $L_\diamond$ without any application of (Cut), then $\Sigma[\alpha] \Rightarrow \beta$ is provable in $L_\diamond$ without any application of (Cut). We proceed by induction on (I) the complexity of (Cut) formula α. In each case we proceed by induction on (II) the sum of the length of two premises of (Cut). Assume that $\Delta \Rightarrow \alpha$ is obtained by (R_l) and $\Sigma[\alpha] \Rightarrow \beta$ is obtained by (R_r). We refer the details to the standard cut elimination proof.

(1) We consider the case that at least one premise in a cut is an axiom or a primary assumption. The cases for axiom is very simple, thus we omit it. We only refer the details for the primary assumptions as follows.

(i) The left premise is a primary assumption. The proof

$$\frac{p_1, ..., p_n \Rightarrow q \qquad \dfrac{\circ \Gamma[q] \Rightarrow \alpha}{\Gamma[q] \Rightarrow \blacksquare \alpha}(\blacksquare R)}{\Gamma[p_1, ..., p_n] \Rightarrow \blacksquare \alpha}(Cut)$$

can be transformed into

$$\frac{\dfrac{p_1, ..., p_n \Rightarrow q \qquad \circ \Gamma[q] \Rightarrow \alpha}{\circ \Gamma[p_1, ..., p_n] \Rightarrow \alpha}(Cut)}{\Gamma[p_1, ..., p_n] \Rightarrow \blacksquare \alpha}(\blacksquare R)$$

Thus the new application of (Cut) has lower length. By induction hypothesis (II), the claim holds.

(ii) The right premise is a primary assumption. The proof

$$\frac{\dfrac{\Gamma[\circ \alpha] \Rightarrow p_i}{\Gamma[\Diamond \alpha] \Rightarrow p_i}(\Diamond L) \qquad p_1, ..., p_n \Rightarrow q}{p_1, ... \Gamma[\Diamond \alpha], ..., p_n \Rightarrow q}(Cut)$$

can be transformed into

$$\frac{\dfrac{\Gamma[\circ \alpha] \Rightarrow p_i \qquad p_1, ..., p_n \Rightarrow q}{\Gamma[\Diamond \alpha] \Rightarrow p_i}(Cut)}{p_1, ... \Gamma[\Diamond \alpha], ..., p_n \Rightarrow q}(\Diamond L)$$

Thus the new application of (Cut) has lower length. By induction hypothesis (II), the claim holds.

(2) α is not introduced by (R_l). We transform the derivation by first applying (Cut) to premises of (R_l) and $\Sigma[\alpha] \Rightarrow \beta$. After that we apply (R_l) to the resulting sequent. Take $(\cdot L)$ as an example to interpret this. The remaining

cases can be treated similarly.

(R_l) is $(\cdot L)$. Then the proof

$$\cfrac{\cfrac{\Gamma[(\alpha,\beta)]\Rightarrow\gamma}{\Gamma[\alpha\cdot\beta]\Rightarrow\gamma}\,(\cdot L)\qquad\Delta[\gamma]\Rightarrow\theta}{\Delta[\Gamma[\alpha\cdot\beta]]\Rightarrow\theta}\,(\text{Cut})$$

can be transformed into

$$\cfrac{\cfrac{\Gamma[(\alpha,\beta)]\Rightarrow\gamma\qquad\Delta[\gamma]\Rightarrow\theta}{\Delta[\Gamma[(\alpha,\beta)]]\Rightarrow\theta}\,(\text{Cut})}{\Delta[\Gamma[\alpha\cdot\beta]]\Rightarrow\theta}\,(\cdot L)$$

Thus the applications of (Cut) in the premises have lower length. Hence by induction hypothesis (II) the claim holds.

(3) α is introduced by (R_l) only. We transform the derivation by first applying (Cut) to the premise of (R_r) and $\Delta\Rightarrow\alpha$. After that we apply (R_r) to the resulting sequent. Take $(\backslash L)$ as an example to interpret this. The remaining cases can be treated similarly.

(R_r) is $(\backslash L)$, then the proof

$$\cfrac{\Pi\Rightarrow\theta\qquad\cfrac{\Gamma[\beta]\Rightarrow\gamma\qquad\Delta[\theta]\Rightarrow\alpha}{\Gamma[\Delta[\theta],\alpha\backslash\beta]\Rightarrow\gamma}\,(\backslash L)}{\Gamma[\Delta[\Pi],\alpha\backslash\beta]\Rightarrow\gamma}\,(\text{Cut})$$

can be transformed into

$$\cfrac{\Gamma[\beta]\Rightarrow\gamma\qquad\cfrac{\Pi\Rightarrow\theta\qquad\Delta[\theta]\Rightarrow\alpha}{\Delta[\Pi]\Rightarrow\alpha}\,(\text{Cut})}{\Gamma[\Delta[\Pi],\alpha\backslash\beta]\Rightarrow\gamma}\,(\backslash L)$$

Thus the new application of (Cut) has lower length of its premise. By induction hypothesis (II), the claim holds.

(4) α is introduced in both premises. We transform the derivation by applying (Cut) to the premise of (R_l) and (R_r). Take $(R_l)=(\Diamond R)$ and $(R_r)=(\Diamond L)$ as an example to interpret this. The remaining cases can be treated similarly. The proof

$$\cfrac{\cfrac{\Gamma\Rightarrow\alpha}{\circ\Gamma\Rightarrow\Diamond\alpha}\,(\Diamond R)\qquad\cfrac{\Delta[\circ\alpha]\Rightarrow\beta}{\Delta[\Diamond\alpha]\Rightarrow\beta}\,(\Diamond L)}{\Delta[\circ\Gamma]\Rightarrow\beta}\,(\text{Cut})$$

can be transformed into

$$\cfrac{\Gamma\Rightarrow\alpha\qquad\Delta[\circ\alpha]\Rightarrow\beta}{\Delta[\circ\Gamma]\Rightarrow\beta}\,(\text{Cut})$$

Thus the complexity of (Cut) formula in the first and second application of (Cut) are lower than the original one. Hence by induction hypothesis (I) the claim holds.

$\square$

Due to the cut eliminiation theorem 2.4 and the fact that the number of connectives in conclusion of any rule is smaller than the one in the premises, one have the following theorem.

Theorem 2.5 $L_\Diamond(\Phi)$ *is decidable.*

Remark 2.6 Note that in the proof of Theorem 2.4 and 2.5, Φ is not required to be tansitive.

3 Context-freeness

Now we consider the type grammars based on $L_\diamond(\Phi)$ where Φ is a set of transitve primary assumptions. A set of primary assumptions Φ is transitive if it satisfies the following conditions:

- Φ is closed under (Cut)

- $p_1, \ldots, p_n, q \Rightarrow q \notin \Phi$ for any $p_1, \ldots, p_n, q$.

Let us recall some basic definitions of type grammars. A type grammar based on a type logic TL(shortly a TL-grammar) is formally defined as a triple $\mathcal{G} = \langle \Sigma, I, D \rangle$ such that Σ is a nonempty finite alphabet, I is a map which assigns a finite set of types to each element of Σ, and D is a designated type. Usually D is an atomic type, often denoted by s. Σ, I, D are called the alphabet(lexicon), the lexical(initial) type assignment and the designated type of $\mathcal{G}$, respectively. Type grammars based on TL are referred to as TL-grammars. We consider type logics enriched with finitely many assumptions Φ. Type grammars based on TL enriched with finitely many assumptions Φ are referred to TL(Φ)-grammars.

The string of formulae obtained from a formula tree Γ by dropping all structure operations and the corresponding parentheses is called the yield of Γ and denoted as $st(\Gamma)$. A language $\mathcal{L}(\mathcal{G})$ generated by a TL(Φ)-grammars $\mathcal{G} = \langle \Sigma, I, D \rangle$ is defined as a set of strings $a_1, \ldots, a_n$, where $a_i \in \Sigma$, for $1 \leq i \leq n$ and $n \geq 1$, satisfying the following condition: there exists formulae $\alpha_1, \ldots, \alpha_n$ and a formulae tree Γ such that for all $1 \leq i \leq n$, $\langle a_i, \alpha_i \rangle \in I$, $\Phi \vdash_{TL} \Gamma \Rightarrow D$ and $st(\Gamma) = \alpha_1 \ldots \alpha_n$.

We consider $L_\diamond(\Phi)$-grammars here. In what follows we show that $L_\diamond(\Phi)$-grammars are context-free. For doing so, we first need to introduce a definition of positive and negative formula. Then we construct a $L_\diamond(\Phi)$-grammars from $L_\diamond$-grammars.

Definition 3.1 The positiveness (negativeness) of an formula α appearing in a sequent $\Gamma \Rightarrow \beta$ is defined recursively by the following rules:

$\alpha = \beta$ is positive, and $\alpha \in \Gamma$ is negative;

if $\alpha = \alpha_1 \cdot \alpha_2$ is positive(negative), then both α_1 and α_2 are positive(negative);

if $\alpha = \alpha_1 \backslash \alpha_2$ is positive(negative), then α_1 is negative(positive) and α_2 is positive(negative);

if $\alpha = \alpha_1 / \alpha_2$ is positive(negative), then α_1 is negative(positive) and α_2 is positive(negative);

if $\alpha = \Diamond \alpha_1$ is positive(negative), then α_1 is positive(negative);

if $\alpha = \blacksquare \alpha_1$ is positive(negative), then α_1 is positive(negative).

Definition 3.2 We define a substitution g for any $p_1, \ldots, p_n \Rightarrow q \in \Phi$ as replacing positive q with a formula $p_1 \cdot \ldots \cdot p_n$ which depands on its antecedent $p_1, \ldots, p_n$.

Theorem 3.3 *If* $\vdash_{L_\diamond(\Phi)} \Gamma \Rightarrow \beta$, *then* $\vdash_{L_\diamond} g(\Gamma \Rightarrow \beta)$.

Proof. For the left to the right, we proceed by induction hypothesis on the length of proof of $\Gamma \Rightarrow \beta$ in $L_\diamond(\Phi)$. For (Id) it is obvious. We consider the case for the primary assumtion i.e. $\Gamma \Rightarrow \beta$ is $p_1, ..., p_n \Rightarrow q$. Then $g(\Gamma \Rightarrow \beta)$ is $p_1, ..., p_n \Rightarrow p_1, ..., p_n$, it obviously holds in $L_\diamond$. Assume that $\Gamma \Rightarrow \beta$ is obtained by rule (R).

(1) (R)=($\cdot$L). Assume $\vdash_{L_\diamond(\Phi)} \Gamma[(\alpha, \beta)] \Rightarrow \gamma$, then by induction hypothesis we have $\vdash_{L_\diamond} g(\Gamma[(\alpha, \beta)] \Rightarrow \gamma)$. By Definition 3.2, for any positive q in $\Gamma[(\alpha, \beta)] \Rightarrow \gamma$ with such $p_1, ..., p_n \Rightarrow q \in \Phi$, we do the substitution. Then we apply ($\cdot$L) to the result. Hereafter we obtain $\vdash_{L_\diamond} g(\Gamma[(\alpha \cdot \beta)] \Rightarrow \gamma)$.

(2) (R)=($\backslash$L). Assume $\vdash_{L_\diamond(\Phi)} \Gamma[\beta] \Rightarrow \gamma$ and $\vdash_{L_\diamond(\Phi)} \Delta \Rightarrow \alpha$, then by induction hypothesis we have $\vdash_{L_\diamond} g(\Gamma[\beta] \Rightarrow \gamma)$ and $\vdash_{L_\diamond} g(\Delta \Rightarrow \alpha)$. By Definition 3.2, for any positive q in $\Gamma[\beta] \Rightarrow \gamma$ and $\Delta \Rightarrow \alpha$ with such $p_1, ..., p_n \Rightarrow q \in \Phi$, we do the substitution. Then we apply ($\backslash$L) to the result. Hereafter we obtain $\vdash_{L_\diamond} g(\Gamma[(\Delta, \alpha\backslash\beta)] \Rightarrow \gamma)$.

(3) (R)=($\diamond$R). Assume the premise is $\vdash_{L_\diamond(\Phi)} \Gamma \Rightarrow \alpha$. Then by induction hypothesis, we have $\vdash_{L_\diamond} g(\Gamma \Rightarrow \alpha)$. By Definition 3.2, for any positive q in $\Gamma \Rightarrow \alpha$ with such $p_1, ..., p_n \Rightarrow q \in \Phi$, we do the substitution. After that we apply ($\diamond$R) to the result. Thus we obtain $\vdash_{L_\diamond} g(\circ\Gamma \Rightarrow \diamond\alpha)$. The remaining cases can be treated similarly.

$\square$

Given a $L_\diamond(\Phi)$-grammar $\mathcal{G} = \langle \Sigma, I, D \rangle$. One construct a $L_\diamond$-grammar $\mathcal{G} = \langle \Sigma', I', \Lambda \rangle$ as follows:

- $\Sigma = \Sigma'$
- If $\alpha \rightharpoonup a \in I$, then $\alpha_i \rightharpoonup a \in I'$ where α_i is obtained from replacing each positive appearing of q_i by $p_1^i, ..., p_m^i$ in α, where $p_1^i, ..., p_m^i \Rightarrow q_i \in \Phi$.
- Λ is the set of all formula D_i such that D_i is obtained from replacing each positive appearing of q_i by $p_1^i, ..., p_m^i$ in α, where $p_1^i, ..., p_m^i \Rightarrow q_i \in \Phi$.

Theorem 3.4 *If in a* $L_\diamond(\Phi)$*-grammar,* Φ *is finite, then there exists a* $L_\diamond$*-grammar generates the same language.*

Theorem 3.5 *(Kanazawa)* $L_\diamond$*-grammars are context-free.*

Theorem 3.6 $L_\diamond(\Phi)$*-grammars are context-free.*

4   Conclusion

In the present paper, we show the categorial grammars based on modal Lambek calculus with transitive primary assumptions are context-free. This result depands on the cut free sequent system discussing in this paper. Since the generative capacity of Lambek calculus with simply assumption $p/q \Rightarrow q$ goes beyond the context-free languages, it is natrual to guess that the generative capacity of Lambek calculus with primy assumptions may beyond the context-free languages. A interesting further work is to find a concrete examlpe of such kind of categorial grammars which is not context-free.

References

[1] Bulińska, M., *The pentus theorem for lambek calculus with simple nonlogical axioms*, Studia Logica **81** (2005), pp. 43–59.

[2] Buszkowski, W., *Some decision problems in the theory of syntactic categories*, Mathematical Logic Quarterly **28** (1982), pp. 539–548.

[3] Buszkowski, W., *Lambek calculus with nonlogical axioms*, Language and Grammar. Studies in Mathematical Linguistics and Natural Language (2005), pp. 77–93.

[4] Buszkowski, W. and Z. PAWLAK, *Generative capacity of nonassociative lambek calculus*, Bulletin of the Polish Academy of Sciences. Mathematics **34** (1986), pp. 507–516.

[5] Dudakov, S. M., B. N. Karlov, S. L. Kuznetsov and E. M. Fofanova, *Complexity of lambek calculi with modalities and of total derivability in grammars*, Algebra and Logic **60** (2021), pp. 308–326.

[6] Foret, A., *On associative lambek calculus extended with basic proper axioms*, Categories and Types in Logic, Language, and Physics: Essays Dedicated to Jim Lambek on the Occasion of His 90th Birthday (2014), pp. 172–187.

[7] Jäger, G., *On the generative capacity of multi-modal categorial grammars*, Research on Language and Computation **1** (2003), pp. 105–125.

[8] Lambek, J., *The mathematics of sentence structure*, The American Mathematical Monthly **65** (1958), pp. 154–170.

[9] Pentus, M., *Lambek grammars are context free*, in: *[1993] Proceedings Eighth Annual IEEE Symposium on Logic in Computer Science*, IEEE, 1993, pp. 429–433.

Part II

LAIL2023

Computing Logic for Initial Probabilities of Bayesian Artificial Intelligence

WENJING DU

Wenbo College, East China University of Political Science and Law,
Shanghai 201620, China.
E-mail: 2178@ecupl.edu.cn

ZIHAN NIU

Institute of Logic and Cognition, Sun Yat-sen University,
Guangzhou 510275, China.
E-mail: niuzh@mail2.sysu.edu.cn

Abstract

With the iterative updates of technology, Bayesian artificial intelligence which is based on probability inference has mighty computational power and can handle a large number of complex uncertain inference problems. Bayesian artificial intelligence uses Bayesian networks as carriers and Bayesian formula as the foundation to visualize and intellectualize probability inference. However, when using Bayesian networks to model specific application domains, "how to calculate initial probabilities" is the biggest challenge. In this paper, we propose that a calculation model needs to be constructed from three dimensions to ensure the accuracy of initial probabilities, namely logical requirements, calculation methods, and rational standards. Logically, initial probabilities is necessary to meet the consistency requirements of probability axioms, and probability rules provide the basis for probability inference. Methodologically, probability interpretations are used to integrate probability information and figure out the numbers of initial probabilities. Rationally, the two reliable standards, that are transparency and consensus, should be required, and argumentation provides a means of proof for reliability. Only through the cooperation and coupling of the three can consistent and accurate initial probability values be obtained, thus ensuring the reliability of the outputs of Bayesian artificial intelligence and providing bases and suggestions for decision-making.

Keywords: Bayesian artificial intelligence, Bayesian formula, initial probabilities, probability interpretations, probability calculation.

1   Introduction: The origin of the problem

The evolution of artificial intelligence has formed two representative paths in the past 60 years: the first path is knowledge-driven symbolism, which constructs artificial intelligence with three elements: knowledge, algorithms, and computing power; the second path is data-driven connectionism, which also constructs artificial intelligence with three elements: data, algorithms, and computing power. These two paths only simulate human intelligent behavior from a particular perspective, and thus have their limitations. However, Bayesian artificial intelligence is driven by both knowledge and data, and constructs artificial intelligence with four elements: knowledge, data, algorithms, and computing power. It belongs to the third generation of artificial intelligence and has stronger robustness and interpretability. Bayesian artificial intelligence [11] uses Bayesian networks to represent knowledge and data, and uses Bayesian formula as a guide for knowledge inference, making probability inference visualized and intellectualized. Bayesian networks were first proposed by Pearl [14] at the University of California in the 1980s. To solve the dilemma of uncertain knowledge representation and inference in symbolism and connectionism, inspired by cognitive science and bionics Pearl combined probability inference with graph theory to form Bayesian networks, which simulate the operating mechanism of neurons in the human brain and become a commonly used method for knowledge inference. The reason why Bayesian networks can model any real-world problems involving uncertainty lies in the probability foundation behind them, i.e., the probability theory based on Bayesian formula. Bayesian probability theory is the most promising formal theory for dealing with uncertain information.

Bayesian formula is the core of the probability theory and the origin of Bayesian philosophy. It refers to the posterior probability $P(H|E)$ of a proposition H conditional on evidence E, which equals the product of the likelihood $P(E|H)$ of the evidence and the prior probability $P(H)$ of the proposition, and then divided by the probability $P(E)$ of the evidence. It can be expressed in the formula as:

$$p(H|E) = \frac{P(E|H)P(H)}{P(E)},$$

where $P(E)$ can be calculated using the law of total probability, which is $P(E) = P(E|H) \times P(H) + P(E|\neg H)P(\neg H)$, here $\neg H$ represents the logical negation of the proposition H. The prior probability and likelihood are both initial probabilities that respectively measure the possibility of proposition H and the occurrence of evidence E before the evidence is received. The posterior probability, as the final probability desired, serves as the basis for decision-making and is the target output of the Bayesian network.

Modeling a specific application domain using a Bayesian network typically involves three steps: first, identifying the key variables and their possible states in the domain; second, clarifying the structural relationships between these variables and representing them graphically; and third, calculating the probability distribution of each variable. The first two steps aim to construct the property of the Bayesian network graph, which is feasible with collaboration and communication with relevant experts despite the operational difficulty. The third step aims to construct the quantity part of the Bayesian network, which is often more challenging. "Where do the initial probability numbers come from?" is a commonly asked puzzle and also the most significant challenge faced by Bayesian artificial intelligence. Without initial probabilities as numerical inputs to the Bayesian network, posterior probabilities cannot be calculated, and hence, advice for decision-making cannot be provided. Of course, any field involving probabilistic reasoning faces the challenge of initial probabilities. This is because there is "no 'true' or 'uniquely/objectively right' probability," and "probability is a very liberal concept in that it does not tell one what their probability should be" [1, p. 13]. Given a situation, probability logic does not provide any standard for determining initial probability; it only provides rules from initial probability to posterior probability [8]. Therefore, no "precise" theory or method for universally calculating initial probabilities exists. Our goal is not to theoretically solve the "problem of initial probability numbers", but to provide method guidelines and evaluation criteria for decision-makers to calculate initial probabilities practically, thereby providing an important guarantee for the wide application of Bayesian artificial intelligence.

From the perspective of practical decision-making, the calculation of initial probabilities as a decision-making behavior depends on the background information [1] that decision-makers have mastered. Rational decision-makers will try their best to ensure that the calculated initial probabilities truly reflect the content expressed by background information and meet their value pursuits, such as truth-seeking, fairness, and justice. However, real people are not entirely rational or omniscient like God. Therefore, the calculated initial probabilities are "imprecise" but can be accurate. This requires decision-makers to shoulder the obligation of calculating initial probabilities and provide evidence for the calculation process while also to bear the risks and responsibilities of inaccurate initial probabilities. How to calculate accurately initial probabilities? To solve this problem, this paper proposes three suggestions. First, based on the requirement of logical consistency, the initial probabilities must satisfy the basic restrictions of probability axioms. Second, at the methodological level, probability interpretations should be used as bases when processing and integrating various probability information and calculating probability values. Finally, the entire calculation process should meet the reliable standards of "transparency" and "consensus," and in order to improve the accuracy and the acceptability of numerical calculation, we advocate argumentation method as the guarantee for justification.

2 Logical requirements: consistency of probability calculations

The first step in calculating initial probability is to ensure no conflict about probabilities of all uncertain events in the same uncertain phenomenon. However, the basis of probability calculation lies in various interpretations of the

---

[1] Background information includes statistical data, literature, expert knowledge, general knowledge about the world, and the empirical knowledge of decision makers.

concept of probability. These interpretations characterize the meaning of probability from different perspectives, which have different advantages and disadvantages respectively. Despite the rapid development of probability theory and its applications, there is still ongoing philosophical debate about what probability is and how to interpret it. No complete consensus has been reached so far.

Probability has a peculiar dual meaning since its birth. Pascal used probability to describe random events in chance games and also evaluated propositions such as "the existence of God" with probability in his famous gambling argument. As described by Hacking, probability is "Janus faced. On the one side it is statistical, concerning itself with the stochastic laws of chance processes. On the other side it is epistemological, dedicated to assessing the reasonable degrees of belief in propositions quite devoid of statistical background" [7, p. 12]. Based on the inherent duality of probability, it can be divided into statistical probability and epistemic probability. Statistical probability, also known as objective probability, mainly includes the frequency interpretation and propensity interpretation. Epistemic probability, also known as subjective probability, mainly includes classical interpretation, personal subjective interpretation, inter-subjective interpretation, and logical interpretation. These interpretations of probability depict the meaning of probability from different perspectives, each with its advantages and limitations. The debate over probability interpretations exists because some scholars insist that objective probability and subjective probability compete each other and are mutually exclusive. This debate can be attributed to the philosophical questioning of "what is probability," and the result of this debate will not have a winner or loser, nor will it reach a cognitive consensus, because the boundaries of various interpretations of probability are inherently fuzzy and can even permeate each other. Specifically, objective probability contains subjective assumptions of reference class selection, while subjective probability entails objective data foundations. According to this, instead of engaging in philosophical debates about objectivity and subjectivity, we advocate that probability interpretations should be viewed as bases for calculating probability under different information conditions.

In fact, probability, as a measurement tool for the uncertainty of unknown events, must satisfy two necessary conditions. First, for all possible events of the same uncertain phenomenon, probability calculation must be consistent and cannot be contradictory. Second, when new evidence occurs, the probability of the uncertain event can be reasonably revised or updated [5, p. 69]. Based on these two conditions, in 1933, Kolmogorov separated the interpretation of probability from its mathematical properties, discarded the philosophical debates on probability interpretation, and proposed a mathematical axiomatic definition of probability. This definition can be regarded as a mathematical response to the question of "what is probability." In his view, each specific uncertain phenomenon can be described as an experiment, which produces various outcomes, and the set of all these outcomes is called the sample space (denoted by Ω). Elements of the sample space is also called sample points. Any subset of the sample space is called an event, and the set consisting of single sample point is called a basic event. Basic events are mutually exclusive; that is, they cannot occur simultaneously. The set of all subsets of the sample space is called the event field (denoted by F), so each element in the event field corresponds to some uncertain event. Kolmogorov defined probability as a set function P from the event field F to the real interval $[0, 1]$ satisfying the following three axioms [5, p. 90-92]:

Axiom 1 (Non-negativity): The probability $P(E)$ of any event E in F is a non-negative number, that is, $P(E) \geq 0$.

Axiom 2 (Normalization): The probability of the entire sample space is 1, that is, $P(\Omega) = 1$

Axiom 3 (Additivity): For mutually exclusive events $E_1, E_2, E_3, \ldots$, the probability of at least one of them occurring is equal to the sum of their probabilities, that is, $P(E_1 \cup E_2 \cup E_3 \cup \ldots) = P(E_1) + P(E_2) + P(E_3) + \ldots$

The probability distribution of an uncertain phenomenon is formed by a numerical assignment function that satisfies the three axioms. Depending on the type of outcomes, uncertain phenomena can be divided into two types: discrete and continuous. The sample space of a discrete uncertain phenomenon can be divided into finite and infinite discrete sample spaces according to the number of basic events it contains, and their corresponding event fields are finite and infinite sets, respectively. The corresponding distributions are also called finite discrete probability distributions and infinite discrete probability distributions, respectively. For example, considering the uncertain phenomenon of "tomorrow's weather", assuming that there are three interesting outcomes: sunny, cloudy, and rainy, then the sample space $\Omega=$ {sunny, cloudy, rainy}, which is a finite discrete sample space. The event field $F =$ {ϕ, {sunny}, {cloudy}, {rainy}, {sunny, cloudy} , {sunny, rainy },{cloudy, rainy}, Ω}. F has eight events, among which {sunny}, {cloudy}, and {rainy} are the three basic events. The empty set ϕ is called an impossible event, and the sample space Ω itself is called a certain event. Now, the following eight numerical assignments are made to these events: $P\{$sunny$\}=$ 0.5, $P\{$cloudy$\}=$ 0.3, $P\{$rainy$\}=$ 0.2, $P\{$sunny, cloudy$\}=$ 0.8, $P\{$sunny, rainy$\}=$ 0.7, $P\{$cloudy, rainy$\}=$ 0.5, $P(\Omega)=$ 1, $P(\phi) = 0$. It is mathematically easy to verify that the assignment function satisfies the requirements of the three probability axioms, thereby forming the probability distribution of "tomorrow's weather", which is a finite discrete probability distribution. However, in practice in order to reduce the number of required probabilities and avoid inconsistent situations when constructing the probability distribution of a finite discrete uncertain phenomenon, it is only necessary to assign probability values to basic events according to the non-negativity and normalization axioms. Probabilities of the other events can be calculated based on the three axioms. In the above example, only the first three probability values need to be determined: $P\{\text{sunny}\}= 0.5$, $P\{\text{cloudy}\}= 0.3$, $P\{\text{rainy}\}= 0.2$, and the remaining five probability can be calculated based on the axioms. It should be noted that in order to satisfy the requirements of the three axioms, people should consider the probabilities of all basic events "simultaneously" and not calculate the probability of each basic event separately in isolation; otherwise, it may lead to inconsistent results. If the probabilities of sunny, cloudy, and rainy are calculated separately at different times, then inconsistent numerical assignments may occur, for example, $P\{\text{sunny}\}= 0.4$, $P\{\text{cloudy}\}= 0.5$, $P\{\text{rainy}\}= 0.3$.

In addition to assigning probability values directly to basic events, predefined probability distributions in mathematics can be used to construct distributions for uncertain phenomena. Examples of known finite discrete probability distributions include the two-point distribution and the binomial distribution, while known infinite discrete probability distributions include the Poisson distribution and the geometric distribution. Each known probability distribution is a model for computing probabilities and has been mathematically proven to satisfy three probability axioms. Depending on the characteristics of the uncertain phenomenon, an appropriate known probability distribution that can describe those characteristics is selected as the probability distribution for that uncertain phenomenon. For example, the Poisson distribution is often used to describe the number of machine failures.

For continuous uncertain phenomena, both the sample space and event domain are infinite sets. Unlike discrete uncertain phenomena, probability distributions for continuous uncertain phenomena are typically described by density functions. Mathematics has pre-defined many density functions that satisfy the logical requirements of the three axioms, such as uniform distribution, exponential distribution, and normal distribution. Each density function is a mathematical model for calculating probabilities. An appropriate density function that can describe the characteristics of the uncertain phenomenon is selected to construct its probability distribution. For example, a normal distribution is often used to describe human height.

In fact, probability theory has a fourth axiom, called the "conditional probability" rule, which is usually expressed as a definition for calculating conditional probabilities in mathematics. The conditional probability rule and Bayes' theorem can be derived from each other and are equivalent, so Bayesianists also consider the latter as the fourth axiom of the probability system [5, p. 110, 135]. The three probability axioms, combined with the conditional probability rule, can be used to derive probability complementary formula, Bayes' formula, probability addition formula, multiplication formula, total probability formula, and other probability rules. They collectively constitute the entire formal system of probability theory and play different roles in the calculation process. The probability axioms require logical consistency for probability calculations. In contrast, the probability rules provide rules and guarantees for further calculations of other probabilities, with Bayes' theorem being the only rational rule for probability revision or updating. The formal system of probability theory is a logically consistent and robust system that provides a solid logical framework for uncertain reasoning.

As a quantitative attribute, probability has to satisfy the logical consistency requirements of probability axioms. All probability distributions must satisfy probability axioms. Prior probability distributions are relative to posterior probability distributions, both must satisfy probability axioms.

It is worth noting that besides satisfying the consistency restrictions of the three probability axioms, the initial probability is free. The formal system of probability theory does not provide any other requirements or instructions for computing the initial probability, nor can it provide specific methods for calculating the initial probability. Therefore, the formal system of probability theory neither specifies the method of calculating the initial probability nor guarantees its reliability. It only accepts the initial probability as input and produces logically consistent output based on it. The rationality of the output depends on the reliability of the initial probability.

3 Method Empowerment: Feasibility of Probability Calculation

Formal probability theory only imposes consistency requirements on the initial probabilities, that is satisfying the three probability axioms, but it does not restrict the methods used to calculate them. Therefore, the next problem is how to compute the initial probabilities at the operational level. In order to obtain methodological guidance, we need to turn to probability interpretations. Unlike the probability axioms, which remain silent on computational methods, the important role of probability interpretations is to provide methods for calculating initial probabilities from different perspectives by incorporating specific background information of practical problems. However, as mentioned earlier, there are multiple versions of probability interpretations, and it is up to decision-makers to consider which interpretation to use for probability calculation.

Probability interpretations should be analyzed independently from two perspectives. One is the probability concept perspective, which provides one answer to "what is probability". The other is the probability calculation perspective, which provides a method for calculating probability. The classical interpretation does not define what probability is but essentially provides a method for computing probability under the conditions of "equiprobability" and "finiteness". The frequency interpretation gives a definition of the probability concept, stating that probability is the limit of infinite frequency sequences, and provides a method for calculating probability by counting frequencies. The propensity and logical interpretations focus on analyzing the probability concept but provide little practical guidance for computing initial probabilities. The subjective interpretation defines the probability as a degree of personal subjective belief, and while it does not provide a clear method for computing probability, it acknowledges multiple ways of computing probability.

Probability concept and probability calculation are two different things, and it is necessary to separate and treat them differently in probability interpretations. The main reason for the controversy surrounding probability interpretations is to lump probability concept and probability calculation together. From the perspective of probability concept, there can only be one definition of probability, so various interpretations are mutually exclusive, which can lead to disputes. However, from the perspective of probability calculation, the methods for calculating probability can be diverse, and probability interpretations provide methods for calculating initial probabilities from different dimensions by incorporating specific background information of practical problems, thus avoiding disputes. On the contrary, at the level of probability calculation, the subjective interpretation accepts multiple methods of probability calculation, so it is an extension of other interpretations. The subjective interpretation broadens the scope of probability application. The critical bottleneck for the application of probability is not the probability concept but the probability calculation. Currently, the focus of this paper is on probability calculation, specifically on the basis for calculating initial probabilities. We will examine the applicability and conditions of the classical interpretation, frequency interpretation, and subjective interpretation in order to provide practical guidance for decision-makers to make rational choices.

3.1   Classical Interpretation Method

The core of the classical interpretation is to provide an epistemological approach to probability calculation under the conditions of finiteness and equiprobability. The classical interpretation is usually attributed to Laplace, who argued that probability is "relative in part to [...] ignorance and in part to our knowledge" [12, p. 2]. In other words, probability is always based on available knowledge, although it may be difficult to obtain all knowledge about the world. Probability can be used to expand the scope of cognition. In this sense, probability provides a rational tool for understanding the world and grasping its regularities, whether in science or in everyday life. Therefore, the classical interpretation is regarded as an epistemological probability that assigns probabilities in the absence of any evidence or the presence of symmetric evidence.

Based on this, Laplace proposed the "principle of indifference," which assumes that the probability of each basic event of an uncertain phenomenon is the same, and the initial probability is calculated accordingly. In this process, the classical interpretation has two underlying assumptions: finiteness and equiprobability. According to the available knowledge, it is assumed that the uncertain phenomenon only includes a finite number of basic events. Based on the principle of indifference, it is assumed that each basic event has an equal probability of occurrence. Therefore, for any event A in an uncertain phenomenon, its probability calculation formula is: the probability of event A = the number of basic events contained in event A/the total number of basic events in the sample space.

It is mathematically easy to verify that the classical probability calculation formula satisfies the three axioms, thus meeting the requirement of consistency.

The classical interpretation is an idealized model for calculating initial probabilities. Many uncertain phenomena in reality, as long as they meet or satisfy the characteristics of "finiteness" and "equiprobability," can use the classical interpretation to calculate initial probabilities.

However, the two underlying assumptions of the classical interpretation limit its applicability. Firstly, equiprobability indicates that the classical interpretation is an idealized mathematical model that is difficult to meet the equiprobability condition in real-life situations precisely. However, from a practical standpoint, assuming equiprobability is acceptable if insufficient evidence suggests that the events being examined do not have equiprobability. Therefore, the classical interpretation is suitable for situations where there is no evidence or where there is symmetrically balanced evidence. For example, when tossing a coin, it is generally assumed that the probability of head and tail is equiprobable unless there is evidence that the coin is biased. Some scholars argue that equiprobability is a circular definition because it actually refers to "equal probability." It should be noted that we do not regard the classical interpretation as a definition of probability but as a method for calculating probability, so there is no problem of the circular definition. Secondly, finiteness indicates that classical probability is only suitable for describing uncertain phenomena with a finite number of basic events. When there are an infinite number of basic events, classical interpretation is powerless. Therefore, when applying classical interpretation to calculate initial probabilities, decision-makers must transparently disclose the reasons for the assumptions of equiprobability and finiteness and provide evidence to support them.

3.2 Frequency Interpretation Method

The core of the frequency interpretation is to provide an ontological approach to probability calculation under the condition that objective attribute and empirical estimation are integrated into each other. The frequency interpretation was originated by John Venn and later modified by Richard von Mises. There are multiple versions of the frequency interpretation, among which the most accepted version by statisticians advocates that probability is the limit of an infinite frequency sequence under specific stochastic assumptions. Empirical evidence indicates that frequency has a significant statistical regularity; namely, the frequency of an uncertain event always oscillates around a fixed constant, which is called "frequency stability." If the same object is measured multiple times, although the results may differ slightly, a certain regularity becomes more apparent as the number of measurements increases: each measurement value fluctuates around a constant and exhibits some symmetry. Moreover, Bernoulli's law of large numbers proves that the limit value of frequency is equal to the probability value. Frequency stability indicates that the magnitude of the possibility of an uncertain event is an inherent objective property that is not subject to volitional changes, and the constant toward which frequency tends is the magnitude of the possibility of the uncertain event occurring. Therefore, frequency interpretation is an objective probability. Based on the

ontological perspective, frequentism claims that each uncertain event has a u-
nique and correct probability, which can be obtained through the limit of an
infinite frequency sequence.

However, this idea, although perfect ideally, is impractical in operational
terms because an infinite sequence cannot be obtained. This is the biggest
challenge faced by the frequency interpretation. Fortunately, Von Mises argues
that the limit of an infinite frequency sequence is an ideal model of probability,
similar to the limits of velocity and density in science, and can be perceived
with finite experience. He believes that "the results of a theory based on the
notion of the infinite collective can be applied to finite sequences of observations
in a way which is not logically definable, but is nevertheless sufficiently exact
in practice" [16, p. 85]. Therefore, the method of calculating probability based
on the frequency interpretation in practical applications is firstly to select an
appropriate and finite reference class, such as a reference class composed of N
trial outcomes or observations. Secondly, count the number of times n that
the event A of interest occurs. Thirdly, calculate the frequency n/N, and this
frequency value is approximately regarded as the probability value of event A
occurring. This method of calculating probability is called finite frequencism.
It is easy to verify from a mathematical perspective that this method satisfies
the three probability axioms and meets the requirement of consistency.

The charm of the frequency interpretation lies in the integration of objectivity and empiricism. The frequency interpretation advocates that uncertain
events objectively exist with correct probabilities, but they are unknown to
humans. However, these right probabilities can be empirically estimated using observed frequencies in samples. Although frequency interpretation has
become the mainstream interpretation in the scientific field, it also has the following problems. Firstly, the limit of an infinite frequency sequence cannot be
obtained in practice. Even if assuming the existence of an infinite frequency
sequence through counterfactuals, it involves the "infinitely repeatable" and
"independence" assumptions, which are difficult to satisfy in practice. Therefore, the limit of an infinite frequency sequence is only an idealized model of
probability without practical operability, and probability calculation can only
be estimated through finite frequency. Secondly, finite frequency is not truly
equal to probability. It is only an approximate value of probability. The finite
frequency is based on the actual outcomes to calculate probability, and the error of this approximation can sometimes be significant. For example, suppose
a "fair" coin is tossed, according to the classical interpretation or the limit of
an infinite frequency sequence by counterfactuals, it is generally accepted that
the probability of "head" is 0.5. However, suppose the coin is tossed ten times
in practice. In that case, the result of "nine heads and one tail" may occur, and
according to finite frequencism, the probability of "head" is 0.9, which is significantly different from 0.5. Therefore, as an appropriate understanding of finite
frequency, it should be considered as evidence for reflecting probability, and
as "non-deterministic" evidence. Finally, calculating initial probabilities based
on the frequency interpretation must depend on a specific reference class, but

there may be more than one feasible reference class. The frequency interpretation faces the problem of reference class selection. Therefore, when estimating probability through frequency, decision-makers must transparently disclose the basis and reasons for selecting a reference class and demonstrate the selected reference class in communication and discussion so that it can be accepted and recognized by others.

3.3   Subjective Interpretation Method

The core of the subjective interpretation method is to provide an epistemological approach to probability calculation under the condition that objective and subjective elements are integrated. Subjective interpretation, also known as Bayesian interpretation, defines probability as the degree of belief of an appropriate subject under given information. Based on this, what is an appropriate subject? A satisfactory answer is that an appropriate subject must be rational. Ramsey pointed out that a rational subject should be logically consistent; that is, the calculation of subjective probability requires the subject to follow the consistency requirement of probability axioms. Following Ramsey's approach, subjective interpretation reaches its peak in de Finetti's theory, where he combines the Bayesian rule with exchangeability, promoting the convergence of belief and observed frequency. This is the famous "de Finetti representation theorem," which ensures the applicability of subjective probability in statistical inference.

Calculating subjective probability not only requires meeting the consistency requirement of probability axioms but also considering all available information, including frequency and equiprobability. "Every probability evaluation essentially depends on two components: (1) the objective component, consisting of the evidence of known data and facts; and (2) the subjective component, consisting of the opinion concerning unknown facts based on known evidence." [3, p. 7] Since collecting factual evidence involves several subjective factors, subjective factors are considered as a prerequisite for evaluating objective factors. In addition, in many cases, the decision-makers' ability and experience will affect probability judgments in various ways. Therefore, how to calculate subjective probability depends on what conditional information the decision-maker obtains and how she utilizes this information. Based on the different types and sources of information, subjective probability can be calculated in the following four situations.

Firstly, if the acquired information can provide reasons and support for the assumption of "equiprobability", then decision-makers can calculate probabilities based on the classical interpretation or the uniform distribution. It is worth noting that classical probability applies to finite equally likely events, while the uniform distribution applies to infinite equally likely situations. In this sense, the uniform distribution is an expansion of classical probability.

Secondly, if the acquired information is frequency data, decision-makers can calculate probabilities based on the frequency method.

Thirdly, if the acquired information is empirical knowledge or expert knowledge, decision-makers can estimate probabilities based on empirical perception or intuition after considering this knowledge.

Fourthly, if there is no information available to estimate the probability, then according to the maximum entropy principle, the assumption of "equiprobability" can be made, and probabilities can be calculated based on the classical interpretation or the uniform distribution. The maximum entropy principle, proposed by Jaynes, states that "...in making inferences on the basis of partial information we must use that probability distribution which has maximum entropy subject to whatever is known. This is the only unbiased assignment we can make; to use any other would amount to an arbitrary assumption of information which by hypothesis we do not have." [9, p. 623] The so-called unbiased distribution is the best probability distribution and the optimal mathematical model for calculating probabilities. It can be seen that the maximum entropy principle not only provides an optimal strategy for probability calculation but also solves the problem of assigning initial probability values in the absence of information. Its advantage is that it can maximize the elimination of subjective interference in the absence of any information and does not favor anyone in order to obtain the fairest probability value.

It should be noted that subjective interpretation claims that there does not exist a unique correct or objective probability. "the subjective theory [...] does not contend that the opinions about probability are uniquely determined and justifiable. Probability does not correspond to a self-proclaimed 'rational' belief, but to the effective personal belief of anyone." [2, p. 218] This indicates that probability is primarily a personal subjective belief, and even under the same information, probabilities calculated by different subjects may differ. Therefore, in order to improve the accuracy of probability calculation, calculators should transparently disclose the calculation process and provide evidence for it so that the obtained probability can achieve consensus and acceptability in communication and discussion.

4   Rational Standards: Reliability of Probability Calculation

Probability logic requires consistency in initial probabilities and provides logical rules for calculating further probabilities based on initial probabilities. However, it does not provide a method for calculating initial probabilities. Probability interpretation provides a method for calculating initial probabilities but does not guarantee the reliability of the calculation process and results. To ensure accuracy and acceptability of initial probabilities, the calculation process should be transparent and consensual. Argumentation can be applied to justify the calculation results.

Firstly, the reliability of initial probabilities must meet the standards of transparency and consensus. Initial probabilities are not right or wrong but can only be evaluated as accurate or inaccurate, which makes the calculation of initial probabilities a decision-making process. Decision-makers must bear the consequences and risks of inaccurate probability calculations. To improve the accuracy of initial probability calculations, decision-makers should not be limited to a single probability interpretation but adopt a diversified perspective and select a suitable probability interpretation based on the probability information they possess. Different decision-makers may have different probability information for the same uncertain phenomenon, resulting in different initial probabilities. However, decision-makers' initial probabilities must truly reflect the probability information they possess, and they must provide justification for the calculation results. "If we cannot require everybody sharing the same likelihoods, we can require everybody having justified likelihoods" [6, p. 1506]. Decision-makers should "take responsibility for their quantifications of uncertainty" and make "the best use of all relevant available data, knowledge and information in a way that can be disclosed and audited" [1, p. 55-57]. What calculation method is disclosed and audited? We advocate that transparency and consensus serve as a basis and guideline for auditing the process and results of initial probability calculations.

On the one hand, uphold openness and transparency. The transparency standard is an important safeguard for probabilistic calculations, primarily because it requires all the choices and considerations involved in the initial probability calculation process to be made public and transparent. First, decision-makers should disclose the sources of data, knowledge, and information and clarify the reasons for selecting these sources. They also should fully consider relevant competitive theories and viewpoints, and eliminate biases and ad hoc interpretations of results. Secondly, the concepts and methods used in the calculation process should have clear and explicit definitions and follow generally accepted rules and procedures. For example, probability distributions should meet the requirements of the three probability axioms of Kolmogorov. Finally, decision-makers should disclose the reasoning process of the calculation and make public the assumptions and potential limitations involved, including those hypotheses that may not be verifiable. They should keep an open attitude towards criticism and communication. In any case, it is essential for decision-makers to explain clearly how they arrived at the initial probability values, so that researchers can decide to what extent they support the probability results.

On the other hand, follow cognitive consensus. The accuracy of the outputs of Bayesian artificial intelligence depends on the reliability of the inputted initial probability. Therefore, decision-makers need to negotiate and reach a consensus when communicating and discussing initial probability calculations. Only by calculating initial probabilities in a transparent and open manner can decision-makers improve their calculation results through communication and discussion with others, ultimately gaining recognition and reaching a consensus. Transparency contributes to achieving consensus. When examining uncertain phenomena, reality and facts can only be understood through personal experience. Different people bring different information and viewpoints, and they also use probability information such as data and knowledge in different ways. Therefore, communication and discussion among decision-makers are necessary to ensure that the probability values are accurate and stable. In communication and discussion, consensus cannot be enforced; as a reliability indicator, it refers to promoting consensus.

Secondly, argumentation provides a means of justification for the reliability of initial probabilities. The rational aim of probability calculation is to achieve stable consensus through explicit or implicit free communication. Transparency and consensus provide specific guidelines for achieving this goal. People no longer need to argue whether initial probabilities are objective or subjective, or the advantages and disadvantages of objectivity and subjectivity. It is sufficient to examine the reliability of the calculation process based on these two indicators. Only initial probabilities that withstand such scrutiny are accurate and acceptable. Transparency provides the basic condition for scrutinizing initial probabilities, and argumentation provides a means of justification. The probabilities that have been justified meet the consensus standard, and thus they are rational and acceptable.

Human reasoning and argumentation often take the form of exchanging arguments. Therefore, argumentation is an important tool for decision-makers to communicate and discuss initial probability calculations. For each initial probability value, argumentation completes the process of justification by dialectically scrutinizing the reasons for or against the value. Argument schemes [17] are the core concept of argumentation, capturing standardized reasoning patterns that can provide the normative and formal basis for justifying initial probability calculations. In 2014, Keppens [10] distilled 20 argument schemes for scrutinizing and justifying initial probability calculations. Prakken [15] also proposed developing argument support tools to support probabilistic analysis of complex criminal cases. Phan Minh Dung's abstract argumentation theory [4] can be seen as a calculus about conflicts. It solves the core problem of reasoning in inconsistent situations and has a milestone role in the field of artificial intelligence and logic.

Whether based on statistical data or expert knowledge or experience, computing initial probabilities involves the subjective judgments and assumptions of decision-makers. What is the rationality of such subjectivity? Argumentation provides a means to justify rationality. For example, in probability analysis of complex criminal cases, decision-makers often need to determine the random matching probability of DNA, that is, the probability that the "defendant's DNA matches the DNA collected at the crime scene" in the case where "the defendant did not appear at the crime scene." To do this, the decision-maker needs to consider a reference class F, such as selecting all citizens as the reference class and then randomly selecting a sufficiently large sample or selecting all individuals in F. Assuming that m citizens are selected, their DNA is compared pairwise with the DNA collected at the crime scene. The number of times that two DNA match, such as n times, is counted. According to the frequency interpretation method, the matching frequency can be calculated as n/m. Therefore, it is inferred that there are individuals in class F whose DNA matches the DNA collected at the crime scene at a proportion of n/m, and it is deduced that the random matching probability between the defendant and

the DNA at the crime scene is n/m. The entire calculation process involves t-
wo argument schemes: statistical induction argument and statistical frequency
argument [15, p. 43-44], which at least contain the following critical questions:

a. Are the samples (i.e., m citizens) selected from F biased? Are they large
enough?

b. Is the defendant an element of class F? Does it have properties that are
obviously different from class F?

c. Is class F an appropriate reference class? Are there other reference classes
with different frequency information?

d. Are there reasons why this frequency cannot be used?

Critical questions have three functions in ensuring the reliability of initial
probabilities. Firstly, expose the sources of suspicion that may arise from the
use of argument schemes and reveal the reasons why argument schemes are not
applicable to specific contexts. Secondly, indicate what relevant information
decision-makers should make clear and transparent in the process of probability calculation. Thirdly, serve as a basis for examing whether initial probability
calculation can be justified. If decision-makers can provide good reasons and
answer critical questions well in communication and discussion, then the probability calculation is justified and should be accepted. If decision-makers cannot
withstand the questioning of critical questions, then the probability calculation
is unreasonable, and its results cannot be accepted.

5   Conclusion

The biggest challenge for Bayesian artificial intelligence is "how to calculate
the initial probability." "What is probability?" has been widely discussed
and extensively debated in philosophy. In this paper, we do not discuss the
controversy related to the concept of probability but adopt a pragmatic perspective to discuss how to calculate accurately the initial probability from the
practical application of probability, because the critical point of probability
application does not lie in the concept of probability, but in the calculation of
probability. Statistical data, literature, and expert knowledge are the primary
sources of probabilistic information [13]. However, how to process and integrate this probabilistic information and calculate probability values relies on
probability interpretations. From the perspective of probability calculation, the
subjective interpretation is an extension of other interpretations. It acknowledges multiple ways of calculating probability and advocates that probability
is subjects' degree of belief under obtained conditional information. When the
obtained conditional information satisfies the assumptions of "equiprobability"
and "finite", the decision-maker can use the classical interpretation method to
calculate probability. When the obtained information is frequency data, the
decision-maker can use the frequency method to calculate probability. When
the obtained conditional information is expert knowledge or experiential knowledge, the probability is estimated based on the intuition method. If there is
no information, based on the principle of maximum entropy, "equiprobability"
can be assumed, and probability can be calculated based on the uniform dis-

tribution. Regardless of which method of probability calculation is used, the initial probability must satisfy the logical consistency requirement proposed by probability axioms, and the probability rules provide a rule guarantee for calculating further probability. In order to improve the accuracy and acceptability of initial probabilities, the probability calculation process needs to meet the reliability standards of transparency and consensus, and argumentation provides a means of justification for reliability. Only the division of labor and cooperation among logical requirements, calculation methods, and rational standards can obtain widely recognized initial probabilities, provide accurate inputs for Bayesian artificial intelligence, ensure the reliability of its output, and expand the application boundary of Bayesian artificial intelligence.

References

[1] Biedermann, A. and J. Vuille, *The decisional nature of probability and plausibility assessments in juridical evidence and proof*, International Commentary on Evidence **16** (2018), p. 20190003.

[2] de Finetti, B., *Recent suggestions for the reconciliation of theories of probability*, in: J. Neyman, editor, *Proceedings of the Second Berkeley Symposium on Mathematical Statistics and Probability* (1951), pp. 93–128.

[3] de Finetti, B., *The value of studying subjective evaluations of probability*, in: C.-A. Stal von Holstein, editor, *The Concept of Probability in Psychological Experiments*, Reidel, 1974 pp. 99–130.

[4] Dung, P., *On the acceptability of arguments and its fundamental role in nonmonotonic reasoning, logic programming, and n-person games*, Artificial Intelligence **77** (1995).

[5] Fenton, N. and M. Neil, "Risk Assessment and Decision Analysis with Bayesian Networks," CRC Press, 2019, 2 edition.

[6] Garbolino, P., *Explaining relevance*, Cardozo Law Review **22** (2001).

[7] Hacking, I., "The Emergence of Probability," Cambridge University Press, Cambridge, UK, 1975.

[8] Howson, C., *Popper, prior probabilities, and inductive inference*, The British Journal for the Philosophy of Science **38** (1987), pp. 207–224.

[9] Jaynes, E. T., *Information theory and statistical mechanics*, Physical Review **106** (1957).

[10] Keppens, J., *On modelling non-probabilistic uncertainty in the likelihood ratio approach to evidential reasoning*, Artificial Intelligence and Law **22** (2014).

[11] Korb, K. B. and A. E. Nicholson, "Bayesian Artificial Intelligence," Chapman & Hall/CRC Press UK, 2010, 2 edition.

[12] Laplace, P. S., "Philosophical Essay on Probabilities," Springer, New York, NY, USA, 1995.

[13] Marek, J. D. and C. G. Linda, *Building probabilistic networks: Where do the numbers come from? — a guide to the literature*, IEEE Transactions on Knowledge and Data Engineering **12** (2000).

[14] Pearl, J., *Fusion, propagation, and structuring in belief networks*, Artificial Intelligence **29** (1986).

[15] Prakken, H., *A new use case for argumentation support tools: supporting discussions of bayesian analyses of complex criminal cases*, Artificial Intelligence and Law **28** (2020).

[16] von Mises, R., "Probability, Statistics and Truth," Dover, New York, NY, USA, 1957, 2 edition.

[17] Walton, D., C. Reed and F. Macagno, "Argumentation schemes," Cambridge University Press, Cambridge, 2008.

Finding Factors in Legal Case-Based Reasoning

Cecilia Di Florio [a], Xinghan Liu [b], Emiliano Lorini [b],
Antonino Rotolo[a], Giovanni Sartor [a]

[a] *Alma Human AI, University of Bologna, Italy*
{*cecilia.diflorio2,giovanni.sartor,antonino.rotolo*} *@unibo.it*

[b] *IRIT-CNRS, University of Toulouse, France*
xinghan.liu@univ-toulouse.fr, Emiliano.Lorini@irit.fr

Abstract

We introduce a language and a formal semantics which support reasoning about factors in case-based reasoning, to determine their direction. We model a case-base through a binary classifier linking cases to decisions. We propose a method, based on counterfactual reasoning, to determine whether, relative to the information provided by the classifier, a feature is a factor, i.e., it always supports a given outcome, or it is irrelevant, i.e., it does not support any outcome, or it is ambiguous, i.e., it supports different outcomes depending on co-occurrent features. We explore the connection between this analysis and the case-based reasoning models adopted in AI and Law research.

Keywords: Case-based reasoning, Modal Logic for classifiers, Explainable AI

1 Introduction

Case-based reasoning (CBR) has played an important role in AI & Law research. Broadly speaking, CBR is engaged when a problem-situation is compared with previous cases in order to reach conclusions and take decisions. Hence, CBR is particularly relevant in the legal domain, especially in contexts where decision-makers rely on a body of precedents to make decisions.

Various approaches to CBR have been developed in the legal domain with different purposes, some approaches being more application-oriented [2,24] and others more formally-oriented [23,11]. Since the seminal work on the HYPO system for US trade secrets law, one basic assumption has become prevalent: legal cases are seen as sets of case features, where a case feature is, according to [21], a "legally relevant fact pattern" favouring one of the two opposing parties (typically either the plaintiff or the defendant), or more generally one of the alternative outcomes in a controversial issue. Particular attention has been paid to Boolean legal features, often called *factors* (while the term "dimension" is used for multi-valued features).

One aspect, however, has remained in the background in factor-based legal approaches to CBR, being adopted as an underlying property of case-based systems, namely, the identification of factors, among the features that are present in a case base, and the determination of their direction. Some investigations in this regard have been conducted in an hybrid approach, handling factor ascription to cases also through machine learning methods [19,20].

In this paper we propose a novel logic-based approach for the identification of factors. The approach is based on the idea that a case base can be modelled through a binary classifier. Each case is an instance of the classifier and is represented by a set of features plus the outcome that the classifier links to the presence of those features and the absence of the omitted features. Actually, our work will go beyond factor extraction: assuming that not all features appearing in the case favour a specific direction, through our model we aim to determine, by observing the behaviour of the classifiers, the nature of the features characterising a case. The following are possible options: the feature is a *factor*, if it consistently supports a single outcome; the feature is *irrelevant*, if it does not support any outcome; the feature is *ambiguous* if, depending on what other features co-occur with it, it supports one outcome or the opposite.

Our approach is based on the analytical model of factor-based reasoning which was proposed in [17], where monotonicity conditions were defined enforcing a fortiori reasoning over classifiers, in connection to the framework by [11], and abductive and contrastive explanations were used to explain the outcomes of classifiers. In this paper we shall connect the notion of a relevant feature, and in particular of a factor, to the concept of counterfactual explanation.

The paper is structured as follows. In Section 2 we briefly discuss legal factor-based case-based reasoning. Section 3 introduces classifier models (CMs) within the logic BCL of binary input classifier. In Section 4 we show how a case base can be modelled through a CM. In Section 5 through the analysis of CMs, we investigate the nature of features within case bases. Section 6 shows the relationship between our approach and some factor-based models developed in the AI&Law domain (particularly, Horty's model). Some conclusions end the paper.

2   Factor-based Case-based Reasoning

In this section we provide a brief introduction to legal factor-case based reasoning. The factor-based line of research has been started with the HYPO system [2,24], which addresses cases concerning the alleged misuse of trade secrets, on the basis of cases including different constellation of pro-misuse (pro-plantiff) or con-misuse (pro-defendant features). In the HYPO framework a factor-based representation enables various patterns of analogical reasoning being defined: citing a precedent, distinguishing it, and reasoning a fortiori. The factor-based approach has been developed in various ways. Hierarchies (ontologies) of factors at different levels of abstraction have been proposed [3] and binary factors have been extended into dimensions, i.e., multi-valued features which support a certain outcome to various extents depending on their specific value in the case at stake [6]. John Horty has developed formally-oriented models aimed at formalising a founding concept in common law, that of precedential constraint, by formally establishing the conditions under which the decision of a new case is forced by a body of previous cases [11]. Further developments are based on distinguishing in a case the reason, namely, the set of elements that the judge considers to sufficiently support a case decision, outweighing the factors to the contrary and on providing new refined logical analysis of dimensions.

We introduce now some preliminary notions pertaining formal factor-based approaches to case-based reasoning. In particular we will refer to John Horty's *result model of precedential constraint*, though departing from the original notation and preferring the one introduced in [17] for uniformity reasons. We will do this briefly, referring a more comprehensive and formal discussion on Horty's models of precedential constraint to Section 6. First of all, we assume to have two possible outcomes for an assessed case, 1 and 0, meaning respectively that the case has been decided in favour of the plaintiff or in favour of the defendant. Then, recalling that factors are Boolean case features, we can think of factors as a set of atomic propositions Atm_0 a priori partitioned between pro-plaintiff factors (Plt) and pro-defendant factors (Dfd). Namely, we have $Atm_0 = Dfd \uplus Plt$.[1] A *fact situation* s, of the kind presented within legal case, is a particular subset of such factors $s \subseteq Plt \uplus Dfd$. Within the result model, all factors appearing in a case are assumed to be significant for its outcome. In this context, a *precedent case* can be seen as a triple $\mathfrak{c} = (s, X, c)$, where s is a fact situation; $c \in \{0, 1\}$ is the outcome; X is the set of all pro-factors for c in s, namely $X = s \cap Plt$ if $c = 1$ and $X = s \cap Dfd$ otherwise. A *Horty case base* is a set of precedent cases.

A key element of Horty's model of precedent is the notion of *a fortiori reasoning*, which we will introduce informally here. According to a fortiori reasoning a new case should have the same outcome of a precedent case, if it includes an equally or more inclusive set of factors for the precedent's outcome, and no additional factor against that outcome.

We remark that the presented model is based on two fundamental assumptions, reflected in the partition $Atm_0 = Dfd \uplus Plt$. Firstly, all features appearing in the case base are factors. Secondly the direction of all factors, i.e. which party they favour, is supposed to be a priori known. Our work aims to avoid both these assumptions (and thus the language-level partition $Atm_0 = Dfd \uplus Plt$). In Section 5 we will show how is possible to determine *which* features of a case base can be considered factors and how to extract their direction.

3 Logic of Binary Classifiers

In this section we introduce the language and semantics of binary-input classifier logic BCL first appeared in [15,16] and use it to formalize a notion of counterfactual conditional.

---

[1] $\uplus$ denotes disjoint union. The choice of notation Atm_0, for propositional atoms, instead of Atm, will become clear in the next section

3.1   Language and classifier model semantics

We denote a set of finite atomic propositions by Atm, which is a disjoint union $Atm_0 \cup Dec$, where the former stands for the set of input variables of the classifier and $Dec = \{t(c) : c \in Val = \{0, 1, ?\}\}$ for the set of all three possible output values of the classifier.[2] In addition, let $Val = \{1, 0, ?\}$ where elements stand for *plaintiff wins*, *defendant wins* and *indeterminacy* respectively. Hence $t(c)$ reads as "the actual decision/outcome (of the judge/classifier) takes value c". For $c \in \{0, 1\}$, the "opposite" $\bar{c}$ is noted for the value $1 - c$. The modal language $\mathcal{L}(Atm)$ of BCL is therefore defined as:

$$\varphi \quad ::= \quad p \mid t(c) \mid \neg\varphi \mid \varphi \wedge \varphi \mid [X]\varphi,$$

where p ranges over Atm_0, $t(c)$ ranges over Dec, and $X \subseteq Atm_0$.[3] Modal operator $\langle X \rangle$ is the dual of $[X]$ and is defined as usual: $\langle X \rangle \varphi =_{def} \neg[X]\neg\varphi$. Their meanings will be revealed after Definition 3.2. Finally, for any $X \subseteq Y \subseteq Atm_0$, the following definition syntactically expresses a valuation on Y s.t. all variables in X are assigned as true, while all the rest in Y are false.

$$\mathsf{cn}_{X,Y} =_{def} \bigwedge_{p \in X} p \wedge \bigwedge_{p \in Y \setminus X} \neg p.$$

The language $\mathcal{L}(Atm)$ is interpreted relative to classifier models defined as follows.

Definition 3.1 [Classifier model] A classifier model (CM) is a pair $C = (S, f)$ where:

- $S \subseteq 2^{Atm_0}$ is a set of states (or fact situations), and

- $f : S \longrightarrow Val$ is a decision (or classification) function.

The class of classifier models is noted **CM**.

A pointed classifier model is a pair (C, s) with $C = (S, f)$ a classifier model and $s \in S$. Formulas in $\mathcal{L}(Atm)$ are interpreted relative to a pointed classifier model, as follows.

Definition 3.2 [Satisfaction relation] Let (C, s) be a pointed classifier model with $C = (S, f)$ and $s \in S$. Then:

$$(C, s) \models p \Longleftrightarrow p \in s,$$
$$(C, s) \models t(c) \Longleftrightarrow f(s) = c,$$
$$(C, s) \models \neg\varphi \Longleftrightarrow (C, s) \not\models \varphi,$$
$$(C, s) \models \varphi \wedge \psi \Longleftrightarrow (C, s) \models \varphi \text{ and } (C, s) \models \psi,$$
$$(C, s) \models [X]\varphi \Longleftrightarrow \forall s' \in S : \text{ if } (s \cap X) = (s' \cap X) \text{ then } (C, s') \models \varphi.$$

---

[2] The logic itself can have as finite many members in Dec as possible. For the purpose of representing a partial binary classifier, here we specify Dec as cardinality three.

[3] Notice p and $t(c)$ have different statuses regarding negation: $\neg p$ means that the input variable p takes value 0, but $\neg t(c)$ merely means the output does not take value c: we do not know which value it takes, since the output is trinary.

A formula φ of $\mathcal{L}(Atm)$ is said to be satisfiable relative to the class $\mathbf{CM}$ if there exists a pointed classifier model (C, s) with $C \in \mathbf{CM}$ such that $(C, s) \models \varphi$. It is said to be valid if $\neg \varphi$ is not satisfiable relative to $\mathbf{CM}$ and noted as $\models_{\mathbf{CM}} \varphi$.

We can think of a pointed model (C, s) as a pair (s, c) in f with $f(s) = c$. The formula $[X]\varphi$ is true at a state s if φ is true at all states that are modulo-X equivalent to state s. It has the *selectis paribus* (SP) (selected things being equal) interpretation "features in X being equal, necessarily φ holds (under possible perturbation on the other features)".[4] Notice when $X = \emptyset$, $[\emptyset]$ is the S5 universal modality since every state is modulo-$\emptyset$ equivalent to all states, viz. $(C, s) \models [\emptyset]\varphi \iff \forall s' \in S, (C, s') \models \varphi$.

3.2   A counterfactual conditional

We introduce a simple notion of counterfactual conditional for binary classifiers in [15]. We start our analysis by defining the following notion of similarity between states in a classifier model.

Definition 3.3 [Similarity between states] Let $C = (S, f)$ be a classifier model, $s, s' \in S$. The degree of similarity between s and s' in S relative to the set of features Atm_0, noted $sim_C(s,s',Atm_0)$, is defined as follows:

$$sim_C(s,s',Atm_0) = |\{p \in Atm_0 : (C, s) \models p \text{ iff } (C, s') \models p\}|.$$

A dual notion of distance between worlds can be defined from the previous notion of similarity:

$$dist_C(s,s',Atm_0) = |Atm_0| - sim_C(s,s',Atm_0).$$

This notion of distance is in accordance with [9] in knowledge revision, i.e. the Hamming distance, which counts the cardinality difference of features' values.

The following definition introduces the notion of counterfactual in Lewis' style of the form $\varphi \Rightarrow \psi$ whose reading is "if it were φ, (with respect to features in Atm_0) it would be ψ."

Definition 3.4 [Counterfactual conditional] Let $C = (S, f)$ be a classifier model, $s \in S$. Then, $(C, s) \models \varphi \Rightarrow \psi$ if and only if $closest_C(s,\varphi,Atm_0) \subseteq \|\psi\|_C$, where

$$closest_C(s,\varphi,Atm_0) = \arg\max_{s' \in \|\varphi\|_C} sim_C(s,s',Atm_0),$$

and for every $\varphi \in \mathcal{L}(Atm)$:

$$\|\varphi\|_C = \{s \in S : (C, s) \models \varphi\}.$$

The underlying idea is that $\varphi \Rightarrow \psi$ holds in a state of a classifier model iff all the closest states to the current one, which make φ true, also make ψ true, where "closest" is defined by the Hamming distance. Formula $\varphi \Rightarrow \psi$ captures the standard notion of conditional logic. One can show that $\Rightarrow$ satisfies all semantic conditions of Lewis' logic of counterfactuals VC [14].

---

[4] $[Atm_0 \setminus X]\varphi$ has the standard *ceteris paribus* (CP) interpretation "features other than X being equal, necessarily φ holds (under possible perturbation of the features in X)".

4   From Classifiers Models to Case Bases

A first connection bewteen CBR and reasoning about classifiers has been already explored in [17], where a representation theorem reveals that every Horty's consistent case bases can be modelled by a set of classifier models. In this section will now show how a "larger class" of case base can be modelled via classifier models. The underlying intuition is that a pointed classifier model, representing a factual situation associated with a decision, can model a a case. Consequently, a cases base can be modelled through a classifier model, i.e. it can be seen as the outcome of a classifier. Actually, in order to relate CBR and reasoning about classifiers we will have to consider classifiers models that satisfy two semantic constraints.

First of all, classifiers should allow *incomplete-knowledge*: not all factual situations are associated with a decision in favour of the defendant or the plaintiff. Namely, a fact situation, besides 0 and 1, can also be classified as ?, where ? means absence of decision. That is, we consider classifier models whose classification function is of the form:

$$f : S \longrightarrow Val \text{ with } Val = \{0, 1, ?\}. \tag{1}$$

Secondly, we have to suppose that the classifier is complete, that is every possible situation is take into account by the classifier. Namely,

$$S = 2^{Atm_0}. \tag{2}$$

Then, $\mathbf{CM}^{inc} = \{C \in \mathbf{CM} : C$ satisfies $(1), (2)\}$, is the class of *incomplete-knowledge classifier models*. A formula φ of $\mathcal{L}$ is said to be satisfiable relative to the class $\mathbf{CM}^{inc}$, if there exists a pointed classifier model (C, s) with $C \in \mathbf{CM}^{inc}$ such that $(C, s) \models \varphi$. It is said to be valid relative to $\mathbf{CM}^{inc}$, and noted as $\models_{\mathbf{CM}^{inc}} \varphi$, if $\neg\varphi$ is not satisfiable relative to $\mathbf{CM}^{inc}$. An incomplete-knowledge classifier model $C \in \mathbf{CM}^{inc}$, models a *incomplete-knowledge case base* defined as $CB_C = \{k \mid k = (s, f(s)) \text{ with } s \in S\}$.

Example 4.1 [Running example]

Our example will deal with hypothetical and simplified cases pertaining compensation claims for trespassing to land. Trespass to land involves the unlawful interference with property rights without permission. In these cases the plaintiffs are the owners of the property. Let us assume that the elements considered in this context are limited to: "the defendant causes property damage" (damage); "the defendant acts in defence of public necessity" (necessity); "the defendant intends to enter the land (intentional)".

Let us assume that the judges ruled in favour of plaintiffs's compensation in the three following hypothetical scenario: 1) the plaintiff claimed his neighbour intentionally occupied part of his land for his own use, causing damage (which we will refer to as case k_2) ; 2) the plaintiff claimed for damages from the municipality for injuries to land caused unintentionally by a policeman in an attempt to catch a criminal, acting in the public interest (case k_3 [5] ; 3) a

---

[5] partially inspired by Wegner v. Milwaukee Mut.Ins.Co., 479 N.W.2d 38 (Minn., 1991)

hunter intentionally entered the plaintiff's property, even if he has not caused damage (case k_5). Instead, let us assume that the judges ruled in favour of defendant when: 1) the defendant accidentally entered the plaintiff's land while attempting to extinguish a fire in the neighbouring land (case k_7); 2) the defendant, while hiking, accidentally entered the plaintiff 's property, without causing damage, because the path was not well signposted (case k_8).

The decisions made by the judges are mapped by the classifier $C_1 = (2^{Atm_0}, f)$, with $Atm_0 = \{\texttt{intentional}, \texttt{damage}, \texttt{necessity}\}$ and $f : 2^{Atm_0} \longrightarrow \{0, 1, ?\}$, s.t. $f(s) = 1$ iff $s \in \{\{\texttt{intentional}, \texttt{damage}\}, \{\texttt{damage}, \texttt{necessity}\}, \{\texttt{intentional}\}\}$ $f(s) = 0$ iff $s \in \{\{\texttt{necessity}\}, \emptyset\}$, $f(s) = ?$ otherwise.[6] The modelled case base is $CB_{C_1} = \{k_i \mid k_i = (s_i, f(s_i))\}$ represented in following table.

	s_i	$f(s_i)$
$k_1 :$	$\{\texttt{intentional}, \texttt{damage}, \texttt{necessity}\}$	?
$k_2 :$	$\{\texttt{intentional}, \texttt{damage}\}$	1
$k_3 :$	$\{\texttt{damage}, \texttt{necessity}\}$	1
$k_4 :$	$\{\texttt{intentional}, \texttt{necessity}\}$	?
$k_5 :$	$\{\texttt{intentional}\}$	1
$k_6 :$	$\{\texttt{damage}\}$	?
$k_7 :$	$\{\texttt{necessity}\}$	0
$k_8 :$	$\emptyset$	0

Intuitively, we expect that in this context **damage** is a factor in favour of the plaintiff. Indeed, it is the only *discriminating feature* between the situation in k_3, classified as 1, and that in k_7, classified as 0; moreover **damage** is not discriminating for any situation classified as 0.

Henceforth, due to the connection between classifier models and case bases, when referring to classifier models we will also mean the modelled case bases.

In the next section we will show how, in a classifier model, it is possible to determine, according to provided classifications, which features can be considered factors and what their direction is. We shall also see that not all features of a case base are factors and in that case they may be irrelevant or ambiguous.

5 Factors, Irrelevant and Ambiguous Features

Intuitively, a factor is a feature that is discriminating in only one direction. A feature is said to be discriminating in one direction if just removing it from a case classified in that direction suffices for having the opposite classification.

The discriminating aspect of features is nicely captured by the notion of strong counterfactual explanation for a decision, which we can express in the language $\mathcal{L}(Atm)$. Indeed, we will say that a formula φ of $\mathcal{L}(Atm)$ *strong counterfactually* explains a decision $c \in \{0, 1\}$ for a certain situation s, if not satisfying φ would lead s to be classified as $\bar{c}$. Formally, we have the following.

---

[6] In this context we are implicitly enforcing the "closed-world assumption", assuming that the absence of a feature means that its negation holds.

Definition 5.1 [Strong counterfactual explanation] We write $\mathsf{SCfXp}(\varphi, c)$ to mean that φ strong counterfactually explains a decision for c and define it as

$$\mathsf{SCfXp}(\varphi, c) =_{def} \mathsf{t}(c) \wedge \left(\neg\varphi \Rightarrow \mathsf{t}(\bar{c})\right).$$

Example 5.2 Consider C_1 of our running example. As already noted, removing damage from the factual situation s_3, classified as 1, we obtain s_7, classified as 0. That is, $(C_1, s_3) \models \mathsf{SCfXp}(\text{damage}, 1)$. Similarly, looking at s_5 and s_8, it can be verified that $(C_1, s_5) \models \mathsf{SCfXp}(\text{intentional}, 1)$.

We speak of strong counterfactual explanation for a decision to distinguish it from a "weaker" counterfactual explanation, according to which, φ explains a decision c for a certain situation s, if not satisfying φ would lead s to be classified *differently* than c (and thus in our framework either as $\bar{c}$ or as ?).

Definition 5.3 [Counterfactual explanation] We write $\mathsf{CfXp}(\varphi, c)$ to mean that φ counterfactually explains a decision for c and define it as

$$\mathsf{CfXp}(\varphi, c) =_{def} \mathsf{t}(c) \wedge \left(\neg\varphi \Rightarrow \neg\mathsf{t}(c)\right).$$

Example 5.4 From cases k_2 and k_6 we have $(C_1, s_2) \models \mathsf{CfXp}(\text{intentional}, 1)$. But $(C_1, s_2) \not\models \mathsf{SCfXp}(\text{intentional}, 1)$, since the classification from s_2 to s_6 changes from 1 to ?.

On the basis of strong counterfactual explanation, we can distinguish three related notion: the notion of factor (unidirectional feature), ambiguous (multi-directional) feature and irrelevant (no-directional) feature. In the following we shall provide definitions and examples for each of these notions.

A feature p is a factor in the direction of c if, as already mentioned, is discriminating only in the direction of c, namely if: (i) there is a fact situation such that p strong counterfactually explains the decision c and (ii) there is no fact situation such that p strong counterfactually explains the opposite decision $\bar{c}$. That is, we have the following definition.

Definition 5.5 [Factor] We write $\mathsf{Factor}(p, c)$ to mean that p is a factor in the direction of c and define it as

$$\mathsf{Factor}(p, c) =_{def} \langle\emptyset\rangle\mathsf{SCfXp}(p, c) \wedge [\emptyset]\neg\mathsf{SCfXp}(p, \bar{c}).$$

Example 5.6 From Example 5.2 and the fact that damage doesn't strong counterfactually explain a decision for 0, we have $(C_1, s) \models \mathsf{Factor}(\text{damage}, 1)$. In the same way we have, $(C_1, s) \models \mathsf{Factor}(\text{intentional}, 1)$.

As already noted, not all features are factors. Indeed, a feature could either not explain any decision strong counterfactually and in this case it is *irrelevant*, or explain strong counterfactually opposite decisions and if so it is *ambiguous*.

Definition 5.7 [Irrelevant feature] We write $\mathsf{Irrelevant}(p)$ to mean that p is irrelevant and define it as

$$\mathsf{Irrelevant}(p) =_{def} \neg\langle\emptyset\rangle\mathsf{SCfXp}(p, 1) \wedge \neg\langle\emptyset\rangle\mathsf{SCfXp}(p, 0).$$

Example 5.8 Within C_1, necessity does not explain any decision strong counterfactually, i.e. we have $(C_1, s) \models \mathsf{Irrelevant}(\texttt{necessity})$ for all $s \in S$.

Definition 5.9 [Ambiguous feature] We write $\mathsf{Amb}(p)$ to mean that p is an ambiguous feature and define it as

$$\mathsf{Amb}(p) =_{def} \langle \emptyset \rangle \mathsf{SCfXp}(p, 1) \wedge \langle \emptyset \rangle \mathsf{SCfXp}(p, 0).$$

Example 5.10 In C_1 there are not ambiguous features. But suppose now that a new case is added to our case base. A private university claims compensation for protesters trespassing and causing damages during an event. The protesters call into question public necessity arguing that their actions are justified as reasonable efforts to prevent immediate threat of greater harm caused by the business of the event's guests. The judge accepts the public necessity defence and, while recognising intentionality, rules in favour of the defendant.[7] So that the "updated" case base is the one modelled by $C_2 = (2^{Atm_0}, g)$ described in the following table.

	s_i	$g(s_i)$
$k_1:$	$\{\texttt{intentional}, \texttt{damage}, \texttt{necessity}\}$	**0**
$k_2:$	$\{\texttt{intentional}, \texttt{damage}\}$	1
$k_3:$	$\{\texttt{damage}, \texttt{necessity}\}$	1
$k_4:$	$\{\texttt{intentional}, \texttt{necessity}\}$	?
$k_5:$	$\{\texttt{intentional}\}$	1
$k_6:$	$\{\texttt{damage}\}$	?
$k_7:$	$\{\texttt{necessity}\}$	0
$k_8:$	$\emptyset$	0

But then, in this context we have both $(C_2, s_5) \models \mathsf{SCfXp}(\texttt{intentional}, 1)$ and $(C_2, s_1) \models \mathsf{SCfXp}(\texttt{intentional}, 0)$. Namely, $(C_2, s) \models \mathsf{Amb}(\texttt{intentional})$ for every $s \in S$.

Two properties concerning features behaviour are established by the following propositions. Firstly, factors are unidirectional: if p is a factor in the direction of c, it cannot be a factor in the opposite direction $\bar{c}$. That is, the following holds.

Proposition 5.11

$$\models_{\mathbf{CM}^{inc}} \bigwedge_{p \in Atm_0, c \in \{0,1\}} \Big(\mathsf{Factor}(p, c) \rightarrow \neg \mathsf{Factor}(p, \bar{c}) \Big).$$

Moreover, the already introduced partition of features between factors, irrelevant and ambiguous is captured by the following validity.

Proposition 5.12

$$\models_{\mathbf{CM}^{inc}} \bigwedge_{p \in Atm_0} \Big(\bigwedge_{c \in \{0,1\}} \neg \mathsf{Factor}(p, c) \rightarrow \big(\mathsf{Irrelevant}(p) \vee \mathsf{Amb}(p) \big) \Big).$$

---

[7] partially inspired by [Commonwealth v. Carter], see N. Y. Times, Apr. 16, 1987

Example 5.10 hints at one aspect that deserves further attention. Namely, suppose that decisions for previously undecided cases are provided. In this case, features extracted as ambiguous will remain so. However, it could happen that: a) features previously labelled as irrelevant may become ambiguous or factors, b) features previously labelled as factors may become ambiguous (as it is the case for **intentional** in the "updating" of Example 5.10). That is, with our method, we identify *pro tanto* irrelevant features and *pro tanto* factors, i.e. features that are irrelevant or factors given the current information. Bearing the latter in mind, a classifier model that does not admit ambiguous features, at least given the given classifications, will be called *consistent*.

Definition 5.13 [Consistency] A classifier $C = (S, f)$ is *consistent given the current classifications* if and only if for every $s \in S$, $(C, s) \models$ Cons, where

$$\text{Cons} =_{def} \bigwedge_{p \in Atm_0} \neg\text{Amb}(p).$$

Example 5.14 Classifier C_1 of Example 4.1 is consistent . Whereas C_2 of Example 5.10 is inconsistent (and will remain so, regardless of later classifications of originally undecided cases).

We now introduce a form of a fortiori reasoning that somehow may remind of the a fortiori reasoning in the "Horty's style" briefly introduced in Section 2, but which differs from the latter. That is, intuitively, according to a form of a fortiori reasoning, a case should have the same outcome of a precedent case, if it includes an equally or more inclusive set of factors for the precedent's outcome, and no additional factor against that outcome. In our framework, however, we take into account also ambiguous and irrelevant feature. More precisely, we expect that if the classifier associates a situation s to an outcome c, then it must assign the same outcome to every situation s' such that: (a) s' includes all factors in the direction of c that are in s (b) s' does *not include* factors in the direction of $\bar{c}$ that are *outside of* s and (c) s' *includes exactly the same ambiguous and irrelevant features that are in s*. We shall call *monotone* those classifiers that satisfy the a fortiori constraint such as the one just described. Formally, we have the following.

Definition 5.15 [Monotonicity] We will say that C is monotone iff $(C, s) \models$ ClMon, where

$$\text{ClMon} =_{def} \bigwedge_{c \in \{0,1\}, X,Y \subseteq Atm_0} \Big(\big(\text{SetFactor}(X, c) \wedge \text{SetFactor}(Y, \bar{c}) \big) \rightarrow$$

$$\bigwedge_{Z \not\subseteq X \cup Y, X' \subseteq X, Y' \subseteq Y} \Big(\langle \emptyset \rangle (\text{cn}_{X' \cup Y' \cup Z, Atm_0} \wedge \text{t}(c)) \rightarrow$$

$$\bigwedge_{X \supset X'' \supseteq X', Y'' \subseteq Y'} [\emptyset](\text{cn}_{X'' \cup Y'' \cup Z, Atm_0} \rightarrow \text{t}(c)) \Big) \Big),$$

where, for every $c \in \{0,1\}$ and $X \subseteq Atm_0$:

$$\mathsf{SetFactor}(X, c) =_{def} \bigwedge_{p \in X} \mathsf{Factor}(p, c) \wedge \bigwedge_{p \in Atm_0 \setminus X} \neg\mathsf{Factor}(p, c).$$

Example 5.16 Classifier C_1 of Example 4.1 is not monotone, i.e $(C_1, s) \not\models$ CIMon. Indeed, for CIMon to be satisfied, the case k_1 should have been classified as 1, since it contains more factors in favour of 1 (i.e. `intentional`) than the factual situation of k_3, classified as 1.

Previous examples show us that the consistency of a classifier (i.e. the absence of ambiguous features) does not imply its monotonicity (think of C_1). However, it is worth noting that the reverse is not true either, i.e. the monotonicity of a classifier does not guarantee its consistency.[8] An "extreme" example of this occurs when all features of a classifier are ambiguous, so that CIMon is vacuously satisfied.

Example 5.17 By way of example only, consider this new simplified scenario: $C_3 = (2^{Atm_0}, h)$, where this time $Atm_0 = \{\texttt{intentional}, \texttt{damage}\}$ and the behaviour of h summarized in following table

	s_i	$h(s_i)$
$k_1:$	$\{\texttt{damage}, \texttt{intentional}\}$	0
$k_2:$	$\{\texttt{intentional}\}$	1
$k_3:$	$\{\texttt{damage}\}$	1
$k_4:$	$\emptyset$	0

It can be verified that both `damage`, `intentional` are ambiguous features so that vacuously $(C_3, s) \models$ CIMon for all $s \in S$.

Remark 5.18 We should conclude this part of the paper by observing that, as it is used in knowledge revision as well as in XAI [10], the Hamming distance is adopted to establish similarity between cases. The reader can likewise observe that, usually, analogy is measured in CBR by shared properties, e.g. in HYPO via subset inclusion relation. We point that, in this context, building similarity via Hamming distance coincides with building it via subset inclusion relation, since the notion of factor only uses counterfactual explanation with an atomic explanans.

6 Relation to Horty's Models of Case Bases

In this section we compare our framework with Horty's reason model previewed in Section 2. Its difference to result model is that in a precedent $\mathfrak{c} = (s, X, c)$, X may be a set of *some* (but not all) pro factors for c in s. Result model can thus be viewed as a special kind of reason model. Therefore we handle both

---

[8] It is worth noting that our definition of consistency differs from that provided by Horty. In following section, we will see that to some extent, both monotonicity and the absence of ambiguous features are necessary to retrieve Horty's consistency.

models under the umbrella term reason model. Nevertheless, whether the set or a subset of pro factors is not the key point of this paper.

In [17] it is shown that given a Horty's reason model of a case base, there are a set of classifier models representing it. What if we start from some classifier model and reverse engineer case bases it can represent? What at stake is that in Horty's theory features are a priori (completely) partitioned as plaintiff- or defendant-side, while here we do not know but rather find a feature as factor, irrelevant or ambiguous via counterfactual reasoning. Actually, a main observation is that, e.g. *a plaintiff-factor in Horty's sense may not have the direction for 1*, i.e. $p \in Plt$ but not $\mathsf{Factor}(p, 1)$. Hence the quadrichotomy of features is more fine-grained than the dichotomy of Plt and Dfd.

Let us begin with briefly formalising the theory of Horty, which we started to discuss in section 2, in the modal logic way. For more details see [11,21,17].

Definition 6.1 [Horty case base] Let $Atm_0 = Plt \uplus Dfd$, the latter two standing for plaintiff- and defendant-side respectively. A Horty precedential case (precedent) is a triple $\mathfrak{c} = (s, X, c)$, where: $s \subseteq Atm_0$ is a state/fact situation; $c \in \{0, 1\}$; X is called the *reason* of the decision, with $X \subseteq s \cap Plt$ if $c = 1$, $X \subseteq s \cap Dfd$ otherwise. A Horty case base CB is a set of precedents.

Note that all precedents in a CB are decided deterministically as 1 or 0. We adopt the notation Atm_0^c, which denotes Plt if $c = 1$, and Dfd if $c = 0$.

The key concern of case base is maintaining its *consistency* with respect to a fortiori reasoning, which we mentioned before and formalise now.

Definition 6.2 [A fortiori reasoning] Let CB be a case base and has a precedent $\mathfrak{c} = (s, X, c)$. Then for any fact situation s' s.t. $X \subseteq s' \cap Atm_0^c$ and $s \cap Atm_0^{\bar{c}} \supseteq s' \cap Atm_0^{\bar{c}}$, s' is forced to be decided as c according to CB.

Definition 6.3 [Consistent case base] A case base CB is consistent, if it follows a fortiori reasoning, and there is no fact situation s s.t. it is forced to be decided as 0 by some $\mathfrak{c}_0 \in CB$ and decided as 1 by some $\mathfrak{c}_1 \in CB$.

The representation theorem in [17] states that every consistent case base can be represented by *a set of* classifier models. Actually we can do better in showing that every consistent case base can be transformed into a *unique* classifier, in which sense we call it "genuine". The basic idea is that the genuine classifier of a case base outputs ? for all and only all fact situations which cannot be forced to take a decision by the case base. [9]

Definition 6.4 [Genuine classifier] Let CB be a case base. The genuine classifier of CB is the function $f : 2^{Atm_0} \longrightarrow \{0, 1, ?\}$, s.t. for any situation s,

$$f(s) = \begin{cases} c \text{ if} & \exists \mathfrak{c} = (s', X, c) \in CB \text{ s.t.} \\ & X \cap Atm_0^c \subseteq s \cap Atm_0^c, s' \cap Atm_0^{\bar{c}} \supseteq s \cap Atm_0^{\bar{c}}; \\ ? & \text{otherwise.} \end{cases}$$

---

[9] It is therefore the "smallest" representation, if we think of every classifier as a partial function by naturally viewing ? as *undefined*.

The genuine classifier model of CB is therefore $C = (2^{Atm_0}, f)$.

That is to say, $f(s') =?$ if and only if a fortiori reasoning fails to force s' to take either decision 0 or 1. We can also define the genuine CM for a case base in terms of a formula.

Proposition 6.5 *Let CB be a Horty case base. Then a classifier model $C = (S, f)$ is the genuine classifier of CB, if $\forall s \in S$, $(C, s) \models \varphi_{CB}$ where*

$$\varphi_{CB} =_{def} \bigwedge_{X \subseteq Atm_0} \langle \emptyset \rangle \mathsf{cn}_{X, Atm_0} \wedge \bigwedge_{c \in \{0,1\}} [\emptyset] \Big(\mathsf{t}(c) \leftrightarrow \bigvee_{\mathsf{c} = (s', X, c) \in CB} \mathsf{cn}_{X, X \cup (Atm_0^c \setminus s')} \Big).$$

The "syntactic" representation φ_{CB} is stronger than the one in [17] consisting of Compl, 2Mon and $tr'(CB)$, due to the bi-conditional in φ_{CB} for $\mathsf{t}(c)$ with $c \in \{0, 1\}$. Hence, fixing the language, φ_{CB} is satisfied in exact one CM.

Proposition 6.6 *Every Horty case base induces exactly one genuine classifier model.*

The inverse, however, does not hold. Not only because two case bases may force exactly the same set of fact situations but with different reasons, but also, more intriguing, that there may exist more than one possible dichotomy, say, $Atm_0 = Plt \uplus Dfd = Plt' \uplus Dfd'$, which both give rise to a consistent case base respectively. That means, reverse engineering usually does not guarantee a unique dichotomy but several admissible ones. That leads us to have the following definition.

Definition 6.7 [Admissible dichotomy] Let CB be a Horty case base whose a priori dichotomy is $Atm_0 = Plt \uplus Dfd$. We say that a dichotomy $Atm_0 = Plt' \uplus Dfd'$ is admissible, if by doing so the a fortiori reasoning will not render CB inconsistent.

Plainly speaking, that means, if we change from letting $Atm_0^1 = Plt$ to $Atm_0^1 = Plt'$ and $Atm_0^0 = Dfd$ to $Atm_0^0 = Dfd'$ and run the a fortiori reasoning, the case base keeps consistent. The simplest example is: let $Plt = \{p, q\}$ and $Dfd = \emptyset$ and a singleton case base $\{(\{p, q\}, \{p\}, 1)\}$. The reader can check that by changing to $Plt' = \{p\}, Dfd' = \{q\}$, the case base remains consistent.

Proposition 6.8 *Let CB be a consistent case base and C its genuine classifier. Then a dichotomy $Plt' \uplus Dfd'$ is admissible, if and only if for any state s,*

(i) $(C, s) \models \neg \mathsf{Amb}(p)$ *for any feature p*

(ii) *if $(C, s) \models \mathsf{SetFactor}(X, 1) \wedge \mathsf{SecFactor}(Y, 0)$ for some $X, Y \subseteq Atm_0$, then $X \subseteq Plt', Y \subseteq Dfd'$.*

The proof relies on an observation regarding counterfactual explanation, namely if $p \in Plt'$ then $(C, s) \models [\emptyset] \neg \mathsf{CfXp}(p, 0)$; and if $p \in Dfd'$ then $(C, s) \models [\emptyset] \neg \mathsf{CfXp}(p, 1)$.

Now we show that the Plt/Dfd dichotomy is not only pre-determined but also too coarse-grained from our lens.

Proposition 6.9 *There exists a Horty case base with $Plt \uplus Dfd$ its dichotomy,*
$C = (2^{Atm_0}, f)$ *its genuine classifier model s.t.* $\exists p \in Plt$ *but* $(C, s) \models \neg \mathsf{Factor}(p, 1)$*; and* $\exists q \in Dfd$ *but* $(C, s) \models \neg \mathsf{Factor}(q, 0)$*.*

Why? Because the dichotomy only guarantees that p on the defendant side shall not support the plaintiff, but can not tell whether p is actually irrelevant, viz supporting the defendant neither. See the following example.

Example 6.10 Let $Plt = \{p\}, Dfd = \{q\}$ and $CB = \{(\{p, q\}, \{p\}, 1)\}$. Then the genuine CM for CB is $C = (2^{Atm_0}, f)$ s.t. $f(s) = 1$ if $p \in s$; $f(s) =?$ otherwise. However, $(C, s) \models \neg \mathsf{Factor}(q, 0)$. Moreover, $(C, s) \models \mathsf{Irrelevant}(q)$, since there are no $s, s \setminus \{q\}$, s.t. $f(s) \neq f(s \setminus \{q\})$. As a result, $Plt' = \{p, q\}$ and $Dfd' = \emptyset$ is another admissible dichotomy.

In light of this it is easy to see that exactly due to irrelevant features do we have not the unique, but many admissible dichotomies.

Our main result of this section shows the necessary and sufficient condition of being a genuine classifier for some Horty case base regardless of which dichotomy the case base actually has.

Proposition 6.11 *Let $C = (S, f)$ be a classifier model. Then C is the genuine CM for some consistent Horty case base with an admissible dichotomy $Atm_0 = Plt \uplus Dfd$, if and only if $C \in \mathbf{CM}^{inc}$, C is monotone, C doesn't have ambiguous features, $f(\emptyset) =?$ and $\forall p \in Atm_0^c$, $f(\{p\}) \neq \bar{c}$.*

7 Discussion and Related Work

Following the extensive AI & Law literature springing from the study of HYPO and CATO, in the last decade a significant effort has been put in investigating axioms as well as formal properties of factor-based CBR, and in providing the logical foundations for such reasoning (see, e.g., [11,21,8,12,13,22,7,17,5,1]).

Also due to development of explainable AI [18,4], the quest for logical foundations of factor-based CBR has been recently focused, e.g., on formal models of argumentative explanation [22] or on logics for classifier systems [17]. As suggested in [5] and investigated in [19,20]—also from a machine learning perspective—one aspect has remained in the background and needs a specific logical inquiry: the identification of factors, among the features that are present in a case base. In this paper we proposed a novel approach to address this issue. As already mentioned, this approach builds on the connection already revealed in [17] between factor-based CBR and reasoning about classifier models. Such a connection allowed us to refine the granularity of the analysis of the features within a case base, enabling us to identify not only factors and their directions, but also irrelevant and ambiguous features.

We believe that some aspects of our work need further discussion.

First, as recalled by [21] in regard to the question of whether the fact that a factor favouring one side does not apply in a case favouring the other side, Horty [11] argued that such a "closed-world assumption" should be adopted on a case-by-case basis. However, [25] argued that the negated factor favouring the other side can only be included if there is a case where judges have stated that the absence of a factor favours the other side. We implicitly use the closed-world assumption when treating a fact situation as a conjunction of positive and negative literals. Also it is responsible for the extra constraint $f(\emptyset) = ?$ needed when relating Horty's models. It is worth exploring the case where negation and absence differ, giving the framework some intuitionistic flavour.

A further consideration concerns monotonicity, that allowed us to deal with a fortiori reasoning in preliminary manner. We have defined monotonicity as condition that may or may not be verified by the classifiers (and thus the case bases modelled by them), depending on whether they satisfy an a fortiori constraint. We are already working on a more general framework in which monotonicity is no longer a condition, but a principle underlying inference mechanisms to classify new cases. Among these inference mechanisms, we are investigating one that would allow us to classify new cases even in case of inconsistencies, that is, even when a case base contains ambiguous features. This approach would be a reply to a legitimate issue: in the context of "real" legal case bases it is plausible that many features would result ambiguous according to our classification. Furthermore, as regards both the condition of monotonicity and the inference mechanisms just mentioned, we recall what has already been emphasised in [11] about models of precedential constraint: our model only captures single step reasoning, leading directly from the features appearing in a case to the outcome of the case. In the future we will study the extension of our approach to multi-step reasoning based on intermediate features. In our future research, we also aim to address scalability, from the computational complexity viewpoint.

Moreover, in this paper we view incomplete-knowledge case bases as partial Boolean functions (pBF) and define a notion of monotonicity. A natural question is its relation to the monotonic partial Boolean function. Different with monotone Boolean functions which are thoroughly studied and widely used, the theory of monotone pBFs is less developed to our knowledge. On this our analysis can shed some light. The first insight is that there are at least two ways to define monotone variable in pBF regarding $\mathsf{CfXp}(p, c)$ and $\mathsf{SCfXp}(p, c)$ respectively. We leave this topic to future work.

Lastly, as we have discussed in Remark 5.18, a more technical point regards how we can establish similarity between cases. We relied on Hamming distance, while usually in CBR analogy is measured by shared properties. While the two approaches in the current framework are mathematically equivalent because the notion of factor only uses counterfactual explanation with an atomic explanans, differences emerge when groups of features are analysed. This issue is left to future research.

A   Proofs

A.1   Proof of Proposition 5.12

Proof. Let (C, s) a pointed model, $C \in \mathbf{CM}^{inc}$. Let p s.t, for all $c \in \{0, 1\}$, $(C, s) \models \neg\mathsf{Factor}(p, c)$. Then, $(C, s) \models \big(\langle\emptyset\rangle\mathsf{SCfXp}(p, c) \wedge \langle\emptyset\rangle\mathsf{SCfXp}(p, \overline{c})\big) \vee$

$\left(\neg\langle\emptyset\rangle\mathsf{SCfXp}(p,c) \wedge \neg\langle\emptyset\rangle\mathsf{SCfXp}(p,\bar{c})\right)$. That is $(C,s) \models \mathsf{Amb}(p) \vee \mathsf{Irrelevant}(p)$.□

A.2 Proof of Proposition 6.8

Proof. We claim that if $p \in Plt$ then $(C,s) \models [\emptyset]\neg\mathsf{CfXp}(p,0)$; and if $p \in Dfd$ then $(C,s) \models [\emptyset]\neg\mathsf{CfXp}(p,1)$. And all the statements in the proposition follow as corollaries. To prove the claim, suppose towards a contradiction that $p \in Plt$ but $\exists s' \in 2^{Atm_0}$ s.t. $(C,s') \models t(1) \wedge \neg p \Rightarrow \neg t(1)$. This apparently violates the a fortiori reasoning: since C is genuine to CB, $f(s') = 1$ because some precedent $(s_1,X_1,1) \in CB$ s.t. $X_1 \subseteq s' \cap Plt$ and $s_1 \cap Dfd \supseteq s' \cap Dfd$. Then a fortiori, $f(s'\backslash\{p\}) = 1$ for the same reason. But this contradicts $(C,s') \models \neg p \Rightarrow \neg t(1)$.□

A.3 Proof of Proposition 6.11

Proof. Let C be the genuine CM for an Horty case base. We show that it satisfies all constraints. Obviously $C \in \mathbf{CM}^{inc}$, since S need equal 2^{Atm_0}; any precedent shall not have an empty fact situation by definition; also for any $p \in Plt$, it shall never be a reason of a defendant case and similarly for a Dfd-factor. Now assume towards a contradiction that C does not satisfy ClMon. Then suppose w.l.o.g., there are two sets $X,Y \subseteq Atm_0$ which $\mathsf{Factor}(X,1) \wedge \mathsf{Factor}(Y,0)$ holds in C. Moreover, there is a state $s_1 = X_1 \cup Y_1 \cup Z$, with $X_1 \subseteq X, Y_1 \subseteq Y, Z \subseteq Atm_0 \setminus (X \cup Y)$. However, there is another state $s_2 = X_2 \cup Y_2 \cup Z$ s.t. $X_1 \subseteq X_2 \subseteq X, Y_2 \subseteq Y_1$ but $f(s_2) = 0$.

Now recall C is the genuine CM for some case base. We know there must be $(s_3,X_3,1),(s_4,Y_4,0) \in CB$ s.t. $X_3 \subseteq s_1 \cap Plt, s_3 \cap Dfd \supseteq s_1 \cap Dfd$, and $Y_4 \subseteq s_2 \cap Dfd, s_4 \cap Plt \supseteq s_2 \cap Plt$, which allow the genuine classifier has $f(s_1) = 1$ and $f(s_2) = 0$ via a fortiori reasoning. Now we consider the state $s_5 = X_3 \cup Y_4 \cup Z$. The a fortiori reasoning forces s_5 to be both 1 and 0, contradicting that CB is consistent.

For the other direction let us build a CB from a C satisfying all constraints, while taking care of the dichotomy of Atm_0 and the consistency of CB. Taking advantage of Proposition 1 in [17], we simply do that for any state s and term λ, if $(C,s) \models \mathsf{AXp}(\lambda,1)$ with $lit^+(\lambda) = X$ and $lit^-(\lambda) = Y$, then put $(s,X,1)$ in CB, put members of X in Plt and members of Y in Dfd. We do the similar for $\mathsf{AXp}(\lambda,0)$. Eventually we obtain CB and let Plt what it is, and simply $Dfd = Atm_0 \setminus Plt$ to have $Plt \uplus Dfd = Atm_0$.

For the consistency of CB, suppose not towards a contradiction, then there are two precedents $(s_1,X,1)$ and $(s_0,Y,0)$ s.t. $X \subseteq s_0 \cap Plt \subseteq Plt$ and $Y \subseteq s_1 \cap Dfd \subseteq Dfd$. We show the supposition impossible by using AXp and PImp. In light of the construction, we have that $X \in lit^+(\lambda_1), Y \in lit^+(\lambda_0)$ with $(C,s_i) \models \mathsf{AXp}(\lambda_i,i)$ with $i \in \{1,0\}$. By definition of AXp, we know also $(C,s_i) \models \mathsf{PImp}(\lambda_i,i)$ with $i \in \{1,0\}$. Consider the term $\mathsf{cn}_{X \cup Y,(s_0 \cap Plt) \cup (s_1 \cap Dfd)}$. Since it propositionally entails both λ_1 (the negative part of λ_1 is always "a subset of" $s_0 \cap Dfd$) and λ_2 (in the same way), by definition of PImp any state verifying it shall have outputs both 0 and 1, contradicting that f is a function.

□

References

[1] Amgoud, L. and V. Beuselinck, *Towards a principle-based approach for case-based reasoning*, in: *Scalable Uncertainty Management* (2022), pp. 37–46.

[2] Ashley, K. D., "Modeling Legal Argument: Reasoning with Cases and Hypotheticals," MIT, 1990.

[3] Ashley, K. D. and S. Brueninghaus, *Automatically classifying case texts and predicting outcomes*, Artificial Intelligence and Law **17** (2009), pp. 125–165.

[4] Atkinson, K., T. Bench-Capon and D. Bollegala, *Explanation in ai and law: Past, present and future*, Artificial Intelligence **289** (2020), p. 103387.

[5] Bench-Capon, T. J. M. and K. Atkinson, *Precedential constraint: the role of issues*, in: *ICAIL '21: Eighteenth International Conference for Artificial Intelligence and Law, São Paulo Brazil, June 21 - 25, 2021* (2021), pp. 12–21.

[6] Bench-Capon, T. J. M. and E. L. Rissland, *A note on dimensions and factors*, Artificial Intelligence and Law **10** (2002), pp. 65–77.

[7] Canavotto, I., *Precedential constraint derived from inconsistent case bases*, in: *Legal Knowledge and Information Systems - JURIX 2022: The Thirty-fifth Annual Conference, Saarbrücken, Germany, 14-16 December 2022* (2022), pp. 23–32.

[8] Canavotto, I. and J. Horty, *Piecemeal knowledge acquisition for computational normative reasoning*, in: *AIES'22*, ACM, 2022 p. 171–180.

[9] Dalal, M., *Investigations into a theory of knowledge base revision: preliminary report*, in: *Proceedings of the Seventh National Conference on Artificial Intelligence* (1988), pp. 475–479.

[10] Guidotti, R., *Counterfactual explanations and how to find them: literature review and benchmarking*, Data Mining and Knowledge Discovery (2022), pp. 1–55.

[11] Horty, J. F., *Rules and reasons in the theory of precedent*, Legal theory **17** (2011), pp. 1–33.

[12] Horty, J. F., *Reasoning with dimensions and magnitudes*, Artif. Intell. Law **27** (2019), pp. 309–345.

[13] Horty, J. F., *Modifying the reason model*, Artif. Intell. Law **29** (2021), pp. 271–285.

[14] Lewis, D. K., "Counterfactuals," Harvard University Press, 1973.

[15] Liu, X. and E. Lorini, *A logic for binary classifiers and their explanation*, in: *Proceedings of the 4th International Conference on Logic and Argumentation (CLAR 2021)*, Springer-Verlag, 2021, pp. 302–321.

[16] Liu, X. and E. Lorini, *A unified logical framework for explanations in classifier systems*, Journal of Logic and Computation **33** (2023), pp. 485–515.

[17] Liu, X., E. Lorini, A. Rotolo and G. Sartor, *Modelling and explaining legal case-based reasoners through classifiers*, in: *Legal Knowledge and Information Systems*, IOS Press, 2022 pp. 83–92.

[18] Miller, T., R. Hoffman, O. Amir and A. Holzinger, *Special issue on explainable artificial intelligence (xai)*, Artificial Intelligence **307** (2022).

[19] Mumford, J., K. Atkinson and T. J. M. Bench-Capon, *Explaining factor ascription*, in: *Legal Knowledge and Information Systems - JURIX 2021: The Thirty-fourth Annual Conference, Vilnius, Lithuania, 8-10 December 2021*, 2021, pp. 191–196.

[20] Mumford, J., K. Atkinson and T. J. M. Bench-Capon, *Reasoning with legal cases: A hybrid ADF-ML approach*, in: *Legal Knowledge and Information Systems - JURIX 2022: The Thirty-fifth Annual Conference, Saarbrücken, Germany, 14-16 December 2022*, 2022, p. 93–102.

[21] Prakken, H., *A formal analysis of some factor- and precedent-based accounts of precedential constraint*, Artificial Intelligence and Law (2021).

[22] Prakken, H. and R. Ratsma, *A top-level model of case-based argumentation for explanation: Formalisation and experiments*, Argument Comput. **13** (2022), pp. 159–194.

[23] Prakken, H. and G. Sartor, *Modelling reasoning with precedents in a formal dialogue game*, in: *Judicial applications of artificial intelligence*, Springer, 1998 pp. 127–183.

[24] Rissland, E. L. and K. D. Ashley, *A case-based system for trade secrets law*, in: *Proceedings of the First International Conference on Artificial Intelligence and Law (ICAIL)*, ACM, 1987 pp. 60–6.

[25] Rissland, E. L. and K. D. Ashley, *A note on dimensions and factors*, Artificial Intelligence and Law **10** (2002), pp. 65–77.

A Logical Approach to Post-Quantum Conditional Digital Signatures

Xiang Li [1]

School of Law and Intellectual Property, Guangdong Polytechnic Normal University, Guangzhou, China.

Xin Sun [2]

Research Center for Basic Theories of Intelligent Computing, Zhejiang Lab, Hangzhou, China

Xingchi Su [3]

Research Center for Basic Theories of Intelligent Computing, Zhejiang Lab, Hangzhou, China

Abstract

A digital signature is an electronic and encrypted authentication on digital messages accomplished by some mathematical scheme. However, the classical digital signature faces the problems of fair exchange and 'valid once signing'. To overcome these challenges, the notion of conditional digital signatures is proposed in the early twenty-first century. However, the existing characterizations on this notion show some weakness in security, power of expression and feasibility, for instances. This paper formalizes conditional digital signatures in a more abstract form and the condition expressions are designed with logical structures. Based on a chosen post-quantum digital signature scheme, e.g. Crystal-Dilithium, we distinguish two validities: cryptographic validity and practical validity. A constructive method is given to find all practically valid signatures from a collection of cryptographically valid conditional signatures. An algorithm can be induced accordingly and makes the application feasible. Moreover, so as to capture multi-party handshaking principles and the reasoning about conditional signatures, a consistent proof system $\mathbb{CSL}$ is established, which renders the conditional signatures non-monotonic.

Keywords: Conditional digital signature, handshaking principle, cryptographic validity, practical validity, non-monotonicity.

---

[1] The first author. lixiang2110001@163.com

[2] xin.sun.logic@gmail.com

[3] The correspondence author. x.su@zhejianglab.com

1   Introduction

A digital signature is an electronic and encrypted authentication on digital messages accomplished by some mathematical scheme, which brings a very high confidence to the receiver that the messages are originated by the signer and are not distorted during the transmission. It is generally implemented by asymmetric encryption scheme where private-public key pairs, e.g. (sk, pk) are generated for signing message (an encryption operation) and verifying the signature (a decryption operation) respectively. The classical asymmetric encryption scheme can be seen as a dual to the digital signature since the public key is used for encryption and the private key is for decryption. A digital signature is valid if the receiver decrypts the signed part of the signature by the signer's public key and the resulting text equals the plaintext part of the signature. This process verifies the authentication of the message sender. The Digital Signature Algorithm (DSA) is the most widely used standard for digital signatures based on the concept of modular exponentiation and the mathematical discrete logarithm problem [4], which is a variant of the famous ElGamal signature scheme [6].

1.1   Conditional Digital Signatures

Digital signatures are essentially bit strings which can be copied easily and spread to many uninvolved parties. In other words, once a digital signature is issued by its originator, he/she loses the control over the information flow of this his/her signature. This uncontrolled information is not expected in many scenarios, e.g. business negotiations, where digital signature is indispensable. Conditional digital signatures are thence designed to deal with the problem. A conditional signature can be treated as a pre-signature for committing a message and only if the condition is met, the signature comes into force as a classical signature. For example, a company A will sign the contract to pay for the bill provided that the company B signs the contract to deliver the product to A. There have witnessed several research on this special concept of digital signatures. B. Lee and K. Kim implements fair exchange of digital signatures in electronic commerce by conditional digital signature [9]. Considering the compatibility with the current social infrastructure, their conditional signatures still utilize the classical signature, but replace the content of the message m with a concatenation with the condition in the form of $m||h(c)$ where $h(c)$ is a Hash value of the condition c. A initiator A starts up a negotiation with a signed information $m_A; c_A$ conditioned by c_A. If the receiver B accepts the negotiation, B replies with a signed information $m_B; c_B$ where $m_B \in \{c_A\}$ and $m_A = c_B$, which represents that both sides agree to trade and the fair change of their unconditional signatures can be fulfilled. I. Z. Berta et al. enable users to generate digital signatures via conditional signatures constructed as B. Lee and K. Kim indicates, in the help of a trusted smart card inserted to an untrusted terminal [3]. M. Klonowski et al. put forward the first conditional digital signatures which only sign the messages themselves instead of the concatenations of messages and conditions [8]. Their approach extends the classical ElGamal scheme to a tuple $(a_1, b_1 S^z, a_2^z)$ where (a_1, b_1) amounts to the classical ElGamal signature on message M_1 by the signer A and (S^z, a_2^z) plays the role of a commitment from B's signature on M_2, which collectively represents a conditional signature of A on M_1 conditioned by B's signature on M_2.

1.2   Handshaking Principle

It worth noting that one of the most interesting problems discussed in the relevant literature is the handshaking principle, which says that if A signs message m_1 provided that B signs message m_2 (this is a conditional signature), and B signs message m_2 provided that A signs message m_1 (a conditional signature as well), then they are supposed to imply that A really signs m_1 and B really signs m_2 (two classical signatures). This principle is not trivially held in the classical logic system since from $p \rightarrow q$ and $q \rightarrow p$, it cannot be inferred that $p \wedge q$. To cope with the subtlety, B. Lee and K. Kim mentioned above provide a characterization within their framework. That is if A signs $m_A; c_A$ conditioned by c_A and B replies with a signed information $m_B; c_B$ where $m_B \in \{c_A\}$ and $m_A = c_B$, then they 'shake hands' and A actually signs m_A and B actually signs m_B. Upon M. Bartoletti and R. Zunino's formal characterization on contract terms, a logic for contract (PCL) is proposed to reason about the terms, contents and even duties in contracts. They specifically investigate handshaking principle and validate it in PCL by constructing a special implication $\twoheadrightarrow$ for characterizing the conditional clauses [1]. They construct a sound and complete axiom system based on intuitionistic logic with addition to axioms about $\twoheadrightarrow$ and an equivalent Gentzen-style sequence calculus is also given.

1.3   Limitations on Current Conditional Signature Schemes

Not quantum secure All the current constructions of conditional digital signatures are based on the classical cryptographic algorithms, whose security relies on some mathematical hardness problem, such as large number factoring and discrete logarithm problem. However, with the rapid development of quantum computers, the security of classical algorithms is very likely to be broken by Shor's algorithm [10]. Taking into account the quantum-security in the future, we set the digital signature algorithm used in our paper as the post-quantum signature algorithm, such as CRYSTALS-Dilithium [5], Falcon [7] or SPHINCS$^+$ [2].

Interactions with other participants In the previous research like M. Klonowski et al.'s approach, a conditional signature of A is constructed successfully only if B makes a commitment and partially calculates his/her own digital signature. It implies that a conditional signature of A is only possible when both sides interact with each other and an agreement needs to be achieved beforehand. Their approach essentially excludes these scenarios where no precommunication exists. For example, the expression 'A promises to sign the payment contract provided that the production company signs a new contract to purchase the best material' does not require the production company to be informed that they are involved in this conditional signature. It is just a unilateral commitment from A.

Incapability in dealing with complex conditions Since the notion of conditional digital signatures is inspired by some electronic commerce, they mostly constraint on the binary cases where only two parties try to achieve some agreements. It results in a very simple form of the condition in a signature, which generally expresses that one party signs something. But there are a lot of scenarios where a complex condition is necessary. For example, 'A signs the contract provided that B signs the contract and C signs the contract' or 'A signs the contract provided that B signs the contract or C signs the contract'. The condition can be in some logical structure which decides when the signature of A can practically come into force.

1.4   Contributions of our work

Based on some given digital signature scheme, this paper extends the structure of the conditions in conditional signatures to capture multi-participant scenarios. We do not specify the concrete digital signature scheme. It can be the classical Digital Signature Algorithm or a even more modern post-quantum scheme based on lattice hardness problem, e.g. Crystals-Dilithium issued by NIST (National Institute of Standard and Technology) [5]. Considering the forthcoming quantum computers which are capable of breaking the classical cryptographic algorithms, we set the digital signature algorithm used in this paper as the post-quantum signature algorithm, such as Crystals-Dilithium [5], Falcon [7] or SPHINCS$^+$ [2].

As we distinguish two notions on validity: cryptographic validity and practical validity, a constructive method on generating unconditional signatures from a finite set of conditional signatures is to be given as an application in encryption practice. In the light of the algorithm induced by the method, any trusted third party that is authorized to collect conditional digital signatures from different participants can deduce whose signature really comes into force.

In addition, inspired by the logic for contract and B. Lee and K. Kim's original idea of conditional signatures, a Hilbert-style proof system is also to be given in order to validate the multi-party handshaking principle in a non-monotonic manner. Besides binary handshaking principle, some new axioms are introduced to strengthen the expressivity power and thence to derive the desired properties. For the lack of semantics, the soundness and completeness will not be discussed in this paper. Consistency is given to show that the theories of the system by no means derive conflicts. However, in terms of application, the proof system does not directly provide an executable algorithm for computers to find all practically valid signatures. For encryption practice, we insist on the constructive method to be given in Section 3.

This paper is organized as follows. We first introduce the basic scheme of digital signatures in Section 2. Then we will give a novel formalization of conditional digital signatures and show a constructive method in deriving all practically valid signatures from a set of cryptographically valid conditional signatures in Section 3. Section 4 a discussed the non-monotonicity of conditional signatures and releases consistent proof system $\mathbb{CSL}$ which validates multi-party handshaking principles.

2 Preliminaries

Any digital signature algorithm deployed nowadays respects a common scheme where asymmetric encryption process is used. We next introduce it briefly.

Definition 2.1 (Digital signature scheme) A digital signature scheme Σ is a triple of algorithms

$$\Sigma = (KeyGen, Sign, Verify)$$

known as the key generation, signing and verification algorithms and satisfying the correctness condition.

- The key generation algorithm takes no input and produces a pair of keys $(sk, pk) \leftarrow KeyGen()$ known as the secret and public keys for a signer a.

- The signing algorithm takes a secret key sk and a message m as inputs and produces a signature $\sigma \leftarrow Sign(sk, m)$. A tuple (m, σ, a) is generated and is called a signed message.

- The verification algorithm takes the public key pk of a, the message m in the signed message and the corresponding signature σ as inputs and returns a Boolean value b, i.e. $b = 1$ or 0. We say that σ is a valid signature for m under key pk if $Verify(pk, m, \sigma) = 1$.

We elaborate on the above scheme a bit here. After a pair of public key and private key of an agent A is generated, A keeps the private key him/herself and distributes the public key to all other participants. The mathematical hardness problem, e.g. LWE (Learning With Error) problem in post-quantum lattice-based cryptography, on which the signature algorithm is based guarantees that the secret key cannot be calculated or induced from the public key. $\sigma \leftarrow Sign(sk, m)$ represents that m is encrypted by secret key sk and a ciphertext σ is obtained. Lastly, all other participants can verify whether the message is really authenticated by A by decrypting σ via public pk and checking if it equals m. $Verify(pk, m, \sigma) = 1$ means that they are equal, 0 otherwise.

A digital signature scheme must satisfy the *correctness condition*:

(Correctness Condition) All correctly generated signatures are always valid, i.e. $b = 1$.

3 Conditional Digital Signature

Given a post-quantum digital signature scheme, a key generation process distributes each party their own secret-public key pairs. Then we can formalize the conditional digital signature as follows.

3.1 Syntax of conditional signatures

Since we intend to formalize these conditions in some logical structure, we first give the syntax of the conditions.

Definition 3.1 (Syntax of conditions Φ) Given a set of messages M and a set of parties/agents G, the conditions of conditional digital signature are defined inductively as following BNF form:

$$\Phi := \epsilon \mid (m, a) \mid (\Phi \wedge \Phi) \mid (\Phi \vee \Phi)$$

where $m \in M$, $s \in G$ and ϵ represents the empty condition.

The atomic condition is (m, a) which simply means that the agent a signs the message m in the classical (unconditional) sense. In other words, a's signature on m comes into force. Definition 3.1 indicates that we only consider these conditions about signatures from other parties, which is different from the normal implication where conditions can be any state of affairs. The negated condition is not considered here since a too complex structure of conditions brings about some inconceivable philosophical confusions.

3.2 Logical form of cryptographic expression

In Section 2, we introduced that (m, σ, a) constitutes a signed message where σ is a ciphertext involving message m and the signer a. Referring to B. Lee and K. Kim's idea, we treat a conditional digital signature as a special form of classical signature where the message to be signed (e.g. according to Dilithium algorithm) is concatenated with a condition, i.e. $(m; \Phi, \sigma, a)$ which represents a practically signs $m; \Phi$. Logically, it is equivalent to that a signs m if Φ is satisfied.

As mentioned before, σ is used for verifying if the message is really signed by a. But it is largely irrelevant to the upcoming discussion on the logical aspect of conditional signatures. We hence omit σ in the following formalization and call $(m; \Phi, \sigma, a)$ the *cryptographic form* of a conditional signature and call $Sig((m, a)|\Phi)$ the *logical form* of a conditional signature. As a reminder, in the cryptography, the standard post-quantum signature algorithm first computes the Hash value of the message m and then signs it, which guarantees the length of the signed message always 512 bits. This is a chosen parameter which not only protects the security of the signature, but makes it not exceeding hard for receivers to verify (compute) the signature. For the sake of simplicity and readability, we omit the Hash function used in algorithm and only keep the most general form of digital signatures: $(m; \Phi, \sigma, a)$ as their cryptographic form.

Definition 3.2 (Conditional signature) Given a message $m \in M$, an agent $a \in G$ and a condition Φ, a conditional signature (logical form) is represented by $Sig((m, a)|\Phi)$.

Depending on different Φ, $Sig((m, a)|\Phi)$ can be read differently respectively. For instances,

- $Sig((m, s)|\epsilon)$: a practically signs m;
- $Sig((m, a)|(m', a'))$: a signs m provided that a' signs m';
- $Sig((m, a)|(m_1, a_1) \wedge (m_2, a_2))$: a signs m provided that a_1 signs m_1 and a_2 signs m_2;

- $Sig((m, a)|(m_1, a_1) \vee (m_2, a_2))$: a signs m provided that a_1 signs m_1 or a_2 signs m_2;

When $\Phi = \epsilon$, we also write $Sig((m, a)|\epsilon)$ as (m, a) since both represent that a practically signs m.

Definition 3.3 (Syntax of conditional signatures) Given a set of messages M, a set of agents G, and a countable set of propositions P, the syntax of conditional digital signatures is defined inductively as following BNF form:

$$\phi := p \mid \neg\phi \mid sig((m, a)|\Phi) \mid (\phi \wedge \phi)$$

where $m \in M$, $s \in G$ and $p \in P$.

Definition 3.3 restricts negation to the outside of conditional signatures. So a formula in the form $Sig(\neg(m, a)|\Phi)$ is not well-defined.

3.3 Binary handshaking

The logic for contract formalizes the binary handshaking principle as below:

$$(p \twoheadrightarrow q) \wedge (q \twoheadrightarrow p) \rightarrow (p \wedge q)$$

It shows a basic property of contractual implication ($\twoheadrightarrow$) that it allows two dual contracting parties to handshake so as to make their agreement effective. Similarly, the binary handshaking in the context of conditional signatures can be captured by the following formula in our syntax:

Binary Handshaking Principle (HS)

$$Sig((m, a)|(m', a')) \wedge Sig((m', a')|(m, a)) \rightarrow (m, a) \wedge (m', a')$$

The formula (HS) can be read as: a signs m provided that a' signs m', and a' signs m' provided that a signs m, then a practically signs m and a' practically signs m'. This represents that a and a' shake hands and really sign the messages of their own parts. It plays an important role in deriving unconditional signatures in the upcoming constructive method. And it also must be included as an axiom or a derived formula in our proof system of the conditional digital signatures in Section 4.

3.4 Constructively find practical validity

In the classical research on the cryptographic aspect of digital signatures, the correctness condition gains the most attention. Once a signature is generated correctly, it comes into force. However, a correctly constructed conditional digital signature by no means implies that the signer practically signs the message. It merely represents that the signer makes some commitment under certain condition. To clarify the discrepancy between them, we distinguish two validity: cryptographic validity and practical validity.

Definition 3.4 (Cryptographic validity) Let $Sig((m, a)|\Phi)$ be a conditional signature and $(m; \Phi, \sigma, a)$ is the corresponding cryptographic form of it. $Sig((m, a)|\Phi)$ is cryptographic valid (c-valid) if $Verify(pk_a, m; \Phi, \sigma) = 1$.

A conditional signature is cryptographically valid if it is generated correctly according to some digital signature algorithm. Upon this notion, practical validity is given based on a set of cryptographically valid conditional signatures.

Definition 3.5 (*W*-validity of conditions) Let W be a set of c-valid conditional digital signatures and Φ is a condition defined as Definition 3.1. W-validity of Φ is defined inductively as follows:

- When $\Phi = \epsilon$, then Φ is W-valid trivially;
- When $\Phi = (m, a)$, Φ is W-valid if $(m, a) \in W$ i.e. $Sig((m, a)|\epsilon) \in W$;
- When $\Phi = \Phi_1 \wedge \Phi_2$, Φ is W-valid if Φ_1 is W-valid and Φ_2 is W-valid;
- When $\Phi = \Phi_1 \vee \Phi_2$, Φ is W-valid if Φ_1 is W-valid or Φ_2 is W-valid;

Definition 3.6 (Practical validity) Let W be a set of c-valid conditional digital signatures. (m, a) is valid in W (notated as $W \vDash (m, a)$) is defined as follows:

- If $Sig((m, a)|\Phi) \in W$ and Φ is W-valid, then $W \vDash (m, a)$;
- If $Sig((m, a)|(m', a')) \in W$ and $Sig((m', a')|(m, a)) \in W$, then $W \vDash (m, a)$ and $W \vDash (m', a')$.

As shown above, practical validity is only defined with respect to unconditional signatures. So $W \vDash (m, a)$ intuitively means that (m, a) is a signature that comes into force induced from the set of conditional signatures W. Considering the application, is there a constructive method or an algorithm to find all practically valid signatures given a set of conditional signatures?

Definition 3.7 (An inductive construction) Given a set of conditional signatures W. We define $PV(W) = \bigcup_{i=0}^{\infty} PV_i(W)$ where

$$
\begin{aligned}
\text{(i) } PV_0(W) \quad = \quad &\{(m, a)|(m, a) \quad \in \quad W\} \quad \cup \\
&\{(m_1, a_1)|Sig((m_1, a_1)|(m_2, a_2)) \quad \in \\
&W \text{ and } Sig((m_2, a_2)|(m_1, a_1)) \in W\} \\
\text{(ii) } PV_{i+1}(W) = \quad &PV_i(W) \cup \{(m, a)|Sig((m, a)|\Phi) \in \\
&W \text{ and } \Phi \text{ is } PV_i(W)\text{-valid}\}
\end{aligned}
$$

Let us explain the inductive construction above. In the first step, we collect all unconditional signatures in W and also derive all unconditional signatures generated by the binary handshaking principle. They constitute $PV_0(W)$. The second step is adding all unconditional signatures detached from those conditional signatures whose conditions are 'satisfied' in $PV_0(W)$, i.e. $PV_0(W)$-valid. So on so forth. We take the union of all $PV_i(W)$ and the final set $PV(W)$ is obtained. Definition 3.7 thence provides a constructive method to derive all practically valid signatures given a set of cryptographically valid conditional signatures. It can be easily transformed to an algorithm to compute $PV(W)$. The whole process for an authority to find all practically valid signatures in application is shown as following steps:

Step 1: All participants submit their conditional digital signatures to the authority. Each conditional signature is formalized as $Sig((m,a)|\Phi)$.

Step 2: After collecting all conditional signatures, the authority verifies all and only retains these cryptographically valid conditional signatures.

Step 3: Let the set of all cryptographically valid conditional signatures be W. The authority computes the set $PV_0(W) = \{(m,a)|(m,a) \in W\} \cup \{(m_1,a_1)|Sig((m_1,a_1)|(m_2,a_2)) \in W$ and $Sig((m_2,a_2)|(m_1,a_1)) \in W\}$

Step 4: Based on $PV_0(W)$, the authority computes $PV_1(W) = PV_0(W) \cup \{(m,a)|Sig((m,a)|\Phi) \in W$ and Φ is $PV_0(W)$-valid$\}$

Step 5: According to the method shown in Step 4, the authority computes $PV_2(W)$, $PV_3(W)$, $\cdots$ and so on so forth. When we meet a i such that $PV_i(W) = PV_{i+1}(W)$, we move to the next step.

Step 6: The authority computes $PV(W) = \bigcup_{i=0}^{\infty} PV_i(W)$

The following proposition shows the equivalence between practical validity and the set $PV(W)$. The proof is not hard and is omitted here.

Proposition 3.8 *Given a set of c-valid conditional signatures W, for arbitrary unconditional signatures,*

$$W \vDash (m,a) \text{ iff } (m,a) \in PV(W).$$

4 A Proof System for Conditional Signatures $\mathbb{CSL}$

A constructive method in finding all practical valid signatures based on binary handshaking principle is sufficient for application in most real-life cases. For further study on the reasoning about conditional signatures and multi-party handshaking principles, a more powerful proof system is required. The logic for contract (PCL) given by M. Bartoletti and R. Zunino [1] seems to be a proper candidate for the logical basis of the conditional digital signatures at first glace. However, their new implication for conditional clauses $\twoheadrightarrow$ is monotonic. In the context of conditional signatures, we cannot infer a conditional signature like 'A signs the contract provided that B signs the contract and C signs the contract' from the conditional signature saying 'A signs the contract provided that B signs the contract' without mentioning C's behavior. Therefore, $\twoheadrightarrow$ does not fit to conditional signatures that we intend to study in this paper. Moreover, even the original definition for conditional signatures in commercial fair exchange given in [9] is monotonic. To characterize the non-monotonicity of conditional signatures and multi-party handshaking principles, a proof system is specifically designed in this section.

Fig. 1. Three-party handshaking

4.1   Non-monotonicity

A monotonic implication $\to$ is monotonic if it satisfies $(\phi \to \psi) \wedge (\phi' \to \phi) \to (\phi' \to \psi)$. It means that if ϕ implies ψ, then a stronger precondition ϕ' should imply ψ as well. In the context of logic of conditionals, it is also called *precondition strengthening*. This property is quite intuitive in most cases. However, with development of logic of conditionals and more non-monotonic conditionals are found, such as counterfactual conditionals and conditional obligations, non-monotonicity gains a lot of attention.

   If we take a closer look at the notion of conditional signatures that we propose to study, it does not satisfy monotonicity. Let us see the following example.

Example 4.1 Agent a signs message m provided that b signs message m_1. But it is by no means that 'a signs message m provided that b signs message m_1 and c signs message m_2' can be derived, even though 'b signs message m_1 and c signs message m_2' trivially implies 'b signs message m_1'.

   The example inspires that in a conditional signature, only if the condition is exactly met, the signature takes effect. Any strengthening or weakening is not compliant. Thus, we should exclude the formula below from our characterization of conditional signatures.

$$(\Phi' \to \Phi) \wedge Sig((m, a)|\Phi) \to Sig((m, a)|\Phi')$$

4.2   Multi-party handshaking principle

As an extension to the binary handshaking principle introduced in Section 3.3, three or more participants can achieve some practically valid signatures when their conditional signatures form a 'cycle', shown as Figure 1.

   In Figure 1, the curved arrow from red node a to red node b represents a conditional signature given by the agent b that b signs a message given that agent a signs some message, similarly for the arrow from b to c and c to a. When three participants' conditional signatures (red nodes) form a cycle, all their conditional signatures come into force and three real signatures (green nodes) are obtained. Accordingly, we can formalize the three-party handshaking principle as the following formula:

Three-party Handshaking Principle

$$Sig((m,a)|(m',b)) \land Sig((m',b)|(m'',c)) \land Sig((m'',c)|(m,a))$$
$$\rightarrow (m,a) \land (m',b) \land (m'',c)$$

However, the above formalization is not satisfactory since if there are n ($n \geq 3$) parties, we need to give a specific formula to characterize the n-party handshaking principle, respectively. That looks very redundant. A proof system that is expressively strong enough to derive multi-party handshaking principles are ready to come out.

In the next part , we show a consistent proof system for conditional digital signatures which can derive multi-party handshaking principles.

4.3 The proof system $\mathbb{CSL}$

Definition 4.2 The proof system $\mathbb{CSL}$ consists of following axiom schemes and inference rules:

(TAUT) All instances of propositional tautologies

(Trans) $Sig((m,a)|(m',a')) \land Sig((m',a')|(m'',a'')) \rightarrow Sig((m,a)|(m'',a''))$
(UHS) $Sig((m,a)|(m,a)) \rightarrow (m,a)$
(Detach) $Sig((m,a)|\Phi) \land \Phi \rightarrow (m,a)$
(MP) From ϕ and $\phi \rightarrow \psi$, infer ψ

In the system $\mathbb{CSL}$, the axiom (Trans) represents a transitivity of conditional signatures. It means that if we have a signs m provided that a' signs m' and a' signs m' provided that a'' signs m'', then it is implied that a signs m provided that a'' signs m''. It might not be trivially valid. But it makes sense if we allow the agent do some very reasonably and logically correct, but easy inference. If a intends to sign m and he/she also knows that $Sig((m',a')|(m'',a''))$, then a completely understand his/her conditional signature is practically equivalent to $Sig((m,a)|(m'',a''))$. The axiom (UHS) is an abbreviation to 'unary handshaking' which is a special form of handshaking principles. The axiom (Detach) shows that the signature of a on m takes effect once the condition Φ is satisfied.

4.4 $\mathbb{CSL}$-Theories And Consistency

As mentioned in Section 4.2, we hope handshaking principles are axioms or can be derived from our proof system, in order to show our characterization validate handshaking in the context of conditional digital signatures.

Lemma 4.3 *The following formulas are derivable in $\mathbb{CSL}$:*

(HS) $Sig((m,a)|(m',a')) \land Sig((m',a')|(m,a)) \rightarrow (m,a) \land (m',a')$
(MHS) for any $n \geq 3$, it holds that
$Sig((m_1,a_1)|(m_2,a_2)) \land Sig((m_2,a_2)|(m_3,a_3)) \land \cdots Sig((m_n,a_n)|(m_1,a_1))$
$\rightarrow (m_1,a_1) \land (m_2,a_2) \land \cdots (m_n,a_n)$

Proof. (HS) is derivable in $\mathbb{CSL}$:

1	$Sig((m,a)|(m',a')) \wedge Sig((m',a')|(m,a)) \rightarrow Sig((m,a)|(m,a))$	(Trans)
2	$Sig((m',a')|(m,a)) \wedge Sig((m,a)|(m',a')) \rightarrow Sig((m',a')|(m',a'))$	(Trans)
3	$Sig((m,a)|(m,a)) \rightarrow (m,a)$	(UHS)
4	$Sig((m',a')|(m',a')) \rightarrow (m',a')$	(UHS)
5	$Sig((m,a)|(m',a')) \wedge Sig((m',a')|(m,a)) \rightarrow (m,a)$	1,3
6	$Sig((m',a')|(m,a)) \wedge Sig((m,a)|(m',a')) \rightarrow (m',a')$	2,4
7	$Sig((m,a)|(m',a')) \wedge Sig((m',a')|(m,a)) \rightarrow (m,a) \wedge (m',a')$	5,6

(MHS) is derivable in $\mathbb{CSL}$:

For each $3 \leq i \leq n-1$,

1	$Sig((m_1,a_1)|(m_2,a_2)) \wedge \cdots Sig((m_{i-1},a_{i-1})|(m_i,a_i)) \rightarrow$ $Sig((m_1,a_1)|(m_i,a_i))$	(Trans)
2	$Sig((m_i,a_i)|(m_{i+1},a_{i+1})) \wedge \cdots Sig((m_n,a_n)|(m_1,a_1)) \rightarrow$ $Sig((m_i,a_i)|(m_1,a_1))$	(Trans)
3	$Sig((m_1,a_1)|(m_2,a_2)) \wedge \cdots Sig((m_n,a_n)|(m_1,a_1)) \rightarrow$ $(m_1,a_1) \wedge (m_i,a_i)$	1,2,(HS)

For each $i = n$,

1	$Sig((m_1,a_1)|(m_2,a_2)) \wedge \cdots Sig((m_{n-1},a_{n-1})|(m_n,a_n)) \rightarrow$ $Sig((m_1,a_1)|(m_n,a_n))$	(Trans)
2	$Sig((m_1,a_1)|(m_n,a_n)) \wedge Sig((m_n,a_n)|(m_1,a_1)) \rightarrow$ $(m_1,a_1) \wedge (m_n,a_n)$	(Trans)
3	$Sig((m_1,a_1)|(m_2,a_2)) \wedge \cdots Sig((m_n,a_n)|(m_1,a_1)) \rightarrow$ $(m_1,a_1) \wedge (m_n,a_n)$	1,2,(HS)

As a summary, (MHS) is derivable in $\mathbb{CSL}$.   □

The formula (HS) is the binary handshaking principle. The formula (MHS) characterizes the multi-party handshaking principles in any group scale.

Lemma 4.4 *The following formulas are not derivable in $\mathbb{CSL}$:*

(MN) $\nvdash_{\text{CSL}} Sig((m,a)|\Phi) \wedge (\Phi' \rightarrow \Phi) \wedge \neg(\Phi \rightarrow \Phi') \rightarrow Sig((m,a)|\Phi')$

(US) $\nvdash_{\text{CSL}} (m,a) \rightarrow Sig((m,a)|\Phi)$ *for* $\Phi \neq \epsilon$

Lemma 4.4 can be proved by showing that none of $\mathbb{CSL}$-axioms replaces the condition part of a conditional signatures with a stronger condition. We do not show the proof in detail here. The formula (US) is a special form of (MN) when $\Phi = \epsilon$.

As mentioned above, $\mathbb{CSL}$ is designed to be non-monotonic. The underivability of (MN) is a witness to this property. The formula (US) says if a really signs m, then he/she signs m under arbitrary condition. (US) is valid in PCL, but it is counter-intuitive in terms of conditional signatures.

The last problem is to prove that $\mathbb{CSL}$ is consistent. For the lack of the semantics, we will prove it with respect to the notion of syntax-consistency. The following definition is necessary. In the remaining part, PL represents the classical propositional logic.

Definition 4.5 (PL-Transformation) For each $\mathbb{CSL}$-formula ϕ, ϕ' denotes its PL-transformation which is defined inductively as follows:

- $p' = p$
- $(\neg\phi)' = \neg\phi'$
- $Sig((m,a)|\Phi)' = p_{m,a}$ where $p_{m,a}$ is a newly added propositional letter.
- $(\phi_1 \wedge \phi_2)' = \phi'_1 \wedge \phi'_2$

Lemma 4.6 *The PL-transformation of each $\mathbb{CSL}$-axiom is also a PL-theory.*

Proof. According to Definition 4.5,

- $(\text{Trans})' = p_{m,a} \wedge p_{m',a'} \to p_{m,a}$, which is a PL-theory;
- $(\text{UHS})' = p_{m,a} \to p_{m,a}$, which is a PL-theory;
- $(\text{Detach})' = p_{m,a} \wedge \Phi' \to p_{m,a}$ where Φ can be arbitrary conjunction or disjunction combination of (m, a). $(\text{Detach})'$ is also a PL-theory.
- For each instance of tautologies ϕ, $\phi' = \beta'[p_1/\gamma'_1, \cdots, p_n/\gamma'_n]$. Since β itself is a PL-theory and each γ'_i is a PL-formula, it is trivial to obtain that ϕ' is a PL-theory as well.

$\square$

Lemma 4.7 *The PL-transformation of each $\mathbb{CSL}$ is also PL-theory.*

Proof. Let ϕ be an arbitrary $\mathbb{CSL}$-theory. There must be a proof sequence to derive ϕ by $\mathbb{CSL}$-axioms and (MP) rule. Do induction on the length n of the proof:

(i) $n = 0$: ϕ is a $\mathbb{CSL}$-axiom. By Lemma 4.6, ϕ' is PL-theory.

(ii) $n = l$: The PL-transformation of each formula in the sequence is PL-theory.

(iii) $n = l + 1$:
  (a) ϕ is a $\mathbb{CSL}$-axiom. By Lemma 4.6, ϕ' is PL-theory.
  (b) ϕ is obtained by substituting the propositional part of a formula β with a $\mathbb{CSL}$-formula. Let $\phi = \beta[p/\gamma]$. By induction hypothesis, β' is a PL-theory. γ is a PL-formula. So $\beta'[p/\gamma']$ is still a PL-theory.
  (c) ϕ is obtained by (MP) rule from $\vdash_{\text{CSL}} \beta \to \phi$ and $\vdash_{\text{CSL}} \beta$. By induction hypothesis, $(\beta \to \phi)'$ is a PL-theory and $beta'$ is a PL-theory. So ϕ' is a PL-theory.

$\square$

Theorem 4.8 *The proof system $\mathbb{CSL}$ is syntax-consistent.*

Proof. Let ϕ be an arbitrary $\mathbb{CSL}$-formula. We need to prove that if $\vdash_{\text{CSL}} \phi$, then $\nvdash_{\text{CSL}} \neg\phi$. Suppose that $\vdash_{\text{CSL}} \phi$ and ϕ' is its PL-transformation. By Lemma 4.7, $\vdash_{\text{CSL}} \phi'$. Since PL is syntax-consistent, it holds that $\nvdash_{\text{CSL}} \neg\phi'$. By Lemma 4.7 again, we have $\nvdash_{\text{CSL}} \neg\phi$. Therefore, $\mathbb{CSL}$ is syntax-consistent. $\square$

5   Comparison And Conclusion

Comparison with M. Klonowski *et al.*'s work M. Klonowski *at al.* designed the very first conditional signature scheme based on ElGamal signature scheme. It goes further than B. Lee and K. Kim who initially put forward the concept of conditional signatures in the sense that M. Klonowski *at al.*'s scheme mathematically separates the message and condition. Their scheme extends the classical ElGamal scheme to a tuple $(a_1, b_1 S^z, a_2^z)$ where (a_1, b_1) amounts to the classical ElGamal signature on message M_1 by the signer A and (S^z, a_2^z) plays the role of a commitment from B's signature on M_2, which collectively represents a conditional signature of A on M_1 conditioned by B's signature on M_2. Once B signs M_2 and produces his/her own signature (a_2, b_2), everyone can transform A's conditional signature tuple $(a_1, b_1 S^z, a_2^z)$ into A's signature by computing $\frac{b_1 S^z}{(a_2^z)^{b_2}} = \frac{b_1 S^z}{(a_2^{b_2})^z} = \frac{b_1 S^z}{(S^z)} = b_1$ where $S = a_2^{b_2}$. So the complete A's signature (a_1, b_1) is obtained. It is obvious that this method requires a pre-communication with B if A wants to establish a conditional signature since B needs to compute (S^z, a_2^z) beforehand. In addition, this condition can only be given by one single participant, like B. More complex conditions are incompatible. Our method deals with condition following B. Lee and K. Kim's way in order not to change the basic signature scheme. It paves the way to make any digital signature scheme feasible, including port-quantum signature schemes. Meanwhile, our approach embeds logical relation into the expression of conditions where any finite combination of conjunction and disjunction of atomic signatures are allowed.

Comparison with PCL The logic for contract given in [1] treats conditional clauses in contracts as a new implication $\twoheadrightarrow$. Their proof system includes the axiom $(p' \to p) \wedge (p \twoheadrightarrow q) \to (p' \twoheadrightarrow q)$ which suggests that $\twoheadrightarrow$ is monotonic. As discussed during the whole paper, conditional signatures should not be monotonic and therefore the proof system of $\mathbb{CSL}$ cannot be inherited directly from PCL. $\mathbb{CSL}$ is non-monotonic since $\nvdash_{\text{CSL}} Sig((m, a)|\Phi) \wedge (\Phi' \to \Phi) \wedge \neg(\Phi \to \Phi') \to Sig((m, a)|\Phi')$.

Conclusion The emergence of conditional digital signatures aims to solve the problem of fair exchange and 'valid once signing'. Many digital commercial scenarios requires conditional signatures for protecting interests of signers and brings more efficiency in contracts coming into force. However, the existing characterizations on this notion face several challenges in security, power of expression and feasibility, for instances. This paper formalized conditional digital signatures in a more abstract form $Sig((m, a)|\Phi)$ where (m, a) represents that a signs m and Φ is a condition expression with a logical structure. Based on a chosen post-quantum digital signature scheme, e.g. Crystal-Dilithium, we distinguish two validities: cryptographic validity and practical validity. A constructive method is designed to find all practically valid signatures from a collection of cryptographically valid conditional signatures. An algorithm can be induced accordingly and makes the application feasible. Moreover, so as to capture multi-party handshaking principles and the reasoning about conditional signatures, a consistent proof system $\mathbb{CSL}$ is established, which renders the conditional signatures non-monotonic.

Acknowledgement

This work was supported by the National Key R&D Program of China (2021YFB2700503), the Research Initiation Project of Zhejiang Lab, the National Natural Science Foundation of China (62071222, U20A20176), the Natural Science Foundation of Jiangsu Province (BK20200418), the Guangdong Basic and Applied Basic Research Foundation (2021A1515012650), and the Shenzhen Science and Technology Program (JCYJ20210324134810028, JCYJ20210324134408023).

References

[1] Bartoletti, M. and R. Zunino, *A calculus of contracting processes*, in: *2010 25th Annual IEEE Symposium on Logic in Computer Science*, IEEE, 2010, pp. 332–341.

[2] Bernstein, D. J., A. Hülsing, S. Kölbl, R. Niederhagen, J. Rijneveld and P. Schwabe, *The sphincs+ signature framework*, in: *Proceedings of the 2019 ACM SIGSAC conference on computer and communications security*, 2019, pp. 2129–2146.

[3] Berta, I. Z., L. Buttyán and I. Vajda, *Mitigating the untrusted terminal problem using conditional signatures*, in: *Proceedings of the International Conference on Information Technology: Coding and Computing (ITCC'04) Volume 2-Volume 2*, 2004, p. 12.

[4] Bruce, S., *Applied cryptography: Protocols, algorthms, and source code in c.-2nd* (1996).

[5] Ducas, L., E. Kiltz, T. Lepoint, V. Lyubashevsky, P. Schwabe, G. Seiler and D. Stehlé, *Crystals-dilithium: A lattice-based digital signature scheme*, IACR Transactions on Cryptographic Hardware and Embedded Systems (2018), pp. 238–268.

[6] ElGamal, T., *A public key cryptosystem and a signature scheme based on discrete logarithms*, IEEE transactions on information theory **31** (1985), pp. 469–472.

[7] Fouque, P.-A., J. Hoffstein, P. Kirchner, V. Lyubashevsky, T. Pornin, T. Prest, T. Ricosset, G. Seiler, W. Whyte, Z. Zhang et al., *Falcon: Fast-fourier lattice-based compact signatures over ntru*, Submission to the NIST's post-quantum cryptography standardization process **36** (2018).

[8] Klonowski, M., M. Kutyłowski, A. Lauks and F. Zagórski, *Conditional digital signatures*, in: *Trust, Privacy, and Security in Digital Business: Second International Conference, TrustBus 2005, Copenhagen, Denmark, August 22-26, 2005. Proceedings 2*, Springer, 2005, pp. 206–215.

[9] Lee, B. and K. Kim, *Fair exchange of digital signatures using conditional signature*, in: *Symposium on Cryptography and Information Security*, 2002, pp. 179–184.

[10] Shor, P. W., *Algorithms for quantum computation: discrete logarithms and factoring*, in: *Proceedings 35th annual symposium on foundations of computer science*, Ieee, 1994, pp. 124–134.

A Deontic Logic for Programming Rightful Machines: Kant's Normative Demand for Consistency in the Law

Ava Thomas Wright [1]

California Polytechnic State University San Luis Obispo

Abstract

In this paper, I set out some basic elements of a deontic logic with an implementation appropriate for handling conflicting legal obligations for purposes of programming autonomous machine agents. Relying on Immanuel Kant's philosophy of law, I argue that a deontic logic of the law should not try to work around such conflicts but, instead, identify and expose them so that the rights and duties that generate inconsistencies can be explicitly qualified and the conflicts resolved. Kantian justice demands that enforceable laws be consistent, precise, and minimally justifiable in a system. I then argue that a credulous, non-monotonic deontic logic can handle legal conflicts to satisfy these normative demands, with appropriate modifications. Finally, I propose an implementation of this logic via a modified form of "answer set programming," which I demonstrate with some simple examples. This proposed implementation helps advance the design of "rightful machines," autonomous machine agents that respect the authority of legitimate law.

Keywords: Conflicts, Deontic Logic, Non-Monotonic Logic, Kant, Law, Answer Set Programming, Logic Programming, Rightful Machines, Consistency, Standard Deontic Logic,

1   Rightful Machines

According to Immanuel Kant, appeals to reason alone cannot completely specify what our rights and duties with respect to each other are in disputed cases [7, 6:312]. In a society of moral equals, each person "has [her] own right to do what seems right and good to [her] and not to be dependent on another's opinion about this," Kant says [7, 6:312]. Hence even if everyone strives to act perfectly ethically with respect to others, rightful relations are impossible in the absence of a legitimate public authority, since "when rights are in dispute (ius controversum), there would be no judge competent to render a verdict having rightful force" [7, 6:312].

What is required, Kant argues, is

---

[1] avwright@calpoly.edu

a system of laws for a people...which because they affect one another, need a
rightful condition under a will uniting them, a constitution (constituto), so
that they may enjoy what is laid down as right. [7, 6:311]

Kant refers to this system of public laws and institutions as "public right" and
a society existing under such a system as one existing in a "rightful" or "civil"
condition, as opposed to a "state of nature." Only by constituting a *united*
will to authoritatively define, enforce, and determine our rights and duties
with respect to each other can we avoid injustice in inevitable cases of conflict
between our rights, Kant argues (see [7, 6:313-14]). The determinations of a
legitimate public authority as to the rights and duties of everyone interacting
in community therefore generally take moral priority over individual ethical
judgments in cases of conflict. To reject the authority of legitimate public
law and institutions and instead use one's own private ethical judgment in
such cases is to act wrongly, indeed, to commit wrong "in the highest degree,"
according to Kant [7, 6:308n].

Kant's philosophy of law is thus neither strictly positivist nor natural-legal.
On the one hand, Kant argues that there is a necessary connection between
positive law and morality because positive laws that violate conditions necessary to constitute the united will are illegitimate and, therefore, lack moral
authority. [2] These conditions include, among others, respect for fundamental
natural rights of freedom, independence, and equality as well as requirements
that the laws be minimally rational and justifiable [10, p. 170-185]. A positive
law that violates such fundamental rights or principles imposes no moral duty
to obey it. But on the other hand, Kant argues that positive laws that do not
violate such rights or principles are in general morally authoritative, even if
those laws may otherwise be unwise or unjust. The positive laws of a legitimate public authority are necessary to cure the problem of indeterminacy of
rights in the state of nature that makes rightful relations impossible. A positive
law that is unjust may still impose a moral duty to obey it.

Kant's hybrid philosophy of law demands that enforceable laws be made
consistent, precise, and minimally rationally justifiable in a system. These normative demands will shape and constrain how a deontic logic of the law should
handle apparent conflicts between legal obligations. My aim in what follows is
to set out the basic elements of such a deontic logic, together with an implementation that could reasonably be part of the programming of an autonomous
machine agent. I first argue that a credulous, non-monotonic deontic logic can
adequately satisfy Kant's normative demand for consistency in the laws, with
appropriate modifications. I then propose an implementation of this logic via
a modified form of "answer set programming," which I demonstrate with some
simple examples. I hope this proposal will help advance the design of "rightful

---

[2] The authority of a law refers to whether citizens have a moral duty to obey it, whereas its
legitimacy refers to the moral permissibility of enforcement of the law, regardless of whether
citizens have a duty to obey it or not. Kant's position is that the law's legitimacy is a
necessary condition for its authority.

machines," autonomous machine agents that properly respect the authority of the law within the bounds of fundamental rights and principles of justice [12] [13].

2 Conflicts Between Legal Obligations

2.1 The Normative Demand for Consistency in the System of Laws

Kant appears to deny that there can ever be conflicts between legal obligations:

> [S]ince duty and obligation are concepts that express the objective practical necessity of certain actions, and two rules opposed to each other cannot both be necessary at the same time—rather if it is one's duty to act according to one of them, to act according to the opposite one is not only no duty, but even contrary to duty—a collision of duties and obligations is not even conceivable (obligationes non colliduntur). [7, 7:224]

Kant argues here that if one were required to perform an action in accordance with an obligation (Oa) that opposed another simultaneous obligation prohibiting the action ($O \sim a$), then acting in accordance with the first obligation (a) would imply acting in a way that violated the second obligation ($\sim a$), a performance that is not even conceivable ($a \wedge \sim a$). One cannot be obligated to perform what is impossible ($O(a \wedge \sim a)$); therefore, Kant concludes, one cannot simultaneously be subject to opposing obligations ($Oa \wedge O \sim a$). (Here "O" is a monadic operator for an obligation one has; "a" is an action one performs.)

Kant's claim that obligations cannot come into conflict ($\sim (Oa \wedge O \sim a)$) may be understood either descriptively or normatively. Understood descriptively, the claim seems false. There seems to me no reason to think that even a thoroughly rational public authority might not create legal obligations that contradict in situations that authority did not foresee. For example, suppose a state authority enacts a traffic law that requires stopping at stop signs and also another law that forbids stopping in front of military bases (see [9] : 179). It is not inconceivable that a local government agency might then erect a stop sign in front of a military base, creating a conflict of narrow legal obligations under applicable enforceable laws for drivers unfortunate enough to encounter the situation. The possibility of such conflicts seems a mundane descriptive fact about any system of laws, and while one might be tempted to assert that the ordinances in question cannot be held to conflict in the case because the driver can have only one true legal obligation, this assertion seems clearly normative rather than descriptive.

Kant's claim that legal duties cannot conflict should be understood as a normative constraint on the prescriptive system of enforceable legal obligations. At the *descriptive* level, law contains contradictory obligations; but at the *prescriptive* level of enforceable obligations, the laws should be completely consistent. What does Kant's normative demand for consistency in the system of enforceable public laws imply for a deontic logic of the law?

2.2 The Inadequacy of Standard Approaches to Legal Conflict

The standard system of deontic logic (SDL) is a normal modal logic with a deontic gloss on the $\Box$ (box) and $\Diamond$ (diamond) operators, interpreted as obligation (O) and permission (P), respectively. The system is a K logic characterized, syntactically, by the D (deontic) axiom, '$\Box p \to \Diamond p$' (that is, if action p is obligatory, then p is permitted, $Op \to Pp$) or the 'D$\Diamond$Introduction' rule in a Fitch-style proof system, and, semantically, by a seriality condition on frames in the Kripkean possible world semantics (that is, for every world, there is at least one accessible world). What SDL amounts to is the rejection of conflicts of obligation ($\sim (\Box p \land \Box \sim p)$), which is just the D axiom.

But since, as we have seen, there is no reason to think that deontic conflicts do not occur in the law as a descriptive matter, an adequate deontic logic of the law should not deny the possibility of such conflicts, as SDL does. Yet if one rejects axiom D so as to admit conflicts of obligation into SDL, then the logic becomes immediately incoherent, since given some standard principles for the inheritance of obligations (RM) (If $\vdash p \to q$, then $\vdash Op \to Oq$) and aggregation (AND) ($\vdash (Op \land Oq) \to O(p \land q)$), one can derive any obligation from the contradiction in accordance with the classical logical principle ex falso quodlibet (EFQ) $((p \land \sim p) \to q)$ [5, p. 463–4]. That is, given a dilemma where simultaneously Op and $O \sim p$, any arbitrary action q can be proven to be obligatory. (For example: 1. Op. assp. 2. $O \sim p$. assp. 3. $O(p \land \sim p)$. 1,2 AND. 4. $(p \land \sim p) \to q$. EFQ. 5. $O(p \land \sim p) \to Oq$. 4 RM. 5. Oq. 3,4 MP.) A number of efforts to weaken one or more of these principles in order to avoid the deontic explosion of arbitrary obligations have therefore been undertaken, though with limited success.

Semi-classical and paraconsistent logics avoid this inferential explosion by replacing the two truth values (true, false) of classical semantics with a semantics of many values (e.g., null, just true, just false, and both true and false) [4, p. 99–105, 195–196]. These logics have been thought too weak to be very useful, however, because they fail to vindicate certain common, intuitively valid deontic arguments. For example: 1. S ought to fight in the war or perform alternative service to his country ($O(f \lor a)$). 2. S ought not fight ($O \sim f$). 3. Therefore, Smith ought to perform alternative service to his country (Oa) [5, p. 467]. This intuitively valid conclusion cannot be derived in most paraconsistent or relevance deontic logic systems because they lack the disjunctive syllogism of propositional calculus needed to make the inference $(f \lor a) \land \sim f \to a$. Such failures are not conclusive, however, and overcoming them continues to be an area of active research.

Other efforts to describe contradictions while avoiding deontic inferential explosion attempt to do so by weakening Aggregation (AND) or Inheritance of Obligations (RM), rather than by rejecting classical EFQ. They typically do so by imposing prior consistency or permissibility checks. For example, Aggregation (AND) may be weakened by requiring that p and q be jointly possible before allowing their aggregation under obligation (CAND: If $\nvdash p \to \sim q$ then $\vdash (Op \land Oq) \to O(p \land q)$), or by requiring that p and q be jointly

211

permissible (PAND: $\vdash P(p \wedge q) \rightarrow ((Op \wedge Oq) \rightarrow O(p \wedge q)))$. Inheritance of Obligations (RM) may be weakened by requiring that p be permissible before allowing q to inherit an obligation from the obligation that p (RPM: If $\vdash p \rightarrow q$ then $\vdash Pp \rightarrow (Op \rightarrow Oq)$) [5, p. 467–473]. Each resulting logic avoids deontic inferential explosion and has its relative advantages and disadvantages in accounting for the intuitive validity of various example deontic arguments.

2.3 What Kant's Normative Demand for Consistency in the Laws Really Requires

The problem with both paraconsistent logics and these other efforts to weaken SDL's axioms for purposes of a deontic logic of the law, however, is that they offend the demand for consistency understood as a *normative* requirement. Contradictory prescriptive legal obligations are admitted as first-class citizens of such logics. In paraconsistent logics, inferences are derived in the face of such contradictions by the alchemy of a non-classical semantics, which often confounds intuitions. In the weakened deontic logics described above, by contrast, contradictions are like icebergs around which reasoning proceeds gingerly, if at all. In neither case does the logic require that one contradictory obligation be defeated, or that rules generating the contradiction be qualified or revised, in order to allow an inference through the other obligation, or vice versa.

For example, suppose that a criminal statute requires the punishment of anyone who intentionally kills a person ($k \rightarrow Op$), while another statute forbids punishing minors ($m \rightarrow O \sim p$), and suppose a court confronts a case where a minor has intentionally killed someone ($k \wedge m$) (see Alchourron 1991). This licenses the inferences Op and also $O \sim p$, so creating a conflict of obligations. The weakened logics above draw both inferences but then limit further inferences that depend directly on one or another of them. For example, suppose that punishment always consists in incarceration ($p \rightarrow c$). RM would license $Op \rightarrow Oc$, and therefore the inference that the killer ought to be incarcerated, despite that she is a minor (Oc) and ought not to be punished ($O \sim p$). The weakened RPM logic above appropriately blocks this inference because Op is impermissible, $O \sim p$ (where $O \sim p \leftrightarrow \sim Pp$). The RPM logic infers that there is a killer who is a minor (k, m), and that the court is obligated to punish her (Op) and obligated not to punish her ($O \sim p$), but then blocks the explosion of further inferences such as that she ought to be incarcerated (Oc). While the RPM logic thus succeeds in admitting conflicts while avoiding a deontic explosion of inferences, which is its goal, its approach to doing so seems to me to miss the point of admitting deontic conflicts in the first place.

Conflicts between deontic obligations should stimulate rational inference rather than shut it down. What conflicts normatively indicate in a legal deontic context is that one must either revise one or the other of the inconsistent formulas, or prioritize one over the other or, semantically, that one must choose between competing consistent models of (revised) rules given the facts of some conflict situation. While a doxastic or epistemic application of modal logic may perhaps not be subject to the same normative demands, a deontic logic of the law must provide some mechanism to adjudicate between consistent sets of formulas. The goal in the case of the killer who is a minor above should be to render a judgment as to whether her punishment is consistent with everyone's obligations and rights in the system of public laws, subject to constitutional constraints. But paraconsistent logics and weakened deontic logics that admit contradictions seem useless for this purpose.

A court might resolve the case by, for example, qualifying the rule against homicide so as not to apply to minors $(k \wedge \sim m) \rightarrow Op$, or, on the other hand, by qualifying the rule barring the punishment of minors so as not to apply in cases of intentional homicide $(m \wedge \sim k) \rightarrow O \sim p$, or the court might articulate some explicit rule of priority [1, p. 423–424]. A deontic logic of the law should be able to admit the conflict descriptively and provisionally generate inferential alternatives, together with further consequences, in order to evaluate each resulting consistent set of rules and require a decision. The weakened deontic logics above instead simply admit the conflict and limit further inferences. What is needed is a deontic logic that admits the presence of contradictions at a descriptive level but whose semantics insists that they be authoritatively resolved at the prescriptive level of enforceable public laws. This resolution should render our legal obligations precise, and, moreover, must be minimally rationally justifiable by reference to authoritative laws, orders, or judgments.

In the next subsection, I argue that a non-monotonic reasoning system with a classical (rather than paraconsistent) base can meet these normative requirements, with appropriate modifications.

2.4 Non-Monotonic Deontic Logics and the "Credulous" Reasoning Semantics

Non-monotonic reasoning systems (NMRs) are able to admit contradictions without igniting a deontic inferential explosion of obligations because they reject monotonicity (i.e., "If $K' \vdash p$ and $K' \subseteq K$, then $K \vdash p$."). What the rejection of monotonicity means is that some inferences might no longer be drawn when new premises are introduced; for example, one might introduce a new fact that directly contradicts some fact upon which an inference depends, so defeating that inference. NMRs thus avoid the deontic inferential explosion of obligations that plagues SDL; at the same time, NMRs insist that the set of consequences inferred be consistent.

Classical logic can be defined as a structure $S = (F, R)$ where F is a set of formulas and R is a set of rules of inference. R defines a classical consequence relation ('$\vdash$') between a set of formulas and a formula of the language (p). A non-monotonic logic can be defined as a structure $S = \{F, K, R\}$ where F is a set of formulas, K is a set of default rules, and R is a set of rules of inference that define a non-monotonic consequence relation ('$\mid\sim$' note the "snake"). This consequence relation may be defined simply as follows [8, p. 66–67]:

> $F, K \mid\sim p$ if and only if $F, K' \vdash p$ for all subsets $K' \subseteq K$ which are maximally consistent with F.

A subset K' of K is maximally consistent with F if and only if it is consistent with F and there is no superset with F that is consistent and also a subset of K.

For example, suppose $K = \{b \to f, p \to \sim f\}$ ("Birds fly; penguins do not fly."). Suppose $F = \{b\}$. Hence $\{b\} \mathrel{\vdash_K} f$ because for all subsets K' of K that are maximally consistent with $\{b\}$, $\{b\}, K' \vdash f$. That is, $\{b, b \to f, p \to \sim f\} \vdash f$. Suppose b: "chilly is a bird;" therefore, f: "chilly flies" because $b \to f$: "birds fly." But now if we also add p (p: "chilly is a penguin") to F, then $\{b, p\} \mathrel{\not\vdash_K} f$ because $\{b, p\} \mathrel{\nvdash_K} f$ for all subsets K' of K that are maximally consistent with $F = \{b, p\}$; that is, while $\{b, p, p \to f\} \vdash f$, the subset $\{b, p, p \to \sim f\} \nvdash f$. (Note that $\{b, p\} \mathrel{\not\vdash_K} \sim f$, either, because $\{b, p, b \to f\} \nvdash \sim f$. We cautiously infer neither that chilly flies nor that chilly does not fly.) This demonstrates that adding p to the premises causes the conclusion f to be withdrawn in the face of contradiction.

What should the semantics of a deontic NMR appropriate for handling conflicts between enforceable legal obligations be? The consequences a NMR draws given some set of rules K and first-order formulas F can be defined in terms of the "extensions" of (K, F), which, informally, are the rational and justifiable sets of conclusions one can draw given (K, F). Extensions are rational in the sense that conclusions are not accepted if they would create inconsistencies, and justifiable in the sense that 1) all the conclusions that one accepts have some justification in (K, F), while 2) adding any further conclusion would create an inconsistency. "Credulous" reasoning defines the consequences as those in exactly one extension; "skeptical" reasoning defines the consequences as those that lie at the intersection of all extensions. "Ideally" skeptical reasoning defines consequences in terms of the intersection of paths of support, where facts and rules are understood to form an inheritance network [11].

I argue that a credulous rather than skeptical reasoning semantics is needed for a deontic logic of the law. Consider again the case of the killer who is a minor $(k \wedge m)$, where killers ought to be punished $(k \to \mathrm{Op})$ and minors ought not to be $(m \to \sim \mathrm{Op})$. The extensions of these rules and facts are $\{k, m, \mathrm{Op}\}$ and $\{k, m, \sim \mathrm{Op}\}$. Skeptical reasoning cautiously infers as consequences only that there is a killer who is a minor (the intersection of the extensions, $\{k, m\}$), but whether punishment is obligatory or not is left undefined. Yet Kantian justice requires an authoritative ruling in the case; otherwise, any enforcement (or lack of it) is wrongfully coercive. Obligations must be determinate at the prescriptive level of enforceable law—punishment either is or is not obligatory in the case.

Credulous reasoning appropriately requires *exactly one* or the other conclusion as a consequence, which will then persist in the knowledge base to guide and constrain further inferences. For example, suppose we again add $(p \to c)$ ("punishment is by incarceration") to our knowledge base of rules. If the NMR credulously concludes that punishment is obligatory (Op) in the case—perhaps on the theory that minors should be punished for crimes that are felonies such as murder—then the obligation to incarcerate (Oc) will also follow as a consequence. If the NMR credulously concludes, on the other hand, that the minor should not be punished ($\sim$ Op)—even in felony cases—then Oc does not follow as a consequence. This correctly reflects whether and how further inferences such as Oc should be reached.

3   Proposal: Answer Set Programming Rightful Machines

In this section, I propose an approach to programming autonomous machine agents to handle conflicts between legal obligations via a form of logic programming referred to as "answer set programming," which can be viewed as a efficient machine implementation of nonmonotonic formalisms. Answer set programming, with some modifications, can therefore capture the deontic logic of the law I proposed in in the previous section.

The two dominant semantics for extended logic programs are the "answer set" / "stable model" semantics and the "well-founded model" semantics. The answer set semantics for logic programs defines a logic program's consequences in terms of the intersection of its answer sets (extensions), however, and the well-founded model will appear as a subset of this intersection. Both semantics thus reflect skeptical reasoning, where the well-founded semantics reflects "ideally" skeptical reasoning. To achieve credulous reasoning, I exploit only a part of the answer set semantics to 1) enumerate answer sets, and then 2) require the selection of exactly one answer set. This credulous semantics will achieve the main aim of admitting conflicts at the descriptive level while normatively requiring their resolution at the prescriptive level of enforceable law.

3.1   Background: Answer Set Semantics for Logic Programs

First, I briefly sketch the answer set semantics for extended logic programs. My aim here is to provide some background material on how answer set programming works, rather than a formal treatment (see [3] for a formal account). A logic program (Π) consists of a set of rules of this form:

$$\texttt{a :- } b_k, \; b_{k+1}, \; \ldots, \; b_m, \; \texttt{not } c_{m+1}, \; \ldots, \; \texttt{not } c_n.$$

where k, m and n are non-negative integers, a, b, and c are atomic formulas or their negations (i.e. p or $-p$), and "not" is "negation-by-failure." a above is referred to as the head of the rule, while formulas following the ":-" symbol make up the body of the rule. A rule with no head is a constraint, while a rule with no body is a fact.

The answer sets of a logic program Π are obtained by the following procedure [3]. I will first provide a general description of the procedure and then apply it to an illustrative example. First, generate a partial interpretation (I) of Π, which is a consistent set of ground literals formed from the rules of Π. Ground rules, literals, and terms of a logic program contain no variables; hence to create a ground instance of a rule of Π, replace all the rule's variables with ground terms of Π. A literal p is true if it is an element of the interpretation and false if its complement is; otherwise, the literal is undefined. Next, obtain the "reduct" of the program Π with respect to the generated partial interpretation I of Π. The reduct of Π is obtained by first deleting every rule from Π with "not p" in its body, where p is a member of I, and then deleting all "not q" from the remaining rules of Π, where q is any literal. Finally, repeatedly obtain the immediate consequences $(T(I))$ of the reduct by applying modus ponens and avoiding contradictions until reaching a fixpoint (Cn) where the set of immediate consequences no longer changes. If the set of immediate conclusions is the same as the partial interpretation I, then the set is an answer set. Repeat this procedure until you have found all answer sets.

3.2 Encoding Alchourron's Example of the Killer who is a Minor

To illustrate, suppose we want to apply an answer set programming approach to the conflict adduced previously between defeasible legal rules that 1) killers should be punished, but 2) minors should not be punished, and the case is such that a minor has killed someone [1]. How best to capture and resolve this case of conflict between legal obligations in answer set programming? We want to know what the rational, justifiable sets of conclusions (answer sets) are in the case, given the facts and applicable legal rules. Intuitively, there are two such sets: either 1) the minor should be punished because she is a killer, despite being a minor, or 2) the minor should not be punished because she is a minor, despite that she is a killer.

The following intuitive encoding of the rules and facts is inadequate because it has just one answer set consisting only in the initial facts ($\{k, m\}$):

```
k.   m.            % a killer who is a minor
p :- k, not m.     % punish killers unless they are minors
-p :- m, not k.    % do not punish minors unless they are killers
```

Encoding the laws in conflict as "normal" default rules, on the other hand, generates the results we want:

```
k.   m.            % a killer who is a minor
p :- k, not -p.    % punish killers unless they should not be
-p :- m, not p.    % do not punish minors unless they should be
```

This encoding yields two rational, justifiable extensions (answer sets). Either 1) the killer who is a minor should be punished ($\{k, m, p\}$), or 2) the killer who is a minor should not be punished ($\{k, m, -p\}$).

The following table displays how the procedure described above finds answer sets for this program (answer sets are starred):

Interpretation (I)	Reduct (P^I)	$\gamma p(I)$
{}	k. m.	k, m
k	k. m. p :- k. -p :- m.	$\perp$
m	k. m. p :- k. -p :- m.	$\perp$
k, m	k. m. p :- k. -p :- m.	$\perp$
k, m, p	k. m. p :- k.	k, m, p*
k, m, -p	k. m. -p :- m.	k, m, -p*

3.3   Regimenting the Proposed Encoding

While encoding legal rules as normal default rules achieves correct results, the encoding should provide some way to make the qualifications on the rules that generated each answer set explicit. This will meet Kant's normative demand that enforceable legal obligations be made precise. The following intuitive encoding of rules with explicit qualifications, however, has no answer sets:

```
k.   m.                % a killer who is a minor
p :- k, not q1.        % punish killers unless this rule is qualified
-p :- m, not q2.       % do not punish minors unless this rule is
       qualified
```

I propose the following encoding, which explicitly tracks the qualifications required as they appear in each answer set and, moreover, leaves room to later make additional qualifications to the rules as needed:

```
k.   m.   a.      % a killer who is a minor (and an action is taken)

p :- k, not q1.   % rule 1: punish killers
q1 :- a, not p.   % unless this rule is qualified

-p :- m, not q2.  % rule 2: do not punish minors
q2 :- a, not -p.  % unless this rule is qualified
```

The following table indicates how answer sets for this program are found. (Partial interpretations are generated and tested here in order by set inclusion; '...' indicates interpretations omitted to save space.)

Interpretation (I)	Reduct (P^I)	$\gamma p(I)$
{}	k. m. a. p :- k. q1 :- a. -p :- m. q2 :- a.	$\perp$
...	...	...
k, m, a, p	k. m. a. p :- k. -p :- m. q2 :- a.	$\perp$
...	...	...
k, m, a, q1, q2	k. m. a. q1:- a. q2 :- a.	k, m, a, q1, q2*
k, m, a, q1, -p	k. m. a. q1:- a. -p :- m.	k, m, a, q1, -p*
k, m, a, q2, p	k. m. a. p :- k. q2 :- a.	k, m, a, q2, p*
k, m, a, q1, q2, p	k. m. a. q2 :- a.	k, m, a, q2
...	...	...

Answer sets for this program are $\{k, m, a, q1, q2\}$, $\{k, m, a, q1, -p\}$, and $\{k, m, a, q2, p\}$. These answer sets reflect explicit qualifications made on both rules $(q1, q2)$, and then on one rule $(q1)$, or the other $(q2)$. In this encoding, qualifications on legal rules are triggered when there is an action (a) and the negation of the head of the rule qualified (p) is by default negation (e.g., not p), rather than classical negation $(-p)$, as in the first attempt at encoding qualifications above. This encoding thus achieves the same behavior as normal default rules but with explicit qualifications on rules. Each answer set reflects a rational, justifiable set of conclusions one might draw, given the facts and applicable rules in the case; moreover, the rule qualifications necessary to construct each answer set are now made explicit in the set.

While qualifying both rules yields a set of consequences that are both consistent and justifiable (i.e., the set stating merely that there is a killer who is a minor, $\{k, m, a, q1, q2\}$), doing so fails to meet the Kantian normative demand that legal rights in conflict cases must be resolved. The main purpose of the law on Kant's account is to rightfully resolve disputes over our rights and duties with respect to each other. To meet this normative demand, a "ruling" predicate is added to the encoding that requires each answer set to provide a definite determination of the rights and obligations in conflict:

```
% a ruling is required
:- not ruling.
ruling :- p.   ruling :- -p.
```

This eliminates the first answer set that qualifies both rules $\{k, m, a, q1, q2\}$, and thus fails to determine an answer to the question whether the killer who is a minor should be punished, or not. The credulous reasoning semantics will now demand that exactly one of the two remaining answer sets ($\{k, m, a, q1, -p\}$, $\{k, m, a, q2, p\}$) be selected as the program's consequences.

3.4   Simple Legal Case: A Shooting in Self-defense

I illustrate the proposed encoding by evaluating a conflict between the legal rule barring murder and one permitting the use of force in self-defense. Imagine a case of first impression that generates a conflict between these rules. In what follows I use the lparse grounder and clingo parser to ground and solve encoded logic programs (see [2] for details).

First, a situation is described with facts that may satisfy elements of various legal theories and rules:

```
%%% conflict situation
intentional(shooting).   act(shooting).   causes_death(shooting,
    someone).   person(someone).
attacked(me).   force(shooting, me).   retreated(me).
```

The facts as encoded in this example are necessarily somewhat stipulative. An actual autonomous machine agent governance system would include perceptual and other subsystems organized to generate new facts (beliefs), which might then generate messages for processing at progressively higher levels in the rational agent hierarchy. My aim here is to isolate and describe only the legal reasoning part of the system.

Applicable legal rules are then extracted and encoded, perhaps by interpreting a semantic legal knowledge base. The head of a legal rule is encoded as a deontic prescription on an action (e.g., that an action is obligatory ($ob(A)$), forbidden ($ob(-A)$), permissible ($pe(A)$), or omissible ($pe(-A)$)). The body of the rule will then invoke legal theories relevant to establishing the deontic status of the action (e.g., that the action constitutes a murder, or is self-defense, an action by omission, an act of necessity, etc.), where elements establishing these theories are define in existing statutory or case law (e.g., the common law rule that a murder is an intentional act that causes the death of a person).

Here is the first part of the body of a simplified rule prohibiting murder:

```
ob(-A)  :- murder(A).
% legal elements of murder: malice killing
murder(A)  :- intentional(A), act(A), causes_death(A, P), person(P).
```

The body of the rule is then completed with a generic defeasible qualification (e.g. $qual(r1(A))$) on the rule tagged with the rule number ($r1$):

```
ob(-A)  :- murder(A), not qual(r1(A)).
qual(r1(A))  :- act(A), not ob(-A).
```

This preserves the possibility that the rule may be defeated for reasons other than those explicitly anticipated in the local context of the system.

Here accordingly are legal rules prohibiting murder ($r1$) and a conflicting rule permitting the use of force in self-defense ($r2$):

```
%%% r1: it is obligatory not to murder
ob(-A)  :- murder(A), not qual(r1(A)).
qual(r1(A))  :- act(A), not ob(-A).

% legal elements of murder: malice killing
murder(A)  :- intentional(A), act(A), causes_death(A, P), person(P).

%%% r2: it is permissible to use force in self-defense
pe(A)  :- self_defense(A), not qual(r2(A)).
qual(r2(A))  :- act(A), not pe(A).

% legal elements of self-defense: use of force by one who is
     attacked and tried to retreat
self_defense(A)  :- force(A, P), attacked(P), retreated(P).
```

The legal rules provided here are obviously simplified in order to isolate how the system handles conflict. For example, the legal intention required to establish the mens rea for murder is "malice aforethought," which implies at least a reckless awareness that one's act will cause the death of a person (see [6]). I omit such details here.[3]

Next, a constraint to avoid conflicts between deontic obligations within answer sets is added as well as standard deontic implication and convenient equivalence relations between obligations and permissions. These reflect axioms and equivalences of SDL that apply within answer sets.

```
% avoid deontic conflict
:- ob(A), ob(-A).

% deontic implication (D)
pe(A)  :- ob(A).    % obligation implies permission

% inheritance of obligation (RM)
ob(A)  :- ob(B), -A.
```

---

[3] I do not mean to imply that these details cannot be supplied, however—indeed, they must be, if the law is to meet the normative requirement that prescriptive obligations be precisely specified.

```
ob(A) :- ob(B), B.
A :- B.

% deontic equivalences
ob(A)  :- -pe(-A).   -pe(-A) :- ob(A).
pe(A)  :- -ob(-A).   -ob(-A) :- pe(A).
ob(-A) :- -pe(A).    -pe(A)  :- ob(-A).
pe(-A) :- -ob(A).    -ob(A)  :- pe(-A).
```

Finally, an updated "ruling" constraint requiring each answer set to provide some deontic resolution of the legal conflict at issue completes the program. Appropriate **#show** directives have also been added to avoid clutter.

```
% a ruling is required: an obligation, prohibition, permission, or
    omission
:- not ruling.
ruling :- ob(A).   ruling :- ob(-A).   % obligation, prohibition
ruling :- pe(A).   ruling :- pe(-A).   % permission, omission
#show pe/1.   #show ob/1.   #show qual/1.
```

The program generates the following answer sets. These answer sets represent the rational and justifiable sets of consequences with associated explicit rule qualifications that one might infer in the case, given the facts and applicable law as encoded.

Answer: 1
```
murder(shooting) self_defense(shooting) ob(-shooting) pe(-shooting)
    qual(r2(shooting))
```

Answer: 2
```
murder(shooting) self_defense(shooting) qual(r1(shooting))
    pe(shooting)
```

Either (Answer: 1) one is *obligated not* to intentionally shoot to kill, ob(-shooting), in the case because the self-defense rule (r2) is qualified not to apply here, qual(r2(shooting)). Or (Answer: 2) one is *permitted* to intentionally shoot to kill, pe(shooting), because the murder rule is qualified not to apply in the case, qual(r1(shooting)). The law as stipulated is thus conflicted with respect to whether the shooting is forbidden or permitted.

Now, it has long been established in the criminal law that a killing that would otherwise be a murder is justified if the killing meets the elements of self-defense. Hence the conflict here should be resolved by explicitly qualifying the murder rule (r1) to permit killing in self-defense (Answer: 2). To do that, the self-defense qualification is made explicit, while retaining the generic defeasible qualification on the rule so that it remains a candidate for defeat in future cases of conflict (e.g., with other defenses such as necessity or excuses such as insanity). The murder rule (r1) is thus adjusted as follows (changes in bold):

```
%%% r1: it is obligatory not to murder
ob(-A) :- murder(A), not qual(r1(A)), not qual(r21(A)).
qual(r1(A)) :- act(A), not ob(-A).

qual(r21(A)) :- self_defense(A).   % unless in self-defense
```

With this adjustment, the program produces only one answer set in the situation described, where the murder rule is qualified, `qual(r21(shooting))`, because the case as stipulated is clearly self-defense.

```
Answer:  1
murder(shooting) qual(r21(shooting)) self_defense(shooting)
   qual(r1(shooting)) pe(shooting)
```

The shooting in this case is therefore permissible. Further qualifications on the updated murder rule may be progressively entertained and accepted or rejected as new cases arise, such as qualifications for excuses or defenses such as insanity, necessity, etc.

The total number of possible answer sets in a case of conflict will be the power set of the available qualifications; in this simple example, $P(\text{qual}(r1()), \text{qual}(r2()))$, or only four total sets, including the empty set. The answer set semantics eliminates inconsistent combinations of rule qualifications and inferences that violate non-contradiction and modus ponens. The `ruling` predicate eliminates the empty set and the set where no decision is made because all rules are qualified. If there are multiple answer sets remaining, then the credulous reasoning semantics will require the selection of exactly one.

The resulting set of enforceable legal obligations will be a rational and justifiable set of applicable laws explicitly qualified to resolve inconsistencies and so to precisely determine legal obligations and rights in the case. So long as the laws in the set are also legitimate, an autonomous machine agent that acts in accordance with them is a rightful machine.

References

[1] Alchourron, C., *Conflicts of norms and the revision of normative systems*, Law and Philosophy **10** (1991), pp. 413–425.

[2] Gebser, M., R. Kaminski, B. Kaufmann and T. Schaub, "Answer Set Solving in Practice," Synthesis Lectures on Artificial Intelligence and Machine Learning, Morgan & Claypool Publishers, 2012, see https://potassco.org/.

[3] Gelfond, M., *Chapter 1: Answer sets*, Foundations of Artificial Intelligence **3** (2008), pp. 285–316.

[4] Girle, R., "Modal Logics and Philosophy, 2d ed." Montreal, MQUP, 2017.

[5] Goble, L., *A logic for deontic dilemmas*, Journal of Applied Logic **3** (2005), pp. 461–483.

[6] Institute, A. L., "Model penal code: official draft and explanatory notes: complete text of model penal code as adopted at the 1962 annual meeting of the American Law Institute at Washington, D.C., May 24, 1962," The Institute, Philadelphia, 1985.

[7] Kant, I., *The metaphysics of morals (1797)*, in: P. Guyer and A. Wood, editors, *The Cambridge Edition of the Works of Immanuel Kant*, Cambridge University Press, Cambridge, 1992 Translated by M. Gregor. Citations to Kant's work are given using standard Academy pagination.

[8] Maranhao, J., *Why was alchourron afraid of snakes?*, Analisis Filosofico **26** (2006), pp. 62–92.

[9] Navarro, P. and J. Rodriguez, "Deontic Logic and Legal Systems," Cambridge University Press, Cambridge, 2014.

[10] O'Neill, O., "Constructing Authorities," Cambridge University Press, Cambridge, 2011.

[11] Touretzky, D., J. Horty and R. Thomason, *A clash of intuitions: The current state of nonmonotonic multiple inheritance systems*, in: *Proc. IJCAI-87*, Morgan Kaufmann, 1987, pp. 476–482.

[12] Wright, A., *Rightful machines*, in: H. Kim and D. Schönecker, editors, *Kant and Artificial Intelligence*, Walter de Gruyter GmbH & Co KG, 2021 .

[13] Wright, A. T., *A kantian course correction for machine ethics*, in: G. J. Robson and J. Y. Tsou, editors, *Technology Ethics: A Philosophical Introduction and Readings*, Routledge, New York, 2023 pp. 141–151.

Multi-criteria Coherence Ranking of Legal Theories: The Aggregation Problem and Possible Solutions

Tianwen Xu[1]

Guanghua Law School, Zhejiang University
Zhejiang University Law & AI Laboratory
Hangzhou, China

Abstract

While coherentist approach to justification has been trending in law and numerous multi-criteria accounts of theory coherence have been established, it mostly remains unknown how a legal decision maker can obtain an overall coherence ranking among legal theories from the multiple criteria, such that he can tell legal theories from better to worse and figure out which legal judgment is best justified. This paper intends to unravel this puzzle. Inspired by social choice theory, the puzzle is presented as a preference aggregation problem in a multi-criteria decision making context. A common problem setting as well as relevant rational conditions are first generalized from three motivating examples. Such generalization gives rise to a formalization in terms of decision matrix. The problem of obtaining an overall coherence ranking is thereby a problem of making appropriate use of the coherence evaluation matrix, and the aggregation function is defined as a coherence evaluation functional (CEFL). Three CEFLs respectively on the basis of simple majority, Borda count and normalized summation are then formulated, with a detailed examination of their strengths and weaknesses for legal decision making.

Keywords: Coherence ranking of legal theories, Multi-criteria decision making, Preference aggregation.

1 Introduction

Philosophy of law, as well as other major fields of philosophy, witnessed a rise of coherentism in the later half of the 20th century. Debunking the robust correspondence theory of truth, meaning and knowledge has led quite a number of legal theorists to adopt a coherentist approach to many essential questions, typical among which are the nature of law, the myth of legitimacy, the semantics of legal statements, and the methods of legal reasoning [25,33]. As regards legal argumentation, the result is a cluster of coherentist accounts of justification (see, e.g., [1,2,3,10,11,13,17,18,23]). In general these accounts amount to

---

[1] tianwen.xu@zju.edu.cn

the following claim: to justify a legal judgment is to show that the judgment 'fits' in a coherent legal theory, and how well the judgment is justified depends on how coherent the theory is. [2] The crucial question then arises as to how a theory's degree of coherence can be properly assessed, such that one can tell if one theory is more coherent than another, and thereby decides which judgment receives the best justification.

Unfortunately, the above question has no easy answer. It is well recognized that the concept *coherence* is more than mere *consistency*, since a consistent theory may fail to 'hang together' or 'make sense as a whole' to qualify as coherent [18, p. 235]. However, it is far less clear what conceptual contents other than consistency are in coherence, let alone to decide which theory is more coherent.

As a response to this puzzle, efforts have been put to dissemble coherence into many sub-concepts, and accordingly propose multiple criteria to measure and compare the degree of coherence. Such is the approach adopted by [1,6,16,23,27,29]. Some of the suggested criteria are structural properties of a legal theory, e.g., 'the number of supportive relations', 'priority orders between reasons' and 'cumulation-netting of reasons'. Others are theoretical virtues, e.g., 'articulateness', 'explanatory power' and 'simplicity'.

Though being philosophically appealing, these multi-criteria accounts are particularly difficult to implement. The major obstacle is the lack of legal methods to synthesize the different coherence rankings of legal theories, each given by a criterion, into an all-things-considered one. Thus as criticized by Amaya [3, p. 24], a multi-criteria coherence theory 'without an account of how the different criteria may be balanced against each other' is incomplete. Therefore it is not surprising that some coherence theorists, though in favor of a multi-criteria conception, actually turned to other coherentist accounts (such as Paul Thagard's *coherence as constraint satisfaction*, see [30,31]) to compute and compare the degree of coherence (see [7], [27, p. 783-784]).

This challenge of obtaining an overall coherence ranking of legal theories from multiple criteria is exactly what this paper intends to deal with. It will be tackled in a formal way, by considering the challenge as a *preference aggregation* problem in a *multi-criteria decision making* context. There is an analogy to the social choice setting. Each criterion of coherence can be considered as an individual (or more specifically, a voter), and its coherence ranking of legal theories can be treated as an individual's preference over alternatives (or more specifically, candidates). Then the problem of obtaining overall coherence ranking from multiple criteria is similar to that of obtaining a collective preference from individuals.

---

[2] This generalization of coherentist justification in law is borrowed from [13]. As regards a legal *theory*, throughout this paper it is understood in a broad sense as 'covering both descriptive, for example empirical theories, and normative or evaluative theories (norm systems or value systems)' [1][p. 132]. This allows for covering most multi-criteria accounts, even those that conceive *theory* in a specific way, e.g., a five-tuple comprised of cases, factors, rules, preferences between rules and between values, as in [6].

To consider multi-criteria decision making via preference aggregation is not entirely novel. In the past decade, some works have been done on associating social choice with theory choice in science, see for instance [19,20,21,22]. There is even a much earlier work in law [14] that explored the possibility of achieving legal coherence by impossibility results developed in social choice literature, although it has a quite different focus on aggregating preferences of legal norms over decision options, rather than preferences of coherence criteria over theories that are supposed to offer legal justification. Operational research is another subject that has long been benefited from applying the social choice framework to its specific multi-criteria decision problem (see e.g., [5]). It just remains to see how such interdisciplinary effort would bring insights to law. Hopefully, this paper will contribute to the following.

(i) *Conceptually*, to make explicit the underlying mathematical structure of multi-criteria accounts of theory coherence in law;

(ii) *Philosophically*, to bring to light the nature of the proposed criteria, as well as the possibility of acquiring from them an overall coherence ranking.

(iii) *Computationally*, to explore the potential of computing the degree of legal coherence by the proposed criteria.

The paper is organized as follows. Section 2 first demonstrates three multi-criteria accounts of legal coherence as motivating examples. Section 3 will then formalize the examples in a preference aggregation framework using *decision matrix*. Section 4 will finally examine some possible solutions to the problem of aggregating an all-things-considered coherence ranking.

2 Three Motivating Examples

2.1 Alexy and Peczenik's criteria

The first group of criteria is proposed by Alexy and Peczenik [1]. Their criteria are centered around the idea of a *perfect supportive structure*, with the over-arching conception of coherence that 'the more the statements belonging to a given theory approximate a perfect supportive structure, the more coherent the theory' [1, p. 131], and every criterion is supposed to enhance a theory's coherence *ceteris paribus* when it is better fulfilled. To summarize, Table 1 lists all criteria in the group, together with an index specified by each criterion to measure a theory's coherence. [3]

Table 1: Alexy and Pecznik's criteria

No.	Criterion	Index of coherence in a theory
1	Supportive relations	Number of statements that are supported
2	Supportive chains	Length of supportive chains of reasons
3	Strong support	Number of statements that are strongly supported

See next page...

---

[3] For brevity, each criterion and its index are described in a much more concise manner.

. . . Continued from previous page

4	Connection between supportive chains	Number of conclusions supported by the same premise
5	Priority orders between reasons	Number of priority relations between principles
6	Reciprocal justification	Number of reciprocal empirical/analytic /normative relations between statements
7	Generality	Number of statements without individual names, number of general concepts, resemblances between concepts
8	Conceptual cross-connections	Number of common or resembled concepts
9	Coverage of cases	Number of individual cases covered
10	Diversity of fields of life	Number of fields of life covered

Since this paper is primarily concerned with the formal aspect, those terms with complicated philosophical contents, e.g., supportive structure, strong support, reciprocal justification and etc., will not be discussed here, but readers may refer to [1], and to [23, chapter 4] for some extended elaborations. Speaking of the formal concern, it is clear that Alexy and Peczenik's criteria are quantitative, as the index of a theory's coherence given by each criterion is numerical. Nonetheless, they did not provide any articulated method for combining those numerical values into an all-things-considered coherence ranking, although they were fairly aware that one theory might be simultaneously superior and inferior to another with regard to different criteria; the only solution they offered was, quite metaphorically, to conduct a 'weighing and balancing' among criteria so as to determine the degree of coherence of a theory [1, p. 143].

2.2 Kress's criteria

The second group of criteria is proposed by Kress [16], which was intended to generalize intuitions about coherence that had appeared in the literature. They are based on the idea that a coherent theory is coherent if it 'hangs or fits together, if its parts are mutually supportive, if it is intelligible, if it flows from or expresses a single unified viewpoint', and the more of them a theory satisfies, the better is the theory [16, p. 521]. Table 2 illustrates each criterion and its index of coherence.

Table 2: Kress's criteria

No.	Criterion	Index of coherence in a theory
1	Consistency	The theory respect logical consistency between principles and propositions
2	Comprehensiveness	The theory provides answers to questions within the theory's scope
3	Completeness	The theory provides single right answers to all questions within the theory's scope
4	Monism	The theory flows from a single principle, or set of principles with a unified spirit
5	Unity	The theory's principles imply, justify or mutually support one another

See next page. . .

... Continued from previous page

6	Articulateness	The theory provides articulated methods for deciding issues, integrating and unifying its principles as well as resolving conflicts in between
7	Justified	The theory resolves conflicts with reasons, or provides normatively intelligible meta-principles and means to resolve conflicts between principles

At first glance, the above criteria appear to have an all-or-nothing character. But this is not true. Instead, Kress argued that some criteria, for instance comprehensive, are 'unduly restrictive' if a coherent theory is required to fully posses; while some others, like completeness, only needs to be manifested to a substantial degree for a theory to be coherent, or is explicitly stated as a matter of degree, e.g., unity; not to mention that even consistency is just a regulative idea, in the sense that it is more desirable in practice to retain some inconsistencies until satisfactory solution is found [16, p. 528, 530]. Anyway, all of his criteria involve properties that are conceptually possible to be more or less satisfied. Even so, Kress did not formulate any procedure regarding how to put together those criteria to give a singel answer of how coherent a theory is, especially when they pull in opposing directions,

2.3 Bench-Capon and Sartor's criteria

The last group is proposed by Trevor Bench-Capon and Giovanni Sartor in [6,27]. Their criteria are subject to a cognitive conception of coherence that aims at enhancing the cognitive vale of a theory, and are explicitly associated with choosing from a range of candidate theories [27, p. 758-760]. Table 3 demonstrates a full list.

Table 3: Bench-Capon and Sartor's criteria

No.	Criterion	Index of coherence in a theory
1	Explanatory power	Number of cases explained
2	Consistency	Free from unsolved internal collisions
3	Simplicity	Number of general rules that consider fewer factors to be relevant
4	Safety	Number of factors taken into account
5	Non-arbitrariness	Less recourse to arbitrary assumptions

By commenting that 'it is not easy to find a way of combining these different criteria into a unique evaluation, given that they may lead to different results' [27, p. 760], they clearly have in mind the difficulty of aggregating the criteria to an overall coherence ranking of candidate theories. But as mentioned earlier, Bench-Capon and Sartor turned to Thagard's constraint satisfaction theory for help, which is in fact not a multi-criteria account, and left as an open question whether this alternative conception matched the shared intuition of legal coherence.

3   Formalizing Multi-criteria Coherence Ranking

3.1   The common setting

Though respectively developed along their distinct conceptions of coherence, the three multi-criteria accounts do have a common setting. At least they agree with the following. [4]

(i) There are some candidate legal theories, whose degree of coherence awaits to be assessed by a specific group of criteria.

(ii) For each candidate theory and its relevant attributes, each criterion will give a unique evaluation, based on which one theory is better or worse than another. In another words, each criterion will give a unique coherence ranking of candidate theories.

(iii) The coherence rankings given by each criterion may pull in opposing directions. That is to say, for any two theory a, b and any two criteria m, n, it is possible that theory a is more coherent than b with regard to criterion m, but less so with regard to n.

(iv) Though may pull in opposing directions, an all-things-considered coherence ranking over candidate theories is supposed to be generated by, and only by, combining the multiple coherence rankings given by all criteria in the specific group.

(v) When combining the multiple rankings, it is possible for them to offset each other. I.e., for any two theories a, b, a being worse than b with regard to some criteria may be compensated by a being better than b with regard to some other.

As the common setting is made clear, let us now fix the terminology. The coherence ranking given by each criterion in a specific group will be called an *individual coherence ranking*, while the all-things-considered ranking will be called the *overall coherence ranking*. Furthermore, though not articulated in the motivating examples, it is fairly reasonable to infer from the common setting another set of rational conditions, which are intuitively necessary for the multi-criteria coherence account to function.

(i) Any individual coherence ranking should satisfy following conditions:
   1a. Every theory in question is taken into account;
   1b. Any two theory is comparable;
   1c. The ranking should consistently determine whether one theory is more coherent than another.

(ii) The overall coherence ranking should satisfy the following conditions:
   2a. Every theory in question is taken into account;
   2b. Any two theory is comparable;
   2c. The overall ranking should be unique;

---

[4] The setting set forth below is very closed to that of the *epistemic virtue account* of theory choice in science, see [22, p. 141-142].

2d. The unique overall ranking should consistently determine whether one theory is more coherent than another.

It is not hard to see how these conditions are related to the common setting. All indexes specified by criteria are properties of a theory, and are either absent or present to a certain degree (therefore, condition 1a). In the meantime, any group of criteria indeed offers shared scales for all theories that fall under the group's target category (condition 1b). And for any scale, it is natural to require that the scale gives a consistent and determinate evaluation, otherwise the scale simply cannot be put to work (condition 1c). Arguments for condition 2a-2d is similar. The following formalization will be based on these conditions and the common setting.

3.2 The Formalization

The formalization given in this section is inspired by [8] and [24], utilizing the typical *decision matrix* in multi-criteria decision making. For a given multi-criteria account, let $\mathcal{C} = \{C_1, C_2, \ldots, C_n\}(n \in \mathbb{N})$ be the set of its criteria, let $\mathcal{T} = \{T_1, T_2, \ldots, T_m\}(m \in \mathbb{N})$ be the set of theories for evaluation (or 'candidate' theories, for some social choice flavor). In this context, any $T_i \in \mathcal{T}$ can be presented as a vector $\overrightarrow{T_i}$ in the following form:

$$\overrightarrow{T_i} = (a_{i1}, a_{i2}, \ldots, a_{ij}, \ldots, a_{in})$$

where a_{ij} is the attribute of T_i relevant to criterion C_j. Intuitively speaking, $\overrightarrow{T_i}$ sequentially enumerates T_i's attribute a_{ij} considered by criterion C_j. For example, a_{11} is the attribute of T_1 considered by C_1, while a_{12} the attribute considered by C_2. Reversely, for any criterion $C_j \in \mathcal{C}$, vector $\overrightarrow{C_j}$ presents the attribute considered by C_j of each $T_i \in \mathcal{T}$, in a form similar to $\overrightarrow{T_i}$:

$$\overrightarrow{C_j} = (a_{1j}, a_{2j}, \ldots, a_{mj})$$

With $\overrightarrow{T_i}$ and $\overrightarrow{C_j}$, a *coherence ranking problem* is defined as follows.

Definition 3.1 [Coherence ranking problem] Let $\overrightarrow{\mathcal{T}} = \{\overrightarrow{T_1}, \overrightarrow{T_2}, \ldots, \overrightarrow{T_m}\}$, let $\overrightarrow{\mathcal{C}} = \{\overrightarrow{C_1}, \overrightarrow{C_2}, \ldots, \overrightarrow{C_n}\}$. A coherence ranking problem is an $m \times n$ matrix $\mathcal{A}$ whose i-th row is comprised of $\overrightarrow{T_i} \in \overrightarrow{\mathcal{T}}$, and j-th column of $\overrightarrow{C_j} \in \overrightarrow{\mathcal{C}}$, i.e.,:

$$\mathcal{A} = \begin{bmatrix} a_{11} & a_{12} & \cdots & a_{1j} & \cdots & a_{1n} \\ a_{21} & a_{22} & \cdots & a_{2j} & \cdots & a_{2n} \\ \vdots & \vdots & \ddots & \vdots & \ddots & \vdots \\ a_{i1} & a_{i2} & \cdots & a_{ij} & \cdots & a_{in} \\ \vdots & \vdots & \ddots & \vdots & \ddots & \vdots \\ a_{m1} & a_{m2} & \cdots & a_{mj} & \cdots & a_{mn} \end{bmatrix}_{m \times n}$$

In this way, $\mathcal{A}$ characterizes a coherence ranking problem as a matrix composed of attributes a_{ij} of T_i considered by C_j. It intuitively dissembles the problem into a number of attributes awaiting evaluation, based on which a criterion C_j will express its preference over $T_i \in \mathcal{T}$ with regard to degree of coherence. For each criterion C_j, let us proceed to define a coherence evaluation.

Definition 3.2 [Coherence evaluation] Given a $C_j \in \mathcal{C}$, a coherence evaluation is a function $U_j : \overrightarrow{C_j} \longrightarrow \mathbb{R}^m$, that assigns to $\overrightarrow{C_j}$ a vector $\overrightarrow{E_j} \in \mathbb{R}^m$, where $\overrightarrow{E_j} = (e_{1j}, e_{2j}, \ldots, e_{mj})$.

Roughly speaking, U_j is associated with the particular *scale* used by C_j to evaluate theories $T_i \in \mathcal{T}$, and e_{ij} as the corresponding value. For instance, a mountain has an attribute of *height*, and measuring its height by a metric system will yield a value, say, 1000 with the unit of *meter*. But readers may not take for granted that e_{ij} must always be the exact value that a given scale yields. Rather, it could be a value given by any U_j' that preserves the information of the kind of scale on which U_j is measured. E.g., if U_j is measured on a purely ordinal scale, then any U_j' as a strictly increasing transformation of U_j is informationally equivalent to U_j. If U_j is measured on a cardinal scale with full comparability, then any U_j' that satisfies $U_j' = kU_j + b$ where $k, b \in \mathbb{R}$ and $k > 0$ is informationally equivalent to U_j. If U_j is measured on a ratio scale with full comparability, then any U_j' that satisfies $U_j' = kU_j$ where $k \in \mathbb{R}$ and $k > 0$ is informationally equivalent to U_j. In utility theory these are called *information invariance*, see generally [8, p. 1115-1126], [24, p. 31-36], [28, Chapter 7].

Either way, when taken together the evaluation functions U_j used by $C_j \in \mathcal{C}$ will form a profile $\mathcal{U} = (U_1, U_2, \ldots, U_j)$. Such a profile will transform the matrix $\mathcal{A}$ to another matrix $\mathcal{A}^e$, where:

$$\mathcal{A}^e = \begin{bmatrix} e_{11} & e_{12} & \cdots & e_{1j} & \cdots & e_{1n} \\ e_{21} & e_{22} & \cdots & e_{2j} & \cdots & e_{2n} \\ \vdots & \vdots & \ddots & \vdots & \ddots & \vdots \\ e_{i1} & e_{i2} & \cdots & e_{ij} & \cdots & e_{in} \\ \vdots & \vdots & \ddots & \vdots & \ddots & \vdots \\ e_{m1} & e_{m2} & \cdots & e_{mj} & \cdots & e_{mn} \end{bmatrix}_{m \times n}$$

Let us call $\mathcal{A}^e$ the *evaluation matrix*. It is easy to see that the individual coherence ranking by C_j on $\mathcal{T}$ is a result of utilizing the j-th column in $\mathcal{A}^e$. Here is an example. Let $\succeq_j$ be the individual coherence ranking of C_j, $\geq$ be the usual *greater or equal to* relation on $\mathbb{R}$. Let $T_i, T_h \in \mathcal{T}$. Then a standard way to define $\succeq_j$ is: $T_i \succeq_j T_h$ iff $e_{ij} \geq e_{hj}$, and abbreviate $T_i \succeq_j T_h$ *but not* $T_h \succeq_j T_i$ as $T_i \succ T_h$, while $T_i \succeq_j T_h$ *and* $T_h \succeq_j T_i$ as $T_i \approx T_h$. [5] It is easy to verify that $\succeq_i$ defined as such satisfies the rational conditions 1a-1c as described in Section 3.1. The overall coherence ranking obtains in a similar way. That is to say, the aggregation problem of generating an overall coherence ranking is about how to make use of the matrix $\mathcal{A}^e$ resulted from a profile $\mathcal{U}$. To deal

---

[5] Unless stated otherwise, this will be the default setting for individual coherence ranking through the rest of the paper.

with it, let us introduce a *coherence evaluation functional* as the aggregation function for multi-criteria coherence ranking.

Definition 3.3 [Coherence evaluation functional, CEFL for short] Let $\mathcal{U}^a$ be the set of admissible profiles, let $\mathcal{R}$ be the set of all possible orderings on $\mathcal{T}$. A coherence evaluation functional is a function $F : \mathcal{U}^a \longrightarrow \mathcal{R}$, that assigns to each $\mathcal{U} \in \mathcal{U}^a$ an ordering on $\mathcal{T}$ as the overall coherence ranking.

What profile is admissible in $\mathcal{U}^a$ depends on the types of scale and the patterns of evaluation allowed by $C_j \in \mathcal{C}$. Meanwhile, F being defined on $\mathcal{U}^a$ rather than on the set of n-tuples of $\overrightarrow{E_j}$ shows that the primary concern of aggregation is with the evaluation function, because the value of e_{ij} can hardly reflect the nature of U_j, i.e., ordinal/cardinal/ratio-scale (and etc.) measurability as well as partial/full comparability, and thus makes it even harder to make appropriately use of $\mathcal{A}^e$. Indeed, the coherence evaluation functional defined as such is an instance of *social welfare functional* (SWFL) proposed by Amartya Sen, see e.g., [28, p. 185]. An obvious merit of applying SWFL to the aggregation problem at hand is that much more informations of individual coherence ranking can be taken into account, besides the mere ordinal properties.

Now question remains as to what forms of CEFL will help to aggregate an overall coherence ranking. This will be addressed in the next section, where three basic methods, namely *simple majority, Borda count* and *normalized summation* are considered. They all present very little technical obstacles for legal practice, and their corresponding CEFLs can be viewed as making increasingly richer use of the evaluation matrix $\mathcal{A}^e$. [6]

4　In Search of Coherence Evaluation Functionals

4.1　Simple majority

The simple majority method is the well-known Condorcet pairwise comparison. It shares an important intuition with multi-criteria accounts of theory coherence, that the more 'coherence criteria' are satisfied, the more coherent a theory is. It realizes this intuition by comparing the relative amount of criteria satisfied between each pair of candidate theories, and by doing so utilizes the evaluation matrix $\mathcal{A}^e$ in a purely ordinal way. Formally, a CEFL based on simple majority is defined as follows. [7]

Definition 4.1 [CEFL based on simple majority] Let $U_j(T_i)$ represent e_{ij}, and let $\geq$ be the *greater or equal to* relation on $\mathbb{R}$. Given a profile $\mathcal{U} \in \mathcal{U}^a$

---

[6] For convenience of demonstration, in constructing the CEFLs we will adopt the following assumptions:

  (i) All coherence criteria are assumed to have equal weight.

  (ii) All coherence criteria are supposed to be maximized.

  (iii) The probability distribution of profiles in $\mathcal{U}^a$ are set to equal.

[7] The definition given below is an application of Arrow's majority decision social welfare function on weak orderings, see [4, p. 46].

and two theories $T_i, T_h \in \mathcal{T}$, let set $\mathcal{N}_{(i,h)} = \{U_j \in \mathcal{U} \mid U_j(T_i) \geq U_j(T_h)\}$, and let $\succeq$ stand for the overall coherence ranking on $\mathcal{T}$, where $T_i \succeq T_h$ reads 'T_i is more coherent than or equally coherent to T_h'. Then a CEFL based on simple majority is a function F that assign to $\mathcal{U}$ an overall coherence ranking $\succeq$ by:

$$T_i \succeq T_h \text{ iff } |\mathcal{N}_{(i,h)}| \geq |\mathcal{N}_{(h,i)}|$$

Accordingly, $T_i \succ T_h$ iff $T_i \succeq T_h$ but not $T_h \succeq T_i$, while $T_i \approx T_h$ iff $T_i \succeq T_h$ and $T_h \succeq T_i$.

As shown by Arrow [4], the simple majority method satisfies condition *unrestricted domain, Pareto principle, independence of irrelevant alternatives* and *non-dictatorship*, [8] meanwhile guarantees an overall weak order (i.e., a complete and transitive relation) when there are two alternatives. Condition 2a-2c as prescribed in Section 3.1 is thus satisfied, and 2d too if $|\mathcal{T}| = 2$. The offset effect as expected in the common setting is modeled by $T_i \approx T_h$ when $|\mathcal{N}_{(i,h)}| = |\mathcal{N}_{(h,i)}|$.

Despite the aforementioned good properties, the simple majority method is not very 'robust' for the ranking task. The overall coherence ranking is only assured to be a weak order when there are two theories under consideration, but this is too restricted since in realistic scenarios the number of candidate theories would easily transcend that amount. For instance, Ronald Dworkin's *law as integrity* reasoning, renown for its highly coherentist character, constantly involves numerous candidate theories. In *McLoughlin* case this specific reasoning yields six candidate interpretations, each of which offers a theory of existing laws and cases [10, p. 240-241]. Bench-Capon and Sartor's coherence evaluation also involves more than two theories, e.g., in the showcase given in [6,7] four theories are put to assess. These examples suffice to reveal the looming threat of the famous *Arrow's impossibility theorem* [4, p. 97], according to which the simple majority method fails to secure a weak order as output without domain restriction. This is a well-known difficulty of pairwise comparison in multi-criteria decision making (see [26, p. 32]).

Another known drawback of simple majority is that it will generate a *non-compensatory preference structure*, which was defined by Fishburn [12] as follows:

- X is a set of n-tuples, where $X = X_1 \times X_2 \times \ldots \times X_n$;
- $(x_i, (a_j)_{j \neq i})$ denotes the n-tuple $(a_1, \ldots, a_{i-1}, x_i, a_{i+1}, \ldots, a_n)$;
- $\succ$ is an asymmetric relation on X;
- For each i and all $x_i, y_i \in X_i$, $\succ_i$ is a binary relation on X_i defined by: $x_i \succ_i y_i$ iff $(x_i, (a_j)_{j \neq i}) \succ (y_i, (a_j)_{j \neq i})$ *for all* $(a_j)_{j \neq i} \in \prod_{j \neq i} X_j$;
- $P(x,y) = \{i \mid x_i \succ_i y_i\}$

---

[8] The result stated here is based on Lemma 3 in [4, p. 48], but slightly modified according to a new set of conditions described in [4, p. 96-97].

Then $(X, \succ)$ is a *noncompensatory preference structure* if and only if, for all $x, y, z, w \in X$:

$$[(P(x,y), P(y,x)) = (P(z,w), P(w,z))] \Rightarrow [x \succ y \text{ iff } z \succ w]$$

This definition reflects that the relation $\succ$ on X between any two n-tuples $x = (x_1, \ldots, x_n)$ and $y = (y_1, \ldots, y_n)$ depends solely on another relation $\succ_i$ between any two coordinates $x_i, y_i \in X_i$ [12, p. 393].

It is clear that the preference structure $(\overrightarrow{\mathcal{T}}, \succ)$ (or equivalently $(\mathcal{T}, \succ)$), since each $\overrightarrow{T_i} \in \overrightarrow{\mathcal{T}}$ is representative of $T_i \in \mathcal{T}$), where $\succ$ results from Definition 4.1, is a noncompensatory preference structure. A sketch of proof: $(P(\overrightarrow{T_x}, \overrightarrow{T_y}), P(\overrightarrow{T_y}, \overrightarrow{T_x})) = (P(\overrightarrow{T_z}, \overrightarrow{T_w}), P(\overrightarrow{T_w}, \overrightarrow{T_z}))$ implies $\mathcal{N}_{(x,y)} = \mathcal{N}_{(z,w)}$ and $\mathcal{N}_{(y,x)} = \mathcal{N}_{(w,z)}$, [9] thereby $|\mathcal{N}_{(x,y)}| = |\mathcal{N}_{(z,w)}|$ and $|\mathcal{N}_{(y,x)}| = |\mathcal{N}_{(w,z)}|$, then by Definition 4.1 follows $T_x \succ T_y$ iff $T_z \succ T_w$, and finally by $\overrightarrow{T_i}$ being representative of T_i yields $\overrightarrow{T_x} \succ \overrightarrow{T_y}$ iff $\overrightarrow{T_z} \succ \overrightarrow{T_w}$. In stark contrast, the next two methods *Borda count* and *normalized summation* will prove to be compensatory (see also [24, p. 119-120]). The implication of simple majority being noncompensatory is that the method only conceives the offset effect in a crude manner, in that each theory's loss and gain in different criteria is not really taken into account, that whether a theory is more coherent than another solely depends on the number of criteria favoring each, thus entirely put aside *how* better a theory is to another with regard to certain criteria. [10]

4.2 Borda count

The second method is Borda count, a specific instance of positional scoring rule. Compared with simple majority, Borda count makes use of the positional information offered by evaluation matrix $\mathcal{A}^e$. There are some preliminary notations and definitions that lead to defining a CEFL based on Borda count:

- Let $\succeq_j$ be the individual coherence ranking of criterion C_j, and let $U_j(T_i)$ represent e_{ij}. Then $\succeq_j$ is defined by $T_i \succeq_j T_h$ iff $U_j(T_i) \geq U_j(T_h)$. Subsequently, $T_i \succ_j T_h$ iff $T_i \succeq_j T_h$ but not $T_h \succeq_j T_i$; $T_i \approx_j T_h$ iff $T_i \succeq_j T_h$ and $T_h \succeq_j T_i$. Note that $\succ_j$ is a strict order and $\approx_j$ is an equivalence relation.

- On the basis of $\succeq_j$, define a maximal $\succeq_j$-chain which arranges $T_{ik} \in \mathcal{T}$ in a sequence $\mu = T_{i1} \succeq_j T_{i2} \succeq_j \ldots \succeq_j T_{im}$. By the definition of $\succ_j$ and $\approx_j$, all $\succeq_j$ in μ can be substituted by $\succ_j$ or $\approx_j$, thus turn μ into $\mu = T_{i1} \succ_j T_{i2} \approx_j T_{i3} \succ_j, \ldots, \succ_j T_{im}$. Further, label μ from m to 1 in a decreasing order, and denote the resulted chain as $\mu^* = T_{i1}^m \succ_j T_{i2}^{m-1} \approx_j T_{i3}^{m-2} \succ_j \ldots \succ_j T_{im}^1$.

- As $\approx_j$ is an equivalence relation, let $\approx_j$ partition $\mathcal{T}$ into equivalence classes $\mathcal{E}_1, \mathcal{E}_2, \ldots \mathcal{E}_n$. Let $\mathcal{E}_k$ be the equivalence class to which T_{ik}^m belongs. Note

---

[9] Note that $\mathcal{N}_{(x,y)} = P(\overrightarrow{T_x}, \overrightarrow{T_y}) \cup \mathcal{Q}_{(x,y)}$ where $\mathcal{Q}_{(x,y)} = \{U_j \in \mathcal{U} \mid U_j(T_x) = U_j(T_y)\}$. Given that $\geq$ on $\mathbb{R}$ is complete and $(P(\overrightarrow{T_x}, \overrightarrow{T_y}), P(\overrightarrow{T_y}, \overrightarrow{T_x})) = (P(\overrightarrow{T_z}, \overrightarrow{T_w}), P(\overrightarrow{T_w}, \overrightarrow{T_z}))$, it follows $\mathcal{Q}_{(x,y)} = \mathcal{Q}_{(y,x)} = \mathcal{Q}_{(z,w)} = \mathcal{Q}_{(w,z)}$. Hence, $\mathcal{N}_{(x,y)} = \mathcal{N}_{(z,w)}$ and $\mathcal{N}_{(y,x)} = \mathcal{N}_{(w,z)}$.
[10] In this sense the crude account set forth here resembles the *cancellation* property of social choice function, see for example [32, p. 45].

that if $T_{ik}^m \approx T_{ik'}^{m'}$, then $\mathcal{E}_k = \mathcal{E}_{k'}$.

- Let m_k stand for the label m of T_{ik}^m. Then $\mathcal{B}_j$ is a *Borda scoring function* based on U_j that assigns to T_{ik}^m a score according to its position in μ^*, in the following manner:

$$\mathcal{B}_j(T_{ik}^m) = \frac{1}{|\mathcal{E}_k|} \sum_{T_{ik}^m \in \mathcal{E}_k} m_k$$

A quick example. Suppose $\mu^* = T_{i3}^5 \succ_j T_{i2}^4 \approx_j T_{i1}^3 \approx_j T_{i5}^2 \succ_j T_{i4}^1$, then $\mathcal{B}_j(T_{i3}^5) = 5$, $\mathcal{B}_j(T_{i2}^4) = \mathcal{B}_j(T_{i1}^3) = \mathcal{B}_j(T_{i5}^2) = \frac{(4+3+2)}{3} = 3$, $\mathcal{B}_j(T_{i4}^1) = 1$. This is the so called *mid-rank method*, whose advantage is that 'the sum of the ranks for all members remains the same as for an untied ranking' [15, p. 34].

For simplicity, let $\mathcal{B}_j(T_i)$ represent $\mathcal{B}_j(T_{ik}^m)$ where $T_i = T_{ik}^m$. A CEFL based on Borda count is defined as follows.

Definition 4.2 [CEFL based on Borda count] Let $\geq$ be the *greater or equal to* relation on $\mathbb{R}$. Given a profile $\mathcal{U} \in \mathcal{U}^a$ and two theories $T_i, T_h \in \mathcal{T}$. Then a CEFL based on Borda count is a function F that assigns to $\mathcal{U}$ an overall coherence ranking $\succeq$ by:

$$T_i \succeq T_h \text{ iff } \sum_j \mathcal{B}_j(T_i) \geq \sum_j \mathcal{B}_j(T_h)$$

Accordingly, $T_i \succ T_h$ iff $T_i \succeq T_h$ *but not* $T_h \succeq T_i$, while $T_i \approx T_h$ iff $T_i \succeq T_h$ *and* $T_h \succeq T_i$.

A great merit of the CEFL so defined is that it guarantees a weak order as output, since the overall coherence ranking takes advantage of the $\geq$ on $\mathbb{R}$. In this way, condition 2a-2d in Section 3.1 are satisfied. Another positive property is that the preference structure $(\mathcal{T}, \succ)$ generated by Borda count is compensatory. To show this, let us look at the following example:

$\mathcal{T}$	C_1	C_2	C_3	Sum
T_1	3	4	2	9
T_2	1	1	3	5
T_3	4	3	1	8
T_4	2	2	4	8

Consider this a Borda score table, where the 'Sum' column gives the total Borda score of T_i. It is clear that $(P(T_1, T_2), P(T_2, T_1)) = (P(T_3, T_4), P(T_4, T_3))$, but by Definition 4.2 the overall coherence ranking is $T_1 \succ T_2$ and $T_3 \approx T_4$, hence the preference structure $(\mathcal{T}, \succ)$ is compensatory.

Nevertheless, a CEFL based on Borda count still has some weaknesses, the most prominent among which is disregarding *independence of irrelevant alternatives* (IIA for short). To see this, below is an example modified from the removal paradox as illustrated in [9, p. 215], showing a paradoxical situation where introducing a new theory will affect the coherence ranking of the existing ones: [11]

---

[11] Technically speaking, this example does not strictly accord with the definition of IIA,

$\mathcal{T}$	C_1	C_2	C_3	Sum
T_1	1	2	3	6
T_2	2	3	1	6
T_3	3	1	2	6

$+T_4 \implies$

$\mathcal{T}'$	C_1	C_2	C_3	Sum
T_1	2	3	4	9
T_2	3	4	1	8
T_3	4	1	2	7
T_4	1	2	3	6

Before introducing T_4, the overall coherence ranking is $T_1 \approx T_2 \approx T_3$. After T_4 joins the evaluation, the overall ranking becomes $T_1 \succ T_2 \succ T_3 \succ T_4$. This result is bizarre, because all the coherence criteria are supposed to measure a legal theory's coherence, and it is reasonable to say that the overall coherence ranking on theories should change only when some theory has its attribute altered, thereby renders the evaluation different. This 'measuring aspect' is perhaps a counterpart of the ordering aspect of IIA ([28, p. 142-143]) in social choice. Nevertheless, there is a crucial distinction: unlike the ordering aspect of IIA, the measuring aspect does not exclude intensity of preference. In another word, measuring degree of coherence allows one to say that theory T_1 is two times better than T_2 with regard to certain criterion, e.g., the strong supports in T_1 has twice the number of that in T_2. This observation makes the foregoing example not as much a paradox as a puzzle. The puzzle is: In what sense is a theory, if deemed so, really irrelevant? Could it be the case that some 'irrelevant' theory is just irrelevant in certain way, but actually become relevant when considered from a different viewpoint? It is this puzzle that lead us to *normalized summation*, where the irrelevant become relevant and the loss of IIA is purchased at a reasonable price.

4.3 Normalized summation

Let us first set forth the philosophical intuition underlying this last method. In real world, the coherence of a legal theory is seldom evaluated in an absolute manner, nor is the legal judgment backed by that theory justified in an absolute sense. Rather, whether a legal theory is coherent and how much if it is, is relative to theories available for evaluation. In a word, it is about making the best of what one has at hand. Such notion is central to major legal coherentism. For example, Dworkin [10] thought of the 'right answer' of legal disputes as the interpretation among the proposed that best served fitness and substantial political virtues, Peczenik [23, p. 106-107] regarded legal reasoning as an argumentative dialogue and the final result as its survival, while Amaya [2] and Thagard [30] considered coherentist reasoning to be an inference to the best (legal) interpretation from piecemeal information. None of them rely on an ideal Platonist entity, to which the word 'coherent' is perfectly entitled, to evaluate and decide a theory's degree of coherence.

---

since the inter-profile comparison in IIA occurs between profiles with the same alternative set. However, it shares with IIA an important intuition, that the overall preference on alternatives should not be affected by irrelevant alternatives. In this respect, the example given here reveals that a Borda-based CEFL yields an overall coherence ranking that somehow depends on the criteria's evaluation of something outside.

Accordingly, the measuring aspect of coherence ranking may be modified to allow the once 'irrelevant' alternatives to be 'relevant'. The point is, every measurement in coherence ranking has a contextual baseline that depends on the status quo of candidate theories. Think of it in this way: if coherence in law is about making the best of things at hand, then the overall ranking will hinge on what are best of the candidate theories. That is to say, when aggregating the individual coherence ranking, the distance of a theory to the best theory available with regard to each criterion should be taken into account. Then, the all-things-considered most coherent theory should be the one that minimizes the distance, and the rest be ranked according to their performances in this respect.

Normalized summation intends to capture this idea. A normalization is a data processing that aims at arranging on a comparable scale those values obtained by coherence evaluation. Speaking formally, a normalization is a function V that assigns to each $\overrightarrow{E_j} = (e_{1j}, e_{2j}, \ldots, e_{mj})$ (recall Definition 3.2) a new vector $\overrightarrow{E_j^v} = (e_{1j}^v, e_{2j}^v, \ldots, e_{mj}^v)$ where $e_{ij}^v \in [0, 1]$. A *normalized coherence evaluation* U_j^v is then a composition of U_j and normalization V, i.e., $U_j^v = U_j \circ V$. Note that the normalization process is not of unique kind, nor is U_j^v. In the literature there are at least three normalization processes that have notable affinity with the idea of 'minimizing distance to best candidate', among others. For a coherence evaluation U_j, let $max\ e_{ij}$ be the maximal value in U_j, and $min\ e_{ij}$ the minimal. Then each e_{ij} in U_j is respectively normalized by the three processes in the following way:

$$(1)\ e_{ij}^v = \frac{e_{ij}}{max\ e_{ij}} \ ; \quad (2)\ e_{ij}^v = \frac{e_{ij} - min\ e_{ij}}{max\ e_{ij} - min\ e_{ij}} \ ; \quad (3)\ e_{ij}^v = \frac{e_{ij}}{\sum_i e_{ij}}$$

Though each minimizes the distance to best candidate in their own way, process (1) and (3) have a particular advantage over(2), in that it preserves the proportionality between pairs of e_{ij} [12] and thus does not alter the extent to which a theory is more coherent than another. Still, process (2) better reflects the status quo of available theories since the range of value from best to worst is included in the calculation. [13] The question as to which kind of normalization is most suitable for coherence ranking of legal theories is reserved for future work. For current purpose, it suffices to proceed with them to construct a CEFL based on normalized summation.

Definition 4.3 [CEFL based on normalized summation] Let $\geq$ be the *greater or equal to* relation on $\mathbb{R}$. Given a profile $\mathcal{U} \in \mathcal{U}^a$, two theories $T_i, T_h \in \mathcal{T}$ and a normalization V, a CEFL based on normalized summation is a function F that assigns to $\mathcal{U}$ an overall coherence ranking $\succeq$ by:

$$T_i \succeq T_h \text{ iff } \sum_j U_j^v(T_i) \geq \sum_j U_j^v(T_h)$$

---

[12] Take e_{1j}^v and e_{2j}^v for example: $\frac{e_{1j}}{max\ e_{ij}} / \frac{e_{2j}}{max\ e_{ij}} = \frac{e_{1j}}{\sum_i e_{ij}} / \frac{e_{2j}}{\sum_i e_{ij}} = e_{1j}/e_{2j}$, no matter what the $max\ e_{ij}$ or $\sum_i e_{ij}$ may be.

[13] These normalization processes as well as their merits can be found in [24].

Accordingly, $T_i \succ T_h$ iff $T_i \succeq T_h$ but not $T_h \succeq T_i$, while $T_i \approx T_h$ iff $T_i \succeq T_h$ and $T_h \succeq T_i$.

Since $\succeq$ takes advantage of the $\geq$ on $\mathbb{R}$, normalized summation always output a weak order as overall coherence ranking. Condition 2a-2d in Section 3.1 is thus satisfied. It does violate IIA, but not always so. If process (1) or (2) is adopted, then normalized summation actually respect IIA to the extent that the change in individual coherence ranking of irrelevant alternatives does not alter $max\ e_{ij}$ (for process (1)) or $min\ e_{ij}$ (both for process (2)); and for those cases where IIA is loss (e.g., $max\ e_{ij}$ is render different), the loss is reasonable and indeed is required, because the once irrelevant alternatives become relevant as the distance to best candidate is altered too.

Another virtue of normalized summation is that it accounts for the offset effect in a way much finer than both simple majority and Borda count. To see this, consider the following example. Suppose $\mathcal{C} = \{C_1, C_2, C_3\}$ where C_1 is 'the cases covered', C_2 is 'strong support' and C_3 is 'reciprocal justification', and $\mathcal{T} = \{T_1, T_2, T_3\}$ is the set of theories for evaluation. Suppose the evaluation is as illustrated in the table below:

$\mathcal{T}$	C_1	C_2	C_3
T_1	16	20	18
T_2	30	18	16
T_3	10	16	13

where T_2 is only slightly inferior to T_1 under C_2 and C_3, but greatly surpasses others with regard to C_1. Intuitively, one can argue that the overall coherence ranking should at least rank $T_2 \succeq T_1$. However, neither simple majority nor Borda count could realize that. For simple majority:

$$|\mathcal{N}_{(1,2)}| = 2,\ |\mathcal{N}_{(2,1)}| = 1;\ 2 > 1,\ \text{there fore } T_1 \succ T_2.$$

As for Borda count:

$$\sum_j \mathcal{B}_j(T_1) = 8,\ \sum_j \mathcal{B}_j(T_2) = 7;\ 8 > 7,\ \text{therefore } T_1 \succ T_2.$$

But for normalized summation, if process (1) is adopted, then:

$$\sum_j U_j^v(T_1) = \tfrac{38}{15},\ \sum_j U_j^v(T_2) = \tfrac{251}{90};\ \tfrac{251}{90} > \tfrac{38}{15},\ \text{therefore } T_2 \succ T_1.$$

Despite these merits, a CEFL based on normalized summation is accompanied by certain drawbacks. First, normalized summation demands a rigorous evaluation to make sure that each e_{ij} is precise. In stark contrast, the precision of evaluation is far less demanding in simple majority and Borda count, for the precision only need to reach a degree that suffices for decision makers to identify the coherence ranking in an ordinal manner. In this regard, normalized summation is indeed the most difficult to apply. Second, normalized summation impose an extra intellectual responsibility on decision makers: those who seek for legal justification in terms of coherence must try their best to find the best available theory, because normalized summation aims at making the best of *status quo* and that *status quo* should be as good as it could be. Otherwise, a critic may argue that the overall coherence ranking so given is only an evidence of irresponsibility and thus not credible, not to mention that failing to take into account some better candidate theory will result in a ranking that does not justify the violation of IIA. Last but not least, the output of normalized summation is vulnerable to the choice of normalization process, which could only be remedied when an appropriate instruction for choosing normalization process is available.

5   Conclusion

So far, to what extent does this paper fulfill the expectations at the outset? The answer is to some extent positive, but not without reservation. The challenge of obtaining an overall coherence ranking of legal theories from multiple criteria is captured in the framework of preference aggregation. The aggregation problem is formulated as finding appropriate coherence evaluation functionals (CEFL) to operate on a coherence evaluation matrix. For this purpose, the paper presents three CEFLs respectively on the basis of simple majority, Borda count and normalized summation, with a detailed examination of their pros and cons. In this way, the multi-criteria accounts of theory coherence in law are equipped with formal tools to realize their ambition.

Nevertheless, this paper has raised lots of puzzles as well. First, to enrich the informational basis for coherence ranking, i.e., to make richer use of the evaluation matrix, is in the meantime to complicate the aggregation problem and render the solution difficult to apply. This is especially the case with normalized summation. It seems that more efforts must be put to model and to make explicit the 'legal theory' so as to catch up with the increasing complexity. This direction invites instructions and contributions from legal knowledge representation. Second, every aggregation procedure has its own weaknesses. To current knowledge, there is no once-and-for-all solution. And special attention should be paid to the fact that none of the proposed CEFLs touch upon the possibility of inter-criteria comparability in evaluation. The normalized summation slightly deals with this issue, through literally forcing the evaluated values obtained by different criteria to be comparable at an interval between 0 and 1. But this is not entirely satisfactory as its own problems have shown. A better way to conceive the comparability of different criteria may be to seriously consider the 'contribution to coherence' of the index used by each criteria in a specific multi-criteria account. In this respect, the constraint satisfaction theory seems to be a good solution, but the exact connection between it and those multi-criteria approaches are yet to discover. All in all, this paper at best takes a starting step. There is still a long way to go.

References

[1] Alexy, R. and A. Peczenik, *The concept of coherence and its significance for discursive rationality*, Ratio Juris **3** (1990), pp. 130–147.
[2] Amaya, A., *Legal justfication by optimal coherence*, Ratio Juris **24** (2011), pp. 304–329.

[3] Amaya, A., "The Tapestry of Reason: An Inquiry into the Nature of Coherence and its Role in Legal Argument," Hart Publishing, Oxford and Portland, Oregon, 2015.

[4] Arrow, K. J., "Social Choice and Individual Values," John Wiley & Sons, New York, London and Sydney, 1963, 2nd edition.

[5] Arrow, K. J. and H. Raynaud, "Social Choice and Multicriterion Decision-Making," MIT Press, Cambridge, Mass. and London, 1986.

[6] Bench-Capon, T. and G. Sartor, *Theory based explanation of case law domains*, in: R. P. Loui, editor, *Proceedings of the 8th international conference on Artificial intelligence and law - ICAIL '01* (2001), pp. 12–21.

[7] Bench-Capon, T. J. M. and G. Sartor, *A quantitative approach to theory coherence*, in: B. Verheij, A. R. Lodder, R. P. Loui and A. J. Muntjewerff, editors, *JURIX 2001: Legal Knowledge and Information Systems* (2001), pp. 53–62.

[8] Bossert, W. and J. A. Weymark, *Utility in social choice*, in: S. Barberà, P. J. Hammond and C. Seidl, editors, *Handbook of Utility Theory (vol.2)*, Springer US, Boston, MA, 2004 pp. 1099–1177.

[9] Brams, S. J. and P. C. Fishburn, *Voting procedures*, in: K. J. Arrow, A. K. Sen and K. Suzumura, editors, *Handbook of Social Choice and Welfare (Vol. 1)*, Elsevier, 2002 pp. 173–236.

[10] Dworkin, R., "Law's Empire," The Belknap Press of Harvard University Press, Cambridge, Mass. and London, 1986.

[11] Dworkin, R., "Justice for Hedgehogs," The Belknap Press of Harvard University Press, Cambridge, Mass. and London, 2011.

[12] Fishburn, P. C., *Noncompensatory preferences*, Synthese **33** (1976), pp. 393–403.

[13] Hage, J., *Law and coherence*, Ratio Juris **17** (2004), pp. 87–105.

[14] Hurley, S. L., *Supervenience and the possibility of coherence*, Mind **94** (1985), pp. 501–525.

[15] Kendall, M. G., "Rank Correlation Methods," Charles Griffin, London, 1970, 4th edition.

[16] Kress, K., *Coherence*, in: D. Patterson, editor, *A Companion to Philosophy of Law and Legal Theory*, John Wiley & Sons, Ltd, 2010 pp. 521–538.

[17] MacCormick, N., "Legal Reasoning and Legal Theory," Oxford University Press, New York, 1978.

[18] MacCormick, N., *Coherence in legal justification*, in: A. Peczenik, L. Lindahl and B. V. Roermund, editors, *Theory of Legal Science. Proceedings of the Conference on Legal Theory and Philosophy of Science* (1984), pp. 235–251.

[19] Morreau, M., *Mr. Fit, Mr. Simplicity and Mr. Scope: From social choice to theory choice*, Erkenntnis **79** (2014), pp. 1253–1268.

[20] Morreau, M., *Theory choice and social choice: Kuhn vindicated*, Mind **124** (2015), pp. 239–262.

[21] Okasha, S., *Theory choice and social choice: Kuhn versus Arrow*, Mind **120** (2011), pp. 83–115.

[22] Patrick, K. and K. Hodesdon, *Is theory choice using epistemic virtues possible?*, in: R. Urbaniak and G. Payette, editors, *Applications of Formal Philosophy*, Springer International Publishing, Cham, 2017 pp. 139–168.

[23] Peczenik, A., "On Law and Reason," Law and Philosophy Library **8**, Springer Netherlands, Dordrecht, 2009.

[24] Pomerol, J.-C. and S. Barba-Romero, "Multicriterion Decision in Management: Principles and Practice," Springer US, Boston, MA, 2000.

[25] Raz, J., *The relevance of coherence*, Boston University Law Review **72** (1992), pp. 273–321.

[26] Roy, B., *Paradigms and challenges*, in: S. Greco, M. Ehrgott and J. R. Figueira, editors, *Multiple Criteria Decision Analysis: State of the Art Surveys*, Springer New York, New York, 2016, 2nd edition pp. 19–39.

[27] Sartor, G., "Legal Reasoning: A Cognitive Approach to Law," A Treatise of Legal Philosophy and General Jurisprudence **5**, Springer, Dordrecht, 2005.

[28] Sen, A., "Collective Choice and Social Welfare: An Expanded Edition," Harvard University Press, Cambridge, Mass., 2017.

[29] Soriano, L. M., *A modest notion of coherence in legal reasoning. a model for the european court of justice*, Ratio Juris **16** (2003), pp. 296–323.

[30] Thagard, P., "Coherence in Thought and Action," MIT Press, Cambridge, Mass. and London, 2000.

[31] Thagard, P. and K. Verbeurgt, *Coherence as constraint satisfaction*, Cognitive Science **22** (1998), pp. 1–24.

[32] Young, H. P., *An axiomatization of Borda's rule*, Journal of Economic Theory **9** (1974), pp. 43–52.

[33] Zipursky, B., *Legal coherentism*, SMU Law Review **50** (1997), pp. 1679–1720.

www.ingramcontent.com/pod-product-compliance
Lightning Source LLC
Chambersburg PA
CBHW071109050326
40690CB00008B/1166